Antonio Córdoba

Suprematism in Harmonic Analysis

Birkhäuser

Antonio Córdoba
Department of Mathematics (UAM)
Autonomous University of Madrid
Madrid, Spain

ISSN 0743-1643 ISSN 2296-505X (electronic)
Progress in Mathematics
ISBN 978-3-031-73736-7 ISBN 978-3-031-73737-4 (eBook)
https://doi.org/10.1007/978-3-031-73737-4

This book is published under the imprint Birkhäuser, www.birkhauser-science.com by the registered company Springer Nature Switzerland AG
The registered company address is: Gewerbestrasse 11, 6330 Cham, Switzerland

Ferran Sunyer i Balaguer (1912–1967) was a self-taught Catalan mathematician who, in spite of a serious physical disability, was very active in research in classical mathematical analysis, an area in which he acquired international recognition. His heirs created the Fundació Ferran Sunyer i Balaguer inside the Institut d'Estudis Catalans to honor the memory of Ferran Sunyer i Balaguer and to promote mathematical research.

Each year, the Fundació Ferran Sunyer i Balaguer and the Institut d'Estudis Catalans award an international research prize for a mathematical monograph of expository nature. The prize-winning monographs are published in this series. Details about the prize and the Fundació Ferran Sunyer i Balaguer can be found at

`https://ffsb.espais.iec.cat/en/the-ferran-sunyer-i-balaguer-prize/`

This book has been awarded the Ferran Sunyer i Balaguer 2024 prize.

The members of the scientific commitee of the 2024 prize were:

Ruth Kellerhals
Université de Fribourg

Jiang-Hua Lu
University of Hong Kong

Marta Sanz-Solé
Universitat de Barcelona

Kristian Seip
Norwegian University of Science and Technology (NTNU)

Yuri Tschinkel
Courant Institute for Mathematical Sciences
New York University

fundació FERRAN SUNYER I BALAGUER

Ferran Sunyer i Balaguer Prize winners since 2014

2014 Veronique Fischer and Michael Ruzhansky
Quantization on Nilpotent Lie Groups,
Open Access, PM 314

2015 The scientific committee decided not to award the prize

2016 Vladimir Turaev and Alexis Virelizier
Monoidal Categories and Topological Field Theory, PM 322

2017 Antoine Chambert-Loir, Johannses Nicaise and Julien Sebag
Motivic Integration, PM 325

2018 Michael Ruzhansky and Durvudkhan Suragan
Hardy Inequalities on Homogeneous Groups, PM 327

2019 The scientific committee decided not to award the prize

2020 Urtzi Buijs, Yves Félix, Aniceto Murillo and Daniel Tanré
Lie Models in Topology, PM 335

Giovanni Catino and Paolo Mastrolia
A Perspective on Canonical Riemannian Metrics, PM 336

2021 Tim Browning
Cubic Forms and the Circle Method, PM 343

2022 Pascal Auscher and Moritz Egert
Boundary Value Problems and Hardy Spaces for Elliptic Systems with Block Structure, PM 346

2023 Xavier Fernández-Real and Xavier Ros-Oton
Integro-Differential Equations, PM 350

Preface

With a little help
from my friends.

Art among theorems

Neoplasticism, whose more conspicuous representative is Piet Mondrian, and suprematism, for which Kazimir Malévich is the more genuine exponent, are abstract pictorial styles which can be often observed in the figures and illustrations appearing in research publications of real variable and harmonic analysis.

Although different, they both advocate for the importance of simple forms, like rectangles with sides parallel to the canvas axes in Mondrian, or with arbitrary directions in the case of Malévich, and the use of primary colors in their compositions.

A typical example is given by the Calderón–Zygmund algorithm. Given a non-negative integrable function f defined on a plane rectangle R, and given m, a fixed number strictly bigger than its average, that is, $m > f_R = \frac{1}{\mu(R)} \int_R f$, let us divide dyadically R into four rectangles and keep those of them for which the average of f is bigger than m. Next, we subdivide the others, also dyadically, and apply the same criterium of maintaining those for which the corresponding average of f is bigger than m, and so on. At the end of the process, we will obtain a sequence of rectangles R_j contained in the original R having the property that $m < f_{R_j} \leq 4m$ and, because of Lebesgue's differentiation theorem, $f(x) \leq m$ almost everywhere in the complement of their union.

If we now paint with identical color (red, blue, yellow, black,. . .) the members of the sequence which belong to the same generation (which have equal sides), the result will look like a Mondrian. And the Mondrian collection obtained with different values of the parameter m will provide us with crucial information about the result of singular integral operators acting on the function f.

There are, however, others important operators in harmonic analysis involving Hilbert transforms in many directions of Euclidean space, whose understanding asks

Neoplasticist painting in the style of Piet Mondrian.

for decomposition in rectangles having those different directions (a good example is the so-called disc multiplier considered in Chapter 5). In those cases, the pictorial images will look more like a composition of Malévich's suprematism.

I hope that these naive digressions help to justify the title of this book. At least of its first part, because the second, Chapters 5 to 9, does not fit so easily into such metaphor and there, the prevalence of oscillatory integrals (stationary phase, Fresnel's diffraction, Gaussian sums, Van der Corput estimates, Cornu's spirals, etc.) would appeal to different pictorial styles. Being audacious, one could associate them with Vincent van Gogh famous "Starry night", as was done in the poster of an analysis meeting held in Madrid in June 2009, with the inclusion in that painting of several suprematistic elements of the proof of the two-dimensional Bochner–Riesz conjecture of Chapter 5.

The heart of the book

Chapter 1 can be considered as a kind of divertimento. Let us define a mathematical haiku as a relevant statement whose description does not need complicated terms and whose proof, sharp and short, can be understood without much effort if one is in possession of the appropriate point of view. Some of the chosen haikus have applications in the following chapters, an example is Fritz John's lemma on convex sets (because it allows to reduce the analysis of the differentiation properties of basis formed by convex sets, to the case of parallelepipeds), but others are there simply because I like them.

Composition following the suprematism of Kazimir Malévich.

In the background of Chapter 2 lies the classical Vitali covering lemma, valid for cubes or balls in Euclidean space and false, in general, for other differentiation basis. But also the relationship between covering properties of families of sets and weak-type estimates for the corresponding maximal operators. A standard linearization reduces the boundedness of the non-linear maximal function to the uniform boundedness of a family of linear operators for which, taking their adjoints or duals, we can read weak-type inequalities in terms of covering properties implied by those adjoints.

The exponential type covering, satisfied by parallelepipeds having sides parallel to the coordinate axes, is also treated in this chapter, together with the positive answer to a conjecture formulated by Antoni Zygmund about monotonic families of parallelepipeds in 3-dimensional Euclidean space. There is also the suprematistic approach to the maximal function associated to parallelepipeds with lacunary directions in the plane.

Chapter 3 is devoted mainly to the theory of Kakeya or Besicovitch sets. That is, compact sets of measure zero containing a unit straight line segment in each direction of the space. Consider in $\mathbb{R}^n$ the collection B_δ, $0 < \delta < 1$, of cylinders (or parallelepipeds) having eccentricity (height/radius, or height/length of the basis cube) equal to $1/\delta$. Given a locally integrable function f, let us define

$$M_\delta f(x) = \sup_{x \in R \in B_\delta} \frac{1}{\mu(R)} \int_R |f(y)| \, \mathrm{d}\mu(y).$$

Suprematistic intervention by Fernando Chamizo in Van Gogh's Starry Night.

About this maximal function there is a conjecture:

Conjecture. *For each $n \geq 2$, there exist constants $C(n)$ and $a(n)$ so that, for each $\delta > 0$, the following estimate holds*:

$$\|M_\delta f\|_n \leq C(n)\,[1 + |\log\delta|]^{a(n)}\,\|f\|_n.$$

The following is the main result of Chapter 3.

Theorem (Kakeya maximal function theorem)**.** *The conjecture above is true in dimension $n = 2$.*

An immediate consequence of this theorem is that every Kakeya set in dimension two must have Minkowski (or box counting) dimension equal to 2, and with some extra arguments it also implies the same conclusion for the Hausdorff dimension.

Its proof goes back to the year 1972, and an outlined version [41] appeared in the Bulletin of the AMS in 1975. It was quoted by C. Fefferman in his plenary talk entitled "Recent progress in classical Fourier analysis" delivered at the International Congress of Mathematicians celebrated in Vancouver in 1974. The paper with a detailed proof was accepted by the American Journal of Mathematics that same year of 1974, but because of its huge backlog, it was published in 1977, see [65]. Several authors have recently written a chronology of results about Kakeya sets giving the date of 1977 for this theorem when, strictly speaking, it was obtained five year earlier, and announced in 1974. It was part of my Ph.D. Thesis at the University of Chicago and, naturally, I was encouraged to try the higher dimensional case, but this turned out to be a very difficult task which today remains as a challenging open problem.

It was clear, nevertheless, that its solution would be implied by an adequate higher dimensional covering lemma for which it was natural to obtain several special cases which, nowadays, are described in the literature under the picturesque names of "bush" and "brush" configurations. That is, when the cylinders (or parallelepipeds) involved share a common point (bush), or when all intersect a given one of them (brush). There are simple geometrical arguments giving the right estimate for the overlapping function in both configurations, bush or brush, but at that time (1974), those observations where not considered to have enough entity for publication. Years later, Jean Bourgain and Tom Wolff rediscovered the properties of those special configurations of fat needles and used them to obtain lower bounds for the fractal dimensions of Kakeya sets in dimension larger than 2.

Being ambitious, we can formulate as a conjecture the following "suprematistic covering lemma":

Definition. A finite family of parallelepipeds $\{R_i\}$ has the *sparseness property* (P) if

$$\mu\Big(R_j \cap \bigcup_{k\neq j} R_k\Big) \leq c\mu(R_j) \quad \text{for each } j,$$

where $c < 1$ is a fixed positive constant.

Definition. A finite family $\{R_\alpha\}$ of parallelepipeds is a *bush* if they have a non-empty intersection and the family satisfies property (P).

Conjecture (Suprematistic covering lemma). *Given a family $\mathcal{B}$ of parallelepipeds in $\mathbb{R}^n$, we have*

$$\sup\left\{\frac{1}{\mu(\bigcup_j R_j)}\Big\|\sum_j R_j\Big\|_p^p : \{R_j\} \subset \mathcal{B} \textit{ satisfies } \text{(P)}\right\}$$
$$\leq C_p \sup\left\{\frac{1}{\mu(\bigcup_\alpha R_\alpha)}\Big\|\sum_\alpha R_\alpha\Big\|_p^p : \{R_\alpha\} \subset \mathcal{B} \textit{ is a bush}\right\},$$

where with abuse of language we have used $R = R(x)$ to denote the indicator function of the set R.

Chapter 4 is about multipliers of the Fourier transform, emphasizing the role of weighted inequalities for singular integrals following the pioneering work of Benjamin Muckenhoup [94] and of Ronald Coifman and Charles Fefferman [82].

Early in 1974, in a joint work with C. Fefferman, published in 1976, see [84], it was observed that given an arbitrary locally integrable function f in $\mathbb{R}^n$ and a positive real number $r > 1$, then $M(f^r(x))^{1/r}$, where M denotes the Hardy–Littlewood maximal function, is a weight in the class A_1 with bounds independent of the function f.

This observation turned out to be crucial for the suprematistic proof of the two-dimensional Bochner–Riesz operators conjecture of Chapter 5, allowing their estimation in terms of the Kakeya maximal function. Being fascinated by the beauty of the method, I used to state in my talks that direct L^p inequalities were not the main issue, being a much better strategy to search for L^2 estimates with appropriate weights.

During the academic year 1976-77, Eli Stein asked me to lecture on his Harmonic Analysis graduate course of Princeton University, covering the, by then, recent results around weighted inequalities for singular integrals. It was a great opportunity for me to organize the material, and very challenging also because Carl Herz, Alex Nagel, Alice Chang and José Luis Rubio de Francia where visitors at the Princeton Mathematics Department that year, and attended my talks. The notes of those lectures were used also in a short course imparted at the Harmonic Analysis Congress celebrated in El Escorial during the summer of 1978, and published in its proceedings. I believe that they have influenced later presentations of the subject by García-Cuerva and Rubio de Francia [4], and Stein [13]. Chapter 4 can be considered an updated version of them.

Chapter 5 begins with the counterexample to the disc multiplier conjecture given by C. Fefferman, where the properties of the Kakeya set were used, and which was at the origin of the approach emphasized in this book. It is followed by the boundedness of Bochner–Riesz operators in dimension two, a result for which there are other two proofs in the literature obtained, respectively, by Lennart Carleson and Per Sjölin [103] and by Lars Hörmander [117], but with methods different to the suprematistic point of view considered here.

Together with the Kakeya maximal theorem, our proof involves several square-functions estimates. Some of them are known to hold in general dimensions: let $\mathcal{L} = \{R_\nu\}$ be a lattice of congruent parallelepipeds in $\mathbb{R}^n$ with pairwise disjoint interiors, and consider the Fourier multipliers P_ν defined by the formula $\widehat{P_\nu f}(\xi) = R_\nu(\xi)\,\widehat{f}(\xi)$. Define the square function $G(f) = (\sum_\nu |P_\nu(f)|^2)^{1/2}$. Then the following inequality holds:

$$\int_{\mathbb{R}^n} |Gf(x)|^2\, w(x)\, \mathrm{d}x \lesssim_r \int_{\mathbb{R}^n} |f(x)|^2\, [M_s w^r(x)]^{1/r}\, \mathrm{d}x,$$

where $r > 1$, M_s denotes the strong maximal function, and f and $w > 0$ are arbitrary functions: f in the Schwartz class, and w locally integrable.

But there are other square-functions whose sharp boundedness is known only in dimension two.

Given $\delta < 1$, let us define a δ-plate to be a cylinder of height δ and radius $\delta^{1/2}$. Suppose that $\{B_j\}$ is a family of pairwise disjoint δ-plates tangent to the unit sphere and that, for each j, f_j is a smooth function whose Fourier transform is supported in B_j.

Conjecture (Suprematistic square-function lemma). *There holds the estimate*

$$\Big\| \sum_j f_j \Big\|_{2n/(n-1)} \lesssim \Big\| \Big(\sum_j |f_j|^2 \Big)^{1/2} \Big\|_{2n/(n-1)}$$

independently of δ.

The fact that the exponent $2n/(n-1)$ is equal to 4 when $n = 2$ helps a lot in the proof of this conjecture in the two-dimensional case, which remains open in higher dimensions.

We can say that the theory in dimension two is by now well understood, and this chapter pretends to be a rather complete exposition of its main features, including the restriction of Fourier transforms to circles, estimates for angular square-functions, a.e. convergence of Bochner–Riesz means, and L^p-estimates for the Hilbert transform along plane curves.

Chapter 6 complements Chapter 5 giving a positive answer to the disproved conjecture about the ball multiplier. If happens that changing $L^p(\mathbb{R}^n)$, for $2n/(n+1) < p < 2n/(n-1)$, by $L^p_{\text{rad}}L^2_{\text{ang}}(\mathbb{R}^n)$, the conjecture holds true and the ball multiplier operator is bounded on that rank of mixed norm spaces. Taking L^2-norm in the angular part helps to defeat the Kakeya set enemy and establishes a theorem.

The methods of proof are now different, and rely on very precise asymptotic formulas for Bessel functions of arbitrary order. The approach was influenced by the study of band limited function in signal theory. In particular, the convergence of series of prolate spheroidal wave functions, as well as the development of functions in series of eigenfunctions of the Laplacian in the unit ball under Dirichlet boundary condition.

Chapter 7 is devoted to the Poisson summation formula, which can be formulated in terms of a lattice $\mathcal{L}$ and its dual $\mathcal{L}^*$: the tempered distribution given by the sum of delta functions placed at the vertices of $\mathcal{L}$ has Fourier transform equal to the sum of deltas supported now at the vertices of $\mathcal{L}^*$ (except for a factor which accounts for the volume of the fundamental cell).

Besides its many important mathematical applications, the formula is also relevant in crystallography, because there the Fourier transform of the electronic density function of a molecule is usually associated to its X-ray spectrum, being historically this observation a crucial step in the understanding of solids. The crystallographic paradigm states that periodic structures produce sharp Bragg peaks, and that the knowledge of the distribution of those peaks would allow the reconstruction of the crystal (modulo perhaps the phase lost, but that is a different problem). The X-ray crystallographers were able to produce a catalogue of crystals and a classification of their symmetry groups. But this great building suffered a seismic attack in the middle-eighties of the past century, due to the discovery of quasicrystals, that is, materials whose X-ray spectra present clear Bragg peaks but showing a pentagonal symmetry not allowed by the theory.

I attended several talks about this issue during my stay at the ETH in Zurich in 1986, where I was participating in a semester in Harmonic Analysis organized by John Garnett. My first reaction was to ask myself about others possible Poisson formulas and, after several weeks of work, I came out with the answer: there are no other formulas. More precisely, if we are given two discrete sets of points $X = \{x_j\}$ and $Y = \{y_k\}$ so that the tempered distribution $\sum \delta_{x_j}$ has Fourier transform equal to $\sum \delta_{y_k}$, then X and Y must be dual lattices having fundamental cell of volume 1.

This is the beginning theorem of this Chapter 7, and was published [154] in the Comptes Rendus of the Académie des Sciences, after several of my friends, all distinguished harmonic analysts, encouraged me to do so because, as they informed me, the proof was new for them. Years earlier, it could have been considered a kind of fundamental theorem of crystallography, but the appearance of quasicrystals changed the classical crystal definition and left the mathematicians with the task of investigate, and classify, the more general Poisson summation formulas needed for the understanding of the X-ray spectrum of the new materials. A task that, as I understand, remains quite open.

Having considered in previous chapters the Fourier transform of balls in $\mathbb{R}^k$, it is unavoidable that a discussion of the Poisson summation formula leads us to lattice point problems in number theory. That is, to estimate the error term committed when we estimate the number of lattice points inside the ball centered at the origin and with radius r by its volume. Following Hardy, the Poisson formula allows us to write an analytic expression for the error term involving Bessel functions and the arithmetical function $r_k(m)$, which counts the number of representations of m as a sum of k squares. There is no pretension of giving here a fair description of such an important and interesting theory, but only to emphasize the role of the Poisson formula in the production of an analytically treatable expression for the error term. Quoting Fernando Chamizo: "Poisson's formula allows the change of counting lattice points to counting waves". A remarkable difference with the Bochner–Riesz problem considered in previous chapters is that, here, the difficulty lies in dimensions two and three, while for $k \geq 4$ we know sharp estimates for the error term due to the more regular behavior of $r_k(m)$.

The chapter contains also an extension of the theory to convex domains in the plane, in the spirit of Feymann's path integration, which helps to understand and delimitate the difficulty of Hardy's conjecture for that error term in the relevant case $k = 2$. There are also several miscellaneous comments about the phase retrieval problem in X-rays spectrographs related to Gaussian sums.

The interface between harmonic analysis and number theory is so vast and rich in results that Chapter 8 can only pretend to point out some of them. Perhaps not the most important, but I hope that somehow significative. It is an area where the names of Riemann and Hardy are ubiquitous, where techniques like "the circle method", "Vinogradov's mean value theorem" or "Weyl estimates for trigonometric sums" are very relevant, and where questions like the previously mentioned about counting lattice points, Waring's problem, or the Goldbach conjecture have been a source of very interesting mathematical interaction between both subjects.

To place in context Chapter 8, a starting point could be the part of Zygmund's book devoted to the study of special trigonometric series. There we can find the properties of lacunary series, that is, series $\sum_k a_k \exp(2\pi i n_k x)$ where the frequencies grow geometrically: $n_{k+1}/n_k > \rho > 1$. It is proved that if such a series is the Fourier series of an integrable function, then any positive power of that function has also to be integrable, and the corresponding L^p-norms are equivalent to its L^2-norm. Nowadays we can precise the statement above in the sense that it belongs also to the space BMO (bounded mean oscillation). A conspicuous reader of Zygmund's treatise misses in that part the corresponding statement when the frequencies grow like k-powers. Preliminary considerations yield the restriction $p < 2k$ for the claim that if the Fourier series of an integrable function has only frequencies in k powers, then that function is necessarily in L^p. But the claim for $p < 2k$ remains as an outstanding and difficult open problem, even in the case $k = 2$, where we have very precise information about Gaussian sums. When we reformulate the problem in terms of multiplier operators of Fourier series, it happens that the corresponding kernels are very different from the singular integrals considered in Chapter 4. Here comes to mind the comment of Henryk Iwaniec: "In harmonic analysis we get nice kernels applied to complicated functions, while in number theory we have difficult kernels applied to very smooth functions."

In this chapter, we find several observations and particular cases of the problem, but not its solution. The frustration of not been able to complete Zygmund's book can be mitigated, nevertheless, by the discovery that one can compute exactly the Minkowski dimension of the graphs of the trigonometric series

$$\sum_{n\geqslant 1} n^{-(1+\delta)}\, e^{2\pi i\, n^k \cdot x},$$

and that those series were already considered in the thesis of Riemann [194], where, among other interesting topics, he generalized the standing notions of integral and derivative.

The chapter ends with an application of number theory to a method of convergence of Fourier series, where the partial sums are now formed taking into account the size of the coefficients. This is a method which is natural in Crystallography by the reasons described in the previous chapter.

Chapter 9 starts with a quick introduction to the Born–Oppenheimer theory of atoms, including an impressionist description of the Thomas–Fermi and Hartree–Fock approximations to the fundamental state of a neutral atom of atomic number Z. The purpose is to illustrate the role that Weyl estimates of trigonometric sums play in the atomic energy oscillations, a phenomenon that has implications for the understanding of the periodic table of chemical elements from the first principles of quantum mechanics. This was an effect discovered by C. Fefferman and Luis Seco [206] on their rigorous treatment of the Dirac–Schwinger correction to the asymptotic formula for the energy of the ground state of large atoms, and that was later analyzed in several publications [200, 201, 207, 209, 223].

It only pretends to suggest further readings to anyone interested seriously in that theory and, in a certain sense, is a counterpart to the mathematical haikus idea of Chapter 1. But it serves to transmit the depth and variety of harmonic analysis: beginning with the elementary geometry of parallelepipeds in Euclidean space and ending with quantum mechanical questions about atomic energy oscillations.

Fortunately, there are outstanding treatises of classical Fourier analysis such as Zygmund's "*Trigonometric series*" [17], Stein's "*Singular integrals and differentiability properties of functions*" [12], or "*Introduction to Fourier analysis on Euclidean spaces*" [14], by Stein and Guido Weiss, and I assume that the reader has an acquaintance with their contents. The main purpose of this book is to present several results which go beyond the scope of those remarkable monographs, a pretension which is more ambitious than what it may seem. A more detailed list of helpful references appears at the beginning of the bibliography.

Some personal remarks

On a certain occasion, I attended a talk where the speaker wrote on the blackboard the sentence "analyst = hunter", which he explained in the following terms: "An analyst, like a hunter, expends most of his time at headquarters improving his tools. Sometimes he goes out hunting, but quickly comes back home and continue sharpening his knives".

Age is not a particularly interesting subject, but usually helps to acquire a certain perspective. As a graduate student at the University of Chicago in the early seventies of the past century, I was a privileged witness of the successful impact that the singular integrals theory developed by the school of Alberto Calderón and Antoni Zygmund had over linear partial differential equations, solving and throwing new light over basic questions like uniqueness for the Cauchy problem of hyperbolic equations; elliptic boundary value problems; characterization of local solvability or the index theorem, etc. Following the indications of my thesis advisor, I got acquainted with the original sources, namely the publications of Calderón–Zygmund, Kohn–Nirenberg, Beals–Fefferman, Atiyah–Singer, and many others, which led me to organize the material in a graduate course delivered at the Universidad Complutense of Madrid in the late seventies, and whose lecture notes were published there under the title "Pseudodifferential operators and applications". Developing the theory of pseudodifferential and Fourier integral operators, and hunting their applications to linear PDEs was, during those years, the main stream of activity in Fourier analysis. That activity originated also the publication of surveys and monographs, which culminated in the *opus monumentalis* [8] of Lars Hörmander which, I think, closed the subject for several decades.

The mainstream changed direction paying now attention to non-linear problems. The seminal work of E. de Giorgi and J. Nash of the late fifties was a kind of Rosetta stone allowing the understanding of elliptic and parabolic non-linear equations, like the one satisfied by minimal surfaces, and opening ways to treat free boundary problems and phase transitions, where the theory of singular integrals was again extremely useful. Then we have also the contributions of J. Leray, and many others, to the equations of fluid mechanics and here, once more, the tools sharpened by Fourier analysts became crucial.

Roughly speaking, we may consider them to be the main hunting areas of harmonic analysts during those years. There were, however, other fields of important activity, like the rigorous mathematical foundation of quantum mechanics by J. von Neumann and H. Weyl, or the outstanding contribution of Hardy–Littlewood–Ramanujan to number theory. But there, and specially in number theory, as H. Iwaniec pointed out, the nature of the kernels involved in the needed estimates deviated from the Calderón–Zygmund paradigm.

Together with several of my distinguished collaborators, I have worked and published articles in those mainstream areas. But I have been interested also in other questions of a more collateral character, such as the mentioned Zygmund's conjecture about differentiation of integrals; Kakeya sets and Bochner–Riesz summability; generalized Poisson summation formulas; or the fractal dimension of Riemann's graphs.

This book is mainly about those collateral subjects, and has its origin in a graduate course given at the Universidad de Murcia during the winter quarter of 2020. The Covid pandemic did not allow the continuation of that course in subsequent years, but although I still do not understand well my reasons to complete the notes, I have tried to present in an orderly manner my thoughts of many years, with the pretension, perhaps exaggerated, that some readers may find inspiration for future research.

Acknowledgements

It is with great pleasure that I extend my gratitude to all my distinguished research collaborators, whose contributions enriched the content of this book. I offer my special thanks to Pablo Fernández Gallardo and Fernando Chamizo for their invaluable insights and suggestions, which played a pivotal role in enhancing the quality of the presentation and coordination of the different chapters.

I would like also to express my appreciation to the Universidad de Murcia, where the graduate course that laid the foundations for this book was conducted, and to the Severo Ochoa Project of the Instituto de Ciencias Matemáticas (ICMAT), for the generous financial assistance.

Lastly, I owe a debt of gratitude to Alfonso Núñez and Irene Quintero for their exceptional professional skills, transforming my original hand-written manuscript into a polished LaTeX format, making it accessible to a wider audience.

Madrid, August 2024 *Antonio Córdoba*

Notation and terminology

- χ_A is the standard notation for the indicator function of the set A. To simplify notation, and when there is no possibility of confusion, it will be simplified writing $A(x) = \chi_A(x)$:

$$A(x) = \chi_A(x) = \begin{cases} 1 & \text{if } x \in A, \\ 0 & \text{if } x \notin A. \end{cases}$$

- Given two real functions f and g defined in a domain D, the symbol $f \lesssim g$ means that there exists a finite positive constant C such that $f(x) \leq Cg(x)$ for every $x \in D$. Similarly, $f \gtrsim g$ means that $f(x) \geqslant Cg(x)$ for $x \in D$.
  With $f \simeq g$, we mean that $f \lesssim g$ and $g \gtrsim f$.
  The symbols $\ll$, $\gg$ are equal, respectively, to $\lesssim$ and $\gtrsim$. They were introduced by I. Vinogradov and are used, mainly, in number theory texts, while $\lesssim$, $\gtrsim$ and $\simeq$ are usual notation in harmonic analysis.
- $[x]$ is the integer part of x; $\{x\}$ is the fractional part of x.
- J_k denotes the Bessel function of the first kind and order k.
- Unless the contrary is explicitly written, $\mu(A)$ is the Lebesgue measure of the set $A \subset \mathbb{R}^n$. On certain occasions, the notation $|A| = \mu(A)$ will also be used.
- We write $⨍_R f$ to denote mean value:

$$⨍_R f = \frac{1}{\mu(R)} \int_R f(x)\, \mathrm{d}\mu(x).$$

- Fourier transform:

$$\widehat{f}(\xi) = \int_{\mathbb{R}^n} e^{-2\pi i x \cdot \xi} f(x)\, \mathrm{d}x.$$

- Mellin transform:

$$Mf(s) = \int_0^\infty t^{-is} f(t) \frac{\mathrm{d}t}{t}.$$

- Convolution:

$$f * g(x) = \int_{\mathbb{R}^n} f(x-y)\, g(y)\, \mathrm{d}y.$$

  The symbol $f *_i g$ means convolution on the i-th variable only.
- $M(f)$ denotes the Hardy–Littlewood maximal function.
- $M_s(f)$ denotes the strong maximal function.
- $M_N(f)$ denotes the Kakeya maximal function (rectangles of eccentricity N).
- Principal value integral:

$$\text{p.v.} \int_{\mathbb{R}^n} g(x)\, \mathrm{d}x = \lim_{\varepsilon \to 0^+} \int_{|x| \geq \varepsilon} g(x)\, \mathrm{d}x.$$

- Hilbert transform in the direction v:

$$H_v f(x) = \lim_{\varepsilon \to 0^+} \int_{|t| \geqslant \varepsilon} \frac{f(x - vt)}{t} \, \mathrm{d}t.$$

- $\mathrm{d}\sigma(\bar{y})$ is the Lebesgue measure in the unit sphere S^{n-1} of $\mathbb{R}^n$.
- $L^p_{\mathrm{rad}} L^2_{\mathrm{ang}}(\mathbb{R}^n)$ is space of functions f satisfying

$$\left[\int_0^\infty \left(\int_{S^{n-1}} |f(r\bar{y})|^2 \, \mathrm{d}\sigma(\bar{y}) \right)^{p/2} r^{n-1} \, \mathrm{d}r \right]^{1/p} < \infty.$$

Contents

Chapter 1
Seven mathematical haikus

In the moonlight,
Color and scent of wisteria
Seems far away.

(Yosa Buson)

Often, in Mathematics, we have to work very hard to prove theorems and, being lucky, one encounters complicated proofs which require also a lot of efforts from their possible readers. It is a rare success to find a clever and short proof of a remarkable mathematical statement. Let us call mathematical haikus those precious gifts from the Muses, of which most mathematicians surely treasure several examples. The following are some of my favorites.

1.1 A geometric proof of Fritz John's lemma on convex sets

Theorem 1.1. *Let $K \subset \mathbb{R}^n$ be a compact and convex set with non-empty interior, and let* E *be an ellipsoid of minimal volume containing K. Then we have the inclusion $\frac{1}{n}E \subset K$ (where $\frac{1}{n}E$ denotes the ellipsoid obtained from E after a contraction with respect to its center of ratio $1/n$).*

An interesting example is the $(n+1)$-simplex $\sum(P_0, P_1, \ldots, P_n)$ of vertices $(P_0, P_1, \ldots, P_n)$, where the vectors $P_1 - P_0, \ldots, P_n - P_0$ are linearly independent:

$$\sum(P_0, P_1, \ldots, P_n) = \left\{P \in \mathbb{R}^n \,\middle|\, P = \sum_{j=0}^{n} t_j P_j, \text{ where } 0 \le t_0, t_1, \ldots, t_n \le 1, \sum_{j=0}^{n} t_j = 1\right\}.$$

The simplex will be called equilateral when $\operatorname{dist}(P_j, P_k)$ is the same for all pairs of different vertices.

Among all the $(n+1)$-simplices inscribed on a sphere in $\mathbb{R}^n$, one with biggest volume must be equilateral, as an straightforward geometrical argument would show. It follows then that given any $(n+1)$-simplex T, there is unique ellipsoid E of minimal volume containing T: this E is obtained as the image of the circumscribed sphere to an equilateral $(n+1)$-simplex T', by the affine transformation which maps T' into T.

A. Córdoba, *Suprematism in Harmonic Analysis*, Progress in Mathematics 356,
https://doi.org/10.1007/978-3-031-73737-4_1

This fact has a dual statement: among all the ellipsoids containing an equilateral $(n + 1)$-simplex, the one with minimum volume is the circumscribed sphere.

Proof. Let us observe first that the existence of an ellipsoid of minimal volume containing K follows easily by a compactness argument, and that the contraction factor $1/n$ is sharp, as the case of the equilateral $(n + 1)$-simplex shows.

We shall proceed by contradiction. Let us assume that there exists a compact and convex set K such that $\frac{1}{n}E$ is not contained in K. Since affine transformations preserve the class of compact, bounded and convex sets, as well as ellipsoids, there is no loss of generality if we assume that E is a sphere. Let P be a point inside $(\frac{1}{n}E) \cap (\mathbb{R}^n - K)$, and denote by H a hyperplane passing throughout P without intersecting K, i.e., K is contained in one of the half-spaces, say H^+, bounded by H (see Figure 1.1, where K is contained in the dashed part of E).

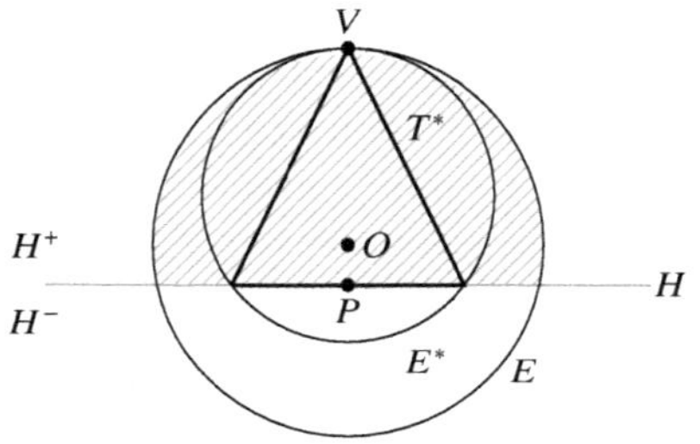

Fig. 1.1: Fritz John's lemma.

Let us draw an equilateral $(n + 1)$-simplex T^* having a vertex at V (placed where the boundary of the dashed region, $E \cap H^+$, intersects the diameter of E throughout P) and so that the remaining n vertices are located on an $(n - 1)$-sphere S^* inside the hyperplane H and centered at P: $S^* = H \cap E^*$, where E^* is the sphere circumscribed to T^*. Let A be the affine transformation which keeps fixed the line OP and maps S^* into $S = E \cap H$. Then $A(E^*)$ is the ellipsoid of minimum volume containing $A(T^*)$. In particular, we have the strict inequality $\mathrm{vol}(E) > \mathrm{vol}(A(E^*))$. But $A(E^*) \supseteq E \cap H^+ \supseteq K$, contradicting the choice of E as an ellipsoid of minimum volume containing K. □

The original proof of Theorem 1.1 given by Fritz John [35] was different and much more complicated. The lemma plays an important role in several theories. For example, it is the adequate tool to reduce maximal functions associated to convex sets to the particular case of those taking averages over parallelepipeds. As an undergraduate student in a real variable course given by M. de Guzmán, circa 1969, I collaborated with my classmate, A. Gallego, answering our teacher request to find an elementary, geometric proof of this interesting lemma. Our proof was presented as a short communication in the meeting of the Spanish Mathematical Society which took place in Murcia (Spain) in the winter of 1970, and it was later recollected in [29, 32]. I believe that nowadays it is just considered a folk result in convex analysis, which has been used, among others, by L. Caffarelli in his work about the Monge–Ampère equations [21], and by C. Fefferman [33] in his Whitney extension results.

1.2 An elementary evaluation of $\zeta(2)$

The famous identity $\sum_{n=1}^{\infty} 1/n^2 = \pi^2/6$ was proved originally by L. Euler using the infinite product expansion of $\sin(z)/z$, but it can be obtained also as a consequence of the Bessel identity applied to the Fourier coefficients of the function $y = x$, $0 < x < 1$. Nowadays one can find in the literature several other proofs which reduce the evaluation of $\zeta(2)$ to a double integral in appropriated plane domains; in the following we present the details of one of them.

Theorem 1.2. *We have*

$$\sum_{n=1}^{\infty} \frac{1}{n^2} = \frac{\pi^2}{6}.$$

Proof. Let us write

$$\zeta(2) = \sum_{n=1}^{\infty} \frac{1}{(2n)^2} + \sum_{n=0}^{\infty} \frac{1}{(2n+1)^2}.$$

That is,

$$\zeta(2) = \frac{4}{3} \sum_{n=0}^{\infty} \frac{1}{(2n+1)^2} = \frac{1}{3} \sum_{n=0}^{\infty} \left(\int_{-1}^{+1} x^{2n}\, \mathrm{d}x \right)^2 = \frac{1}{3} \sum_{n=0}^{\infty} \int_{-1}^{+1} \int_{-1}^{+1} (x^2y^2)^n\, \mathrm{d}x\, \mathrm{d}y$$
$$= \frac{1}{3} \int_{-1}^{+1} \int_{-1}^{+1} \frac{\mathrm{d}x\, \mathrm{d}y}{1 - x^2y^2}.$$

The change of variables

$$\left.\begin{aligned} x = \tanh(s) &= \frac{\exp(s) - \exp(-s)}{\exp(s) + \exp(-s)} \\ y = \tanh(t) &= \frac{\exp(t) - \exp(-t)}{\exp(t) + \exp(-t)} \end{aligned}\right\} \quad -\infty < s, t < +\infty,$$

yields

$$\zeta(2) = \frac{1}{3} \int_{-\infty}^{\infty} \int_{-\infty}^{\infty} \frac{\mathrm{d}s\, \mathrm{d}t}{\cosh(s)^2 \cosh(t)^2 - \sinh(s)^2 \sinh(t)^2}.$$

But since the hyperbolic functions

$$\cosh(s) = \frac{\exp(s) + \exp(-s)}{2} \quad \text{and} \quad \sinh(s) = \frac{\exp(s) - \exp(-s)}{2}$$

satisfy the identities

$$\cosh(s)\cosh(t) - \sinh(s)\sinh(t) = \cosh(s-t),$$
$$\cosh(s)\cosh(t) + \sinh(s)\sinh(t) = \cosh(s+t),$$

we get

$$\zeta(2) = \frac{1}{3}\int_{-\infty}^{\infty}\int_{-\infty}^{\infty} \frac{\mathrm{d}s\,\mathrm{d}t}{\cosh(s-t)\cosh(s+t)}.$$

Then, the change of variables $u = s - t$, $v = s + t$ gives us

$$\zeta(2) = \frac{1}{6}\int_{-\infty}^{\infty}\int_{-\infty}^{\infty} \Big(\frac{\mathrm{d}u\,\mathrm{d}v}{\cosh(u)\cosh(v)}\Big) = \frac{2}{3}\Big(\int_{0}^{\infty}\frac{\mathrm{d}z}{1+z^2}\Big)^2 = \frac{\pi^2}{6},$$

where we have used now the change $x = \log z$, $z > 0$. □

Remark 1.3. As it has been mentioned above, there are many different proofs of this famous formula. The original one, given by Euler, has the advantage that also yields the values of ζ at all the even positive integers $2k$, but the one presented here ranks among the more direct and simpler of all. The idea of the change of variables associated to the hyperbolic tangent was a curious case of serendipity, because I was writing an article about phase transitions in the Ginzburg–Landau model when, at the same time, was asked by the editors of the Spanish Mathematical Society journal, La Gaceta de la Real Sociedad Matemática Española, to contribute with an expository paper [26] about my 1979 unpublished version of R. Apery's proof of the irrationality of $\zeta(3)$. So I was confronted with the evaluation of $\zeta(2)$ and $\zeta(3)$ at the same time that the graph of $y = \tanh(x/\varepsilon)$ was present in my other work, as an example of phase transition between the states $+1$ and -1.

1.3 The irrationality of $\zeta(3)$

The formula

$$\zeta(3) = \sum_{n=1}^{\infty}\frac{1}{n^3} = -\frac{1}{2}\int_0^1\int_0^1 \frac{\log(st)}{1-st}\,\mathrm{d}s\,\mathrm{d}t$$

follows from the well-known identity

$$\int_0^1\int_0^1 \frac{s^{\alpha+\gamma}\,t^{\alpha+\gamma}}{1-st}\,\mathrm{d}s\,\mathrm{d}t = \sum_{n=1}^{\infty}\frac{1}{(\alpha+\gamma+n)^2}$$

taking derivatives with respect to γ and evaluating at $\gamma = 0$:

$$\int_0^1\int_0^1 \frac{s^{\alpha}t^{\alpha}\log(st)}{1-st}\,\mathrm{d}s\,\mathrm{d}t = -2\sum_{n=1}^{\infty}\frac{1}{(\alpha+n)^3}.$$

Therefore, if α is a non-negative integer, we get

$$\int_0^1\int_0^1 \frac{s^\alpha t^\alpha \log(st)}{1-st}\,\mathrm{d}s\,\mathrm{d}t = -2\Big\{\zeta(3) - \sum_{n=1}^{\alpha}\frac{1}{n^3}\Big\},\quad \text{if } \alpha \geq 1,$$

and

$$\int_0^1\int_0^1 \frac{\log(st)}{1-st}\,\mathrm{d}s\,\mathrm{d}t = -2\zeta(3),\quad \text{if } \alpha = 0.$$

Similarly, if $\alpha > \beta$ are integers, we have

$$\int_0^1\int_0^1 \frac{s^{\alpha+\gamma}t^{\beta+\gamma}}{1-st}\,\mathrm{d}s\,\mathrm{d}t = \frac{1}{\alpha-\beta}\sum_{n=1}^{\infty}\Big\{\frac{1}{n+\beta+\gamma} - \frac{1}{n+\alpha+\gamma}\Big\},$$

and taking derivatives with respect to γ and evaluating at $\gamma = 0$,

$$\int_0^1\int_0^1 \frac{s^\alpha t^\beta \log(st)}{1-st}\,\mathrm{d}s\,\mathrm{d}t = -\frac{1}{\alpha-\beta}\sum_{n=\beta+1}^{\alpha}\frac{1}{n^2}.$$

Therefore, if $Q(s,t)$ is a polynomial with integer coefficients and degree n on each variable separately, we have

$$\int_0^1\int_0^1 \frac{Q(s,t)\log(st)}{1-st}\,\mathrm{d}s\,\mathrm{d}t = A_n\,\zeta(3) + B_n,$$

where A_n is an integer and B_n is a rational number whose denominator is a divisor of d_n^3, where

$$d_n = \operatorname{lcm}(1, 2, \ldots, n).$$

The objective, in order to prove the irrationality of $\zeta(3)$, is to find a sequence of polynomials $P_n(s,t)$ of degree $\leq n$ on each variable and such that the integrals

$$\int_0^1\int_0^1 \frac{P_n(s,t)\log(s\cdot t)}{1-st}\,\mathrm{d}s\,\mathrm{d}t$$

are non-zero but very small. It turns out that a good choice is

$$P_n(s,t) = P_n(s)P_n(t),$$

where

$$P_n(s) = \frac{1}{n!}\left(\frac{d}{ds}\right)^n [s^n(1-s)^n]$$

is a Legendre polynomial.

To compute

$$\int_0^1\int_0^1 \frac{P_n(s)P_n(t)\log(st)}{1-st}\,\mathrm{d}s\,\mathrm{d}t,$$

it is convenient to use the formula

$$\frac{\log(st)}{1-st} = \int_0^1 \frac{\mathrm{d}u}{1-(1-st)u}.$$

We obtain

$$\begin{aligned}\int_0^1\int_0^1 \frac{P_n(s)P_n(t)\log(st)}{1-st}\,\mathrm{d}s\,\mathrm{d}t &= \int_0^1\int_0^1\int_0^1 \frac{P_n(s)P_n(t)}{1-(1-st)u}\,\mathrm{d}s\,\mathrm{d}t\,\mathrm{d}u \\ &= \int_0^1\int_0^1\int_0^1 \frac{s^n t^n \cdot u^n (1-s)^n P_n(t)}{[1-(1-st)u]^{n+1}}\,\mathrm{d}s\,\mathrm{d}t\,\mathrm{d}u\end{aligned}$$

after integration by parts with respect to the s-variable. Then the change of variables

$$u = \frac{1-v}{1-(1-st)v}, \qquad \mathrm{d}u = \frac{-st}{[1-(1-st)v]^2}\,\mathrm{d}v$$

yields

$$\int_0^1\int_0^1\int_0^1 \frac{(1-s)^n(1-v)^n P_n(t)}{1-(1-st)v}\,\mathrm{d}s\,\mathrm{d}t\,\mathrm{d}v,$$

and integrating by parts once more respect to t,

$$\begin{aligned} 0 &< \int_0^1\int_0^1 \frac{P_n(s)P_n(t)\log(st)}{1-st}\,\mathrm{d}s\,\mathrm{d}t \\ &= \int_0^1\int_0^1\int_0^1 \frac{s^n(1-s)^n t^n(1-t)^n v^n(1-v)^n}{[1-(1-st)v]^{n+1}}\,\mathrm{d}s\,\mathrm{d}t\,\mathrm{d}v \\ &\le C\Big[\max_{0\le s,t,v\le 1} \frac{s(1-s)t(1-t)v(1-v)}{1-(1-st)v}\Big]^n = C\,(\sqrt{2}-1)^{4n}.\end{aligned}$$

But

$$\begin{aligned} d_n = \operatorname{lcm}(1,2,\ldots,n) &= \prod_{p\text{ prime}\le n} p^{[\log n/\log p]} \le \prod_{p\le n} p^{\log n/\log p} = n^{\pi(n)} \\ &\le n^{(1+\varepsilon)n/\log n} = e^{n(1+\varepsilon)}\end{aligned}$$

for every $\varepsilon > 0$, where we have used the prime number theorem to estimate

$$\pi(n) = \operatorname{card}\{p \text{ prime } \le n\}.$$

Theorem 1.4. *$\zeta(3)$ is irrational.*

Proof. Since $(\sqrt{2}-1)^4\,e^3 < 1$, the construction above yields a sequence of approximations of $\zeta(3)$ by rational numbers E_n/D_n so that

$$0 < \Big|\zeta(3) - \frac{E_n}{D_n}\Big| = o\Big(\frac{1}{D_n}\Big). \qquad \square$$

The first proof of the irrationality of $\zeta(3)$ was obtained by R. Apéry [18], and it was presented at the ICM celebrated in Helsinki the year 1978. The proof given above was inspired on the work of C.L. Siegel, which I learnt on a seminar organized and directed by E. Bombieri and B. Dwork at the Institute for Advanced Studies (Princeton) during the fall of that year.

Siegel [39] considered the following problem:

Find two polynomials $A(x)$ and $B(x)$ of degree n with integer coefficients so that the difference $e^x - A(x)/B(x)$ vanishes at 0 of order $n+1$.

Such condition implies that

$$B(x)\, e^x - A(x) = R(x) = cx^{2n+1} + \cdots ,$$

where $R(x)$ is a power series starting with a term of order $2n+1$.

Writing A and B with undetermined coefficients, and equalizing the terms of order $0, \ldots, 2n$, we get $2n+1$ linear homogeneous equations for the $2n+2$ coefficients of A and B. That shows that the system has a non-trivial solution with the coefficient $c \neq 0$. Taking successive derivatives, we obtain

$$D^{n+1}R = D^{n+1}(Be^x) = e^x(1+D)^{n+1}B = c_0\, x^n + \cdots ,$$

where $c_0 = (2n+1)\cdot(2n)\cdots(n+1)c$. Therefore,

$$(1+D)^{n+1}B = e^{-x}\,(c_0\, x^n + \cdots),$$

and since B is a polynomial of degree n, we get

$$(1+D)^{n+1}B = c_0\, x^n.$$

Similarly,

$$(-1+D)^{n+1}A = c_0\, x^n,$$

and we have freedom to choose c so that $c_0 = 1$. That is,

$$B = (1+D)^{-n-1}x^n, \quad A = (-1+D)^{-n-1}x^n \quad \text{and} \quad D^{n+1}R = x^n e^x.$$

Next, let us consider

$$J^{k+1}\phi(x) = \int_0^x \frac{(x-t)^k}{k!}\,\phi(t)\,dt$$

Note that

$$D^k J^k\phi = \phi \quad \text{and} \quad J^k D^k\phi = \phi$$

so long as $\phi(0) = D\phi(0) = \cdots = D^{k-1}\phi(0)$. Notice also that

$$D^k(e^{\lambda x}\phi) = e^{\lambda x}(\lambda + D)^k\phi.$$

These formulas give us the operators $(\pm 1 + D)^{-k}$, and prove that if Q is a polynomial with integer coefficients, then $(\pm 1+D)^{-k} Q$ also has integer coefficients. In particular,

$$R(x) = \frac{x^{n+1}}{n!} \int_0^1 (1-t)^n\, t^n\, e^{xt}\, dt = \frac{x^{2n+1}}{n!} \int_0^1 P_n(t)\, e^{xt}\, dt,$$

where

$$P_n(x) = \frac{1}{n!} \frac{d^n}{\mathrm{d}x^n} \{x^n (1-x)^n\}$$

is the Legendre polynomial of order n.

Siegel observes that

$$0 < |R(x)| \le \frac{|x|^{2n+1}}{n!}\, e^{|x|},$$

implying that

$$0 < |B_n(m)\, e^m - A_n(m)| \le \frac{m^{2n+1} e^m}{n!},$$

from which the irrationality of any integer powers e^m follows easily. These estimates inspired the proof of Apery's theorem described above.

During several months, I tried unsuccessfully to extend the method to other values of $\zeta(2m+1)$; then I learned that, independently, F. Beukers had sent to publish a manuscript [19] with a proof very similar or almost identical and I decided not to publish the proof although, at that time, I had delivered public lectures about it at the Universities of Wisconsin (Madison) and Princeton. The method, nevertheless, has room to estimate many linear combinations, with rational coefficients, of different $\zeta(2m+1)$. This was carried out by T. Rivoal [38], who showed that the vector space over the rationals spanned by $1, \zeta(3), \ldots, \zeta(2m+1)$ has dimension $\ge \frac{1}{2}\log(m)$. In particular, this implies that the sequence $\{\zeta(2m+1)\}$ contains infinitely many irrationals.

1.4 A pointwise inequality for fractionary powers of the Laplacian

The Fourier transform

$$\widehat{f}(\xi) = \int_{\mathbb{R}^n} e^{-ix\cdot\xi} f(x)\, \mathrm{d}x,$$

applied to the Laplacian $\Delta u = \sum_{i=1}^n u_{x_i x_i}$ in $\mathbb{R}^n$, allows the representation

$$\widehat{\Delta u}(\xi) = -|\xi|^2\, \widehat{u}(\xi)$$

for rapidly decreasing smooth functions u. Then one can define fractional powers of $-\Delta$ by the formula

$$\widehat{\Lambda^\alpha u}(\xi) = |\xi|^\alpha\, \widehat{u}(\xi).$$

When $0 < \alpha < 2$, we have the representation

$$\Lambda^\alpha u(x) = C_{\alpha,n} \int_{\mathbb{R}^n} \frac{u(x) - u(y)}{|x-y|^{n+\alpha}}\, dy,$$

where

$$C_{\alpha,n} = 2^{\alpha-2}(\alpha^2 + \alpha(n-2))\,\Gamma\Big(\frac{n-2+\alpha}{2}\Big)\Gamma\Big(1-\frac{\alpha}{2}\Big) > 0,$$

from which is clear the non-local character of the operator Λ^α. However, given a C^1 convex function ϕ, the following pointwise inequality is satisfied.

Theorem 1.5. *Let ϕ be a C^1 convex function. Then*

$$\Lambda^\alpha(\phi(f))(x) \le \phi'(f)(x) \cdot \Lambda^\alpha f(x).$$

Proof. We have that

$$\begin{aligned}\Lambda^\alpha(\phi(f))(x) &= C_{\alpha,n} \int_{\mathbb{R}^n} \frac{\phi(f)(x) - \phi(f)(y)}{|x-y|^{n+\alpha}}\, dy \\ &\le C_{\alpha,n} \int_{\mathbb{R}^n} \frac{\phi'(f)(x)(f(x) - f(y))}{|x-y|^{n+\alpha}}\, dy = \phi'(f)(x)\, \Lambda^\alpha f(x).\end{aligned}$$ □

Independently of the simplicity of its proof [27, 30], its validity is somehow surprising given the non-local character of the involved operators Λ^α. It has also several interesting applications to non-lineal and non-local evolution problems. Let us describe briefly one of them.

The analysis of water waves, or the Muskat and Hele–Shaw flows inside a porous medium, ask for the evolution of the free boundary between two fluids or a fluid and the vacuum. In both cases, there is a curve (in $\mathbb{R}^2$) or a surface (in $\mathbb{R}^3$) whose time evolution has to be controlled from the fundamental laws (Bernoulli's law for water waves, D'Arcy's law for porous media), and one can assign a velocity field to the moving boundary, obtaining a closed system of somehow complicated non-linear differential equations [28].

However, it was early discovered that with such generality the problem is not well posed. In the porous media case, first Rayleigh and later Taylor, observed that the linearized problem was unstable if a certain quantity σ becomes negative. Therefore, to have a consistent theory, one needs also control the evolution of σ assuming its positivity at time $t = 0$. Choosing an adequate (isothermal) parametrization of the free boundary $\psi : \mathbb{R} \to \mathbb{R}^2$ (or $\psi : \mathbb{R}^2 \to \mathbb{R}^3$), one is led to estimate the evolution of the Sobolev's norms: $\frac{d}{dt}\|\Lambda^\alpha \psi\|^2_{L^2}$.

It happens that, after certain algebraic manipulations, the task is reduced to the following situation:

$$\frac{d}{dt}\|\Lambda^\alpha\psi\|^2_{L^2} \le -\sum_j \sum_{|\alpha|=k} \int \sigma(x,t)\, D^\alpha_x\psi(x,t)\, \Lambda D^\alpha_x \psi(x,t)\, dx + O\big(\|\Lambda^k\psi\|^p_{L^2}\big),$$

for some $p > 0$.

Then the particular case

$$u(x)\Lambda u(x) \geq \frac{1}{2}\,\Lambda u^2(x)$$

of our inequality, together with the positivity of the Rayleigh–Taylor term σ, allow us to dispose of the dangerous term in the inequality above and conclude well posedness.

1.5 Generalized Laplacians and minimal surfaces

Let f be a locally integrable function defined on an open domain of the Euclidean space $\mathbb{R}^n$. Its generalized Laplacian $\Delta^\star f$ is given by the following limit (provided it exists):

$$\Delta^\star f(x) := \lim_{r\to 0^+} \frac{2(n+2)}{r^2} \fint_{B_r^{(n)}(x)} (f(y) - f(x))\,\mathrm{d}y.$$

Here the symbol $\fint_{B_r^n(x)}$ denotes the average on the ball $B_r^{(n)}(x) \subset \mathbb{R}^n$ centered at x and with radius r, $\mathrm{d}y = \mathrm{d}\mu_n(y)$ is the Lebesgue measure in $\mathbb{R}^n$, and $\mu_n(A) = |A| = \mathrm{vol}(A)$ is the measure of the measurable set A.

It is well known that a continuous generalized harmonic function (i.e., $\Delta^\star f(x) = 0$ for every x) must be smooth and, therefore, harmonic in the ordinary sense: $\Delta f = 0$.

There are, however, discontinuous functions which are harmonic in the generalized sense. An example is given by

$$f(x', x_n) = \begin{cases} \alpha_+ & \text{if } x_n > 0, \\ \frac{1}{2}\,(\alpha_+ + \alpha_-), & \text{if } x_n = 0, \\ \alpha_- & \text{if } x_n < 0, \end{cases}$$

where $x = (x', x_n)$ denotes a point in Euclidean space $\mathbb{R}^n$, $n \geq 2$, which is clearly discontinuous when $\alpha_+ \neq \alpha_-$ and, nevertheless, satisfies $\Delta^\star f(x) = 0$ everywhere.

Suppose that S is a smooth surface separating the domain Ω into two non-empty components:

$$\Omega = \Omega^+ \cup \Omega^- \cup S, \quad \text{so that} \quad S = \partial\Omega^+ \cap \Omega = \partial\Omega^- \cap \Omega.$$

Let the function f_S be equal to α_+ inside Ω^+, equal to α_- inside Ω^-, and equal to $\frac{1}{2}\,(\alpha_+ + \alpha_-)$ in S.

Theorem 1.6. *The function f_S (with $\alpha_+ \neq \alpha_-$) is generalized harmonic if and only if S is minimal.*

In the proof, we will make use of the modern notion of viscosity solution of uniformly elliptic equations. Namely, we will show that if $\Delta^\star f_S = 0$, then, locally, S will be given as the graph of a viscosity solution of the minimal surface equation and, therefore, it has to be smooth. We shall use also several well-known properties of minimal surfaces and elliptic equations; [22] and [34] are appropriate references.

A basic calculus lemma

Let S be a smooth (C^4) hypersurface in $\mathbb{R}^n$ with unit normal vector field $\nu(x)$. Given $x \in S$ and $r > 0$, small enough, the ball $B_r^{(n)}(x)$ is separated by S into two connected components, $S_r^+(x)$ and $S_r^-(x)$, where $S_r^+(x)$ (respectively, $S_r^-(x)$) consists of the points inside $B_r^{(n)}(x)$ which are placed above (respectively, below) S in the given normal direction.

Then we have the following.

Lemma 1.7 *As $r \to 0^+$,*

$$\mathrm{vol}(S_r^+(x)) - \mathrm{vol}(S_r^-(x)) = -c_n\, H_S(x) \cdot r^{n+1} + O(r^{n+3}),$$

where $H_S(x)$ denotes the mean curvature of S at the point x, and $c_n > 0$ is a universal constant that only depends on the dimension n.

Proof. Without loss of generality, we can assume that $x = 0$ is the origin of a coordinate system such that the tangent space of S at x is horizontal, i.e., the normal vector is $\nu(0) = (0, \ldots, 0, 1)$. Hence, near $x = 0$, S is the graph of a smooth (C^4) function $x_n = \varphi(x_1, \ldots, x_{n-1})$ satisfying

$$\varphi(0) = 0 \quad \text{and} \quad \nabla\varphi(0) = 0.$$

Then, inside the cylinder $B_r^{(n-1)}(0) \times \mathbb{R}$, for r small enough, we have the inclusion

$$S \cap (B_r^{(n-1)}(0) \times \mathbb{R}) \subset \{x \in \mathbb{R}^n : |x_n| \leq c_1 r^2\}$$

for a positive constant c_1 depending upon the size of the second derivatives of φ.

An elementary computation shows that the vertical projection of $S \cap B_r^{(n)}(0)$ onto $B_r^{(n-1)}(0)$ must contain the ball

$$B^{(n-1)}_{r-c_2 r^3}(0)$$

for a fixed constant c_2. Let

$$D_r^+ = B_r^{(n)}(0) \cap \Omega^+ \cap \{|x_n| \leq c_1 r^2\} \quad \text{and} \quad D_r^- = B_r^{(n)}(0) \cap \Omega^- \cap \{|x_n| \leq c_1 r^2\}.$$

Then, since S is contained in the strip, by symmetry we get the first equality

$$\begin{aligned} \mathrm{vol}(S_r^+(0)) - \mathrm{vol}(S_r^-(0)) &= \mathrm{vol}(D_r^+) - \mathrm{vol}(D_r^-) \\ &= \int_{B^{(n-1)}_{r-c_2 r^3}(0)} (c_1 r^2 - \varphi(x))\, \mathrm{d}x - \int_{B^{(n-1)}_{r-c_2 r^3}(0)} (c_1 r^2 + \varphi(x))\, \mathrm{d}x + I, \end{aligned}$$

where I is the correction due to the restriction of the domain of integration.

A direct computation of the term I shows that

$$|I| \lesssim r^2 \cdot \left(r^{n-1} - (r - c_2 r^3)^{n-1}\right) \lesssim r^{n+3} = O(r^{n+3}).$$

Hence

$$\mathrm{vol}(S_r^+(0)) - \mathrm{vol}(S_r^-(0)) = -2 \int_{B^{(n-1)}_{r-c_2 r^3}(0)} \varphi(x)\, \mathrm{d}x + O(r^{n+3}).$$

Another computation of the volume of the cylinders gives

$$\left| \int_{B_r^{(n-1)}(0)} \varphi(x)\, \mathrm{d}x - \int_{B^{(n-1)}_{r-c_2 r^3}(0)} \varphi(x)\, \mathrm{d}x \right| \lesssim r^{n+3} = O(r^{n+3}).$$

Then Taylor's expansion yields

$$\fint_{B_r^{(n-1)}(0)} \varphi(y)\, \mathrm{d}y = \frac{1}{2(n+1)} \Delta\varphi(0) \cdot r^2 + O(r^4).$$

This allows us to finish the proof of the lemma, because we know that

$$H_S(x', \varphi(x')) = \frac{1}{n-1} \operatorname{div} \left(\frac{\nabla\varphi}{\sqrt{1 + |\nabla\varphi|^2}} \right)(x'),$$

and since $\nabla\varphi(0) = 0$, we have

$$H_S(0) = \frac{1}{n-1} \Delta\varphi(0).$$

□

End of the proof

First, without loss of generality, one can assume that $\alpha_+ = +1$ and $\alpha_- = -1$. Next, let us observe that one of the two implications of the theorem follows immediately: namely, if S is minimal and C^1 then, by the classical theorem of de Giorgi–Nash, S has to be smooth and we can apply Lemma 1.7 to observe that at any point $x \in S$, we have

$$\begin{aligned} \Delta^\star f(x) &= \lim_{r \to 0^+} \frac{2(n+2)}{r^2} \fint_{B_r^{(n)}(x)} (f(y) - 0)\, \mathrm{d}y \\ &= \lim_{r \to 0^+} \frac{2(n+2)}{|B_r^{(n)}(x)|\, r^2} [\mathrm{vol}(S_r^+(x)) - \mathrm{vol}(S_r^-(x))] = 0, \end{aligned}$$

because of the minimality condition $H_S(x) = 0$.

Note that a similar argument with Lemma 1.7 also works to prove that f_S being generalized harmonic implies that S is minimal. Therefore, to finish the proof we just need to prove the regularity C^4 of S.

To continue the proof, recall now that a real continuous function F defined on $\Omega \times \mathbb{R} \times \mathbb{R}^n \times M^n$, where M^n denotes the vector space of $n \times n$ symmetric matrices, yields an elliptic equation $F(x, u, Du, D^2u) = 0$ if

$$F(x, u, \eta, \delta) \leq F(x, u, \eta, \delta + \sigma)$$

for all $(x, u, \eta, \delta) \in \Omega \times \mathbb{R} \times \mathbb{R}^n \times M^n$ and $\sigma \in M^n$ non-negative.

The elliptic equation is called uniformly elliptic if there exist positive constants λ and Λ satisfying the estimate

$$0 < \lambda \|\sigma\| < F(x, u, \eta, \delta + \sigma) - F(x, u, \eta, \delta) \leq \Lambda \|\sigma\|,$$

where $\|\sigma\|$ denotes the (L^2, L^2)-norm (i.e., $\|\sigma\| = \sup_{\|x\|=1} \|\sigma x\|$ = maximum of the eigenvalues of σ).

Definition 1.8. A continuous function u is called a viscosity subsolution (respectively, supersolution) of $F(x, u, Du, D^2u) = 0$ if, for any quadratic polynomial $\Psi \in C^2(\Omega)$ and local maximum (respectively, local minimum) x_0 of $u - \Psi$, we have

$$F(x_0, u(x_0), D\Psi(x_0), D^2\Psi(x_0)) \geq 0 \quad (\text{respectively, } \leq 0).$$

Finally, u is a viscosity solution if it is both a viscosity subsolution and supersolution.

Reference [22] contains the result about regularity (see Corollary 5.7) and uniqueness (see Corollary 5.4) of viscosity solutions of uniformly elliptic equations, which we shall invoke to conclude the proof.

More precisely, under the hypothesis that

$$\Delta^\star f \equiv 0,$$

let P be a paraboloid tangent from below to our hypersurface $S = \{x_n = \varphi(x_1, \ldots, x_{n-1})\}$ at a point $x = (x_1, \ldots, x_{n-1}, x_n) \in S \cap P$.

Let f_P be its corresponding function defined in the introduction. We have the inequality

$$\fint_{B_r^{(n)}(x)} f_P \geq \fint_{B_r^{(n)}(x)} f_S.$$

On the other hand, Lemma 1.7 applied to the hypersurface P yields

$$\fint_{B_r^{(n)}(x)} f_P = -c_n \, H_P(x) \cdot r^{n+1} + O(r^{n+2}),$$

which together with the hypothesis

$$\lim_{r \to 0^+} \frac{1}{r^2} \fint_{B_r^{(n)}(x)} f_S = 0$$

implies that

$$H_P(x) \leq 0.$$

Similarly, if P is now a paraboloid tangent to S from above at the point x, then we must have $H_P(x) \geq 0$. Therefore, φ is a viscosity solution of the equation

$$\operatorname{div}\left(\frac{\nabla\varphi}{\sqrt{1+|\nabla\varphi|^2}}\right) = 0,$$

whose uniform ellipticity is ensured by the hypothesis that S is of class C^1.

The regularity theory of such solutions [34] allows us to conclude the smoothness of φ (and S) and, therefore, its minimality.

1.6 Lattice points on arcs

The function

$$r(n) = \#\{n = a^2 + b^2 : a, b \in \mathbb{Z}\},$$

that counts the number of representations of an integer n as a sum of two squares, is an important character in number theory.

One can find a well-known formula in terms of the decomposition of n into prime factors. Let

$$n = 2^\nu \prod_{p_j \equiv 1 \pmod 4} p_j^{\alpha_j} \prod_{q_k \equiv 3 \pmod 4} q_k^{\beta_k}.$$

Then we have that

(i) $r(n) = 0$ if one of the β_k is odd,
(ii) $r(n) = 4\prod_j (1+\alpha_j)$ if $\beta_k = 2\beta'_k$ is even for every k,

implying that

$$\limsup_{n\to\infty} \frac{r(n)}{(\log n)^p} = \infty, \quad \text{for every } p > 0, \tag{1.1}$$

$$\limsup_{n\to\infty} \frac{r(n)}{n^\varepsilon} = 0, \quad \text{for every } \varepsilon > 0. \tag{1.2}$$

Since $r(n)$ represents the number of lattice points $(a, b) \in \mathbb{Z}^2$ on the circle centered at the origin and radius $R = \sqrt{n}$, then (1.1) shows that one can find as many lattice points as we wish on such a circle for certain values of R big enough.

On the other hand, if $v_j = (a_j, b_j)$, $j = 1, 2, 3$, are three lattice points on the circle of radius $R = \sqrt{n}$, a celebrated theorem attributed to Hero of Alexandria gives the identity

$$\|v_1 - v_2\|\,\|v_2 - v_3\|\,\|v_3 - v_1\| = 4\Delta \cdot R,$$

where Δ is the area of the triangle of vertices $\{v_1, v_2, v_3\}$, which is at least 1/2. Therefore, the arc C containing the three points v_1, v_2, v_3 must satisfy

$$\text{length}\,(C) \geq (2R)^{1/3}.$$

Theorem 1.9. *For every $1/3 \leq \alpha < 1/2$, there exists a finite constant C_α such that on a circle centered at the origin and radius R, any arc of length R^α contains at most C_α lattice points (uniformly in R).*

Proof. (a) It is well known that the properties of $r(n)$ are better uncovered within the ring of Gaussian integers. For an odd prime $p \equiv 1 \pmod 4$, the representation is unique except for the obvious symmetries $p = (\pm a)^2 + (\pm b)^2 = (\pm b)^2 + (\pm a)^2$, and we will adopt the convention of choosing

$$a + bi = \sqrt{p}\, e^{2\pi i \Phi}, \quad 0 < b < a, \quad 0 < \Phi = \frac{1}{2\pi} \arctan \frac{b}{a} < \frac{1}{4}.$$

If

$$n = 2^\nu \prod_{p_j \equiv 1 \,(\mathrm{mod}\, 4)} p_j^{\alpha_j} \prod_{q_k \equiv 3 \,(\mathrm{mod}\, 4)} q_k^{2\beta_k},$$

an elementary argument yields that Gaussian integers corresponding to the $r(n) = 4\prod_j (1 + \alpha_j)$ representations of n as a sum of two squares are given by the formula

$$\sqrt{n}\, e^{2\pi i (\Phi_0 + \sum_j \gamma_j \Phi_j + t/4)}, \tag{1.3}$$

where

- Φ_j is the angle corresponding to the prime p_j,
- γ_j describes the set $\{\gamma \in \mathbb{Z} : |\gamma| \leq \alpha_j, \gamma \equiv \alpha_j (\mathrm{mod}\, 2)\}$,
- $t = 0, 1, 2, 3$,
- $\Phi_0 = 0$ if ν is even,
- $\Phi_0 = 1/8$ if ν is odd.

(b) A direct consequence of (1.3) is that the factors 2^ν and $q_k^{2\beta_k}$, for $q_k \equiv 3 \pmod 4$, do not change the relative position of those points of integer coordinates contained on the circle of radius $R = \sqrt{n}$. Therefore, without loss of generality, one may assume that n has the prime factorization

$$n = \prod_{p_j \equiv 1 \,(\mathrm{mod}\, 4)} p_j^{\alpha_j}.$$

Let us assume that we have, at least, $m + 1$ points placed on an arc of length $\sqrt{2}\, R^\alpha$, and that they corresponds to complex numbers of the form

$$\sqrt{n}\, e^{2\pi i (\sum_j \gamma_j^\sigma \Phi_j + t^\sigma/4)}$$

with

- $\sigma = 1, 2, \ldots, m + 1$,
- $\gamma_j^\sigma \in \{\gamma \in \mathbb{Z} : |\gamma| \leq \alpha_j, \gamma \equiv \alpha_j (\mathrm{mod}\, 2)\}$,
- $t^\sigma \in \{0, 1, 2, 3\}$,
- Φ_j is the angle corresponding to the prime p_j.

For each pair σ, σ', consider

$$\Psi^{\sigma,\sigma'} = \sum_j \Phi_j(\gamma_j^{\sigma} - \gamma_j^{\sigma'}) + \frac{t^{\sigma} - t^{\sigma'}}{4} = 2\Big(\sum_j \Phi_j \frac{\gamma_j^{\sigma} - \gamma_j^{\sigma'}}{2} + \frac{t^{\sigma} - t^{\sigma'}}{8}\Big),$$

and observe that $\gamma_j^{\sigma,\sigma'} = (\gamma_j^{\sigma} - \gamma_j^{\sigma'})/2$ takes only integer values due to the congruence $\gamma_j^{\sigma} \equiv \gamma_j^{\sigma'} \pmod 2$.

There are two cases:

Case (1). $t^{\sigma} \equiv t^{\sigma'} \pmod 2$. Then we have that

$$(t^{\sigma} - t^{\sigma'})/8 = t^{\sigma,\sigma'}/4, \qquad \text{for a certain integer } t^{\sigma,\sigma'}.$$

Case (2). $t^{\sigma} \not\equiv t^{\sigma'} \pmod 2$. Then we have that

$$(t^{\sigma} - t^{\sigma'})/8 = 1/8 + t^{\sigma,\sigma'}/4, \qquad \text{for a certain integer } t^{\sigma,\sigma'}.$$

The angle $\frac{1}{2}\Psi^{\sigma,\sigma'}$ corresponds in the first case to a representation of the integer $\prod_j p_j^{|\gamma_j^{\sigma,\sigma'}|}$ as a sum of two squares, while the second gives $\frac{1}{2}\Psi^{\sigma,\sigma'}$ as the angle of a representation of $2\prod_j p_j^{|\gamma_j^{\sigma,\sigma'}|}$.

If $\|x\|$ denotes the distance to the nearest integer, then we have

$$\sigma \neq \sigma' \quad \Longrightarrow \quad \Big\|\frac{1}{2}\Psi^{\sigma,\sigma'}\Big\| > \frac{1}{2\pi(2\prod_j p_j^{|\gamma_j^{\sigma,\sigma'}|})^{1/2}},$$

where the worst configuration in the inequality above occurs when we have a representation of the form

$$2\prod_j p_j^{|\gamma_j^{\sigma,\sigma'}|} = m^2 + 1.$$

On the other hand, our hypothesis about the location of those $m+1$ lattice points implies

$$\Big\|\frac{1}{2}\Psi^{\sigma,\sigma'}\Big\| \le \frac{\sqrt{2}\,R^{\alpha-1}}{4\pi}.$$

Therefore, for each pair σ, σ' we have

$$R^{\alpha-1} > \prod_j p_j^{-|\gamma_j^{\sigma,\sigma'}|/2}.$$

Since the number of such pairs σ, σ' is $m(m+1)/2$, multiplying together all the inequalities above we get

$$R^{(\alpha-1)m(m+1)/2} > \Big(\prod_{\sigma\neq\sigma'}\prod_j p_j^{|\gamma_j^{\sigma,\sigma'}|}\Big)^{-1/2} = \Big(\prod_j p_j^{\sum_{\sigma\neq\sigma'}|\gamma_j^{\sigma,\sigma'}|}\Big)^{-1/2}.$$

To conclude, let us estimate the maximum of the sum

$$\sum_{\sigma\neq\sigma'} |\gamma_j^{\sigma,\sigma'}| = \frac{1}{2}\sum_{\sigma\neq\sigma'} |\gamma_j^{\sigma} - \gamma_j^{\sigma'}|$$

taking into account that γ_j^σ takes values in the set $\{\gamma \in \mathbb{Z} : |\gamma| \leq \alpha_j, \gamma \equiv \alpha_j (\mathrm{mod}\ 2)\}$, it is easy to see that

- if $m+1$ is even, the maximum is attained when $(m+1)/2$ of the γ_j^σ are equal to α_j and the $(m+1)/2$ remaining are $-\alpha_j$;
- if $m+1$ is odd, then the maximum is attained when $(m+2)/2$ of the γ_j^σ are α_j and the remaining $m/2$ are equal to $-\alpha_j$.

That is,

$$\sum_{\sigma\neq\sigma'} |\gamma_j^{\sigma,\sigma'}| \leq \alpha_j \frac{(m+1)^2-\delta(m)}{4}, \qquad \text{where } \delta(m) = \begin{cases} 1 & \text{if } m \text{ is even,} \\ 0 & \text{if } m \text{ is odd.} \end{cases}$$

Therefore,

$$R^{(\alpha-1)m(m+1)/2} > \left(\prod_j p_j^{\alpha_j((m+1)^2-\delta(m))/4}\right)^{-1/2} = R^{-((m+1)^2-\delta(m))/4},$$

implying

$$\alpha > 1 - \frac{(m+1)^2-\delta(m)}{4} = \frac{1}{2} - \frac{1}{4[m/2]+2},$$

and this allows us to conclude the proof. □

The result was part of a project trying to connect two restriction theorems in harmonic analysis, one due to A. Zygmund for Fourier series in $\mathbb{T}^2$, and the other to C. Fefferman and E. Stein for the Fourier transform in $\mathbb{R}^2$. What happens for $1 > \alpha \geq 1/2$ is an interesting open question, whose answer has aroused a certain amount of interest due to the work of J. Bourgain and Z. Rudnick [20].

There are two other different proofs of Theorem 1.9, obtained respectively by J. Cilleruelo and G. Tenembaum [25] and by D. S. Ramada [37].

1.7 Sumsets and irrationality

In the d-dimensional torus $\mathbb{T}^d$, we define $x = (x_1, \ldots, x_d)$ to be irrational if the set $\{1, x_1, \ldots, x_d\}$ is linearly independent over the rationals $\mathbb{Q}$.

Let us consider the Lebesgue (Haar) measure normalized so that $\mu(\mathbb{T}^d) = 1$.

Theorem 1.10. *If A is a Borel set in $\mathbb{T}^d$ of measure $\mu(A) = 1/2$, then $A - A$ contains every irrational element.*

Proof. Consider the Fourier series of the indicator function of A,

$$\chi_A \sim \sum_{\nu \in \mathbb{Z}^d} \hat{\chi}_A(\nu)\, e^{2\pi i\,\nu\cdot x}, \qquad \text{where} \qquad \hat{\chi}_A(\nu) = \int_{\mathbb{T}^n} \chi_A(x)\, e^{-2\pi i \nu\cdot x}\, dx,$$

and observe that for $-A$ we have $\hat{\chi}_{-A}(\nu) = \overline{\hat{\chi}_A(\nu)}$. Notice also that $\hat{\chi}_A(0) = \mu(A)$.

Bessel's identity yields

$$\mu(A) = \mu(A)^2 + \sum_{\nu\neq 0} |\hat{\chi}_A(\nu)|^2,$$

and therefore,

$$\frac{1}{4} = \sum_{\nu\neq 0} |\hat{\chi}_A(\nu)|^2.$$

The convolution $\chi_A * \chi_{-A}$ is a continuous function whose Fourier series converges uniformly:

$$\widehat{\chi_A * \chi_{-A}}(\nu) = \hat{\chi}_A(\nu)\cdot\hat{\chi}_{-A}(\nu) = \hat{\chi}_A(\nu)\cdot\overline{\hat{\chi}_A(\nu)} = |\hat{\chi}_A(\nu)|^2.$$

Therefore, we can write

$$\chi_A * \chi_{-A}(x) = \int_{\mathbb{T}^d} \chi_A(x-y)\,\chi_{-A}(y)\,\mathrm{d}y = \mu(A)^2 + \sum_{\nu\neq 0} |\hat{\chi}_A(\nu)|^2\, e^{2\pi i\nu\cdot x}$$

and

$$x \notin A - A \quad\Longrightarrow\quad \chi_A * \chi_{-A}(x) = 0 \quad\Longrightarrow\quad \sum_{\nu\neq 0} |\hat{\chi}_A(\nu)|^2 e^{2\pi i\nu x} = -\frac{1}{4},$$

implying that $e^{2\pi i\nu\cdot x} = -1$ for every $\nu \in \mathbb{T}^d$, i.e., x cannot be irrational. □

Clearly, the condition $\mu(A) = 1/2$ cannot be relaxed in the statement above. At this respect, there is a well-known result in the area that can be also considered a mathematical haiku:

Theorem 1.11. *Let $A, B \in \mathbb{T}^d$. Then*

$$\min(1, \mu(A) + \mu(B)) \leq \mu(A+B).$$

In particular, if $\mu(A) + \mu(B) > 1$, we have $A + B = \mathbb{T}^d$.

Proof. We have

$$0 \leq \int_{\mathbb{T}^d} (1-\chi_A(x-y))(1-\chi_B(y)) = \chi_A * \chi_B(x) - \mu(A) - \mu(B) + 1.$$

Hence, when $\mu(A) + \mu(B) > 1$, we have $\chi_A * \chi_B(x) > 0$ for every $x \in \mathbb{T}^d$, and this implies that $x \in A + B$, giving the last part of the theorem.

For the first part, we can assume $\mu(A) + \mu(B) \le 1$. Since

$$\int_{\mathbb{T}^d\times\mathbb{T}^d} \chi_{-x+A}(t)\,\chi_{x+B}(t)\,\mathrm{d}x\,dt = \int_{\mathbb{T}^d\times\mathbb{T}^d} \chi_A(t+x)\,\chi_B(t-x)\,\mathrm{d}x\,dt = \mu(A)\,\mu(B),$$

there exists $x_1 \in \mathbb{T}^d$ such that

$$\mu((-x_1 + A) \cap (x_1 + B)) \le \mu(A)\,\mu(B).$$

Let us define

$$A_1 = (-x_1 + A) \cup (x_1 + B) \quad \text{and} \quad B_1 = (-x_1 + A) \cap (x_1 + B).$$

We have

$$A + B = (-x_1 + A) + (x_1 + B) \supset A_1 + B_1,$$

and by the previous inequality,

$$\mu(B_1) \le \mu(A)\,\mu(B) \le \mu(B)(1 - \mu(B)).$$

On the other hand,

$$\mu(A_1) + \mu(B_1) = \mu(-x_1 + A) + \mu(x_1 + B) = \mu(A) + \mu(B).$$

Iterating this procedure, we find sequences $(A_r)_{r=1}^{\infty}$ and $(B_r)_{r=1}^{\infty}$, with $A_0 = A$ and $B_0 = B$, such that for $r \ge 1$,

$$A_r + B_r \supset A_{r-1} + B_{r-1}, \quad \mu(A_r) + \mu(B_r) = \mu(A_{r-1}) + \mu(B_{r-1})$$

and

$$\mu(B_r) \le \mu(B_{r-1})(1 - \mu(B_{r-1})).$$

This last condition implies $\mu(B_r) \to 0$, and from

$$\mu(A + B) \ge \mu(A_r + B_r) \ge \mu(A_r) = \mu(A) + \mu(B) - \mu(B_r),$$

we conclude the expected formula. □

Chapter 2
All those rectangles

In mathematics, as in life itself, often it is more convenient to integrate, only, the positive part.

Antoni Zygmund

2.1 General facts

Given a measure space $(X, d\mu)$ and a family $\mathcal{B} = \{(E_\alpha; R_\alpha)\}$ of pairs of measurable subsets of finite measure, there is a maximal function operator, acting on the class of locally integrable function, defined by the formula

$$M_{\mathcal{B}} f(x) = \sup_{x \in E_\alpha} \frac{1}{\mu(R_\alpha)} \int_{R_\alpha} |f(y)| \, d\mu(y) \quad \text{when } x \in \bigcup_\alpha E_\alpha,$$

and $M_{\mathcal{B}} f(x) = 0$ otherwise.

In their seminal paper [48] entitled "A maximal theorem with function-theoretic applications" Hardy and Littlewood considered the case where $E_\alpha = R_\alpha$ are the intervals of the real line, that is, the so-called *Hardy–Littlewood maximal function*,

$$(2.1) \qquad Mf(x) = \sup_{x \in I} \frac{1}{|I|} \int_I |f(y)| \, dy,$$

and proved the following inequality: for $\lambda > 0$,

$$(2.2) \quad \mu\{x : Mf(x) > \lambda\} \leq C\lambda^{-1} \|f\|_1, \quad \text{where } C < \infty \text{ is an absolute constant.}$$

This result admits extension to higher dimensions if we take balls or cubes as substitutes for intervals of the real line, and it provides a quantitative version of the so-called *fundamental theorem of calculus* within the theory of Lebesgue, which is usually quoted as Lebesgue's differentiation theorem. The original proof of Hardy and Littlewood was based on the properties of rearrangements of measurable functions, and was expressed in the picturesque language of cricket. But it can also be obtained from elementary geometrical considerations.

A version of the so-called Vitali's covering lemma is the following.

Lemma 2.1. *There exists a positive constant $C_n < \infty$ such that, from any finite family $\{B_\alpha\}$ of balls in Euclidean space $\mathbb{R}^n$, one can select a pairwise disjoint*

A. Córdoba, *Suprematism in Harmonic Analysis*, Progress in Mathematics 356,
https://doi.org/10.1007/978-3-031-73737-4_2

subfamily $\{B_1, \ldots, B_m\}$ *so that*

$$\mu\Big(\bigcup_\alpha B_\alpha\Big) \le C_n\,\mu\Big(\bigcup_{k=1}^m B_k\Big).$$

The works [56] and [46] are good references. The inequality (2.2) is quoted as the Hardy–Litllewood maximal theorem, and it provides a quantitative version of the differentiation theorem:

$$\lim_{B\Rightarrow x} \frac{1}{\mu(B)} \int_B f(y)\,\mathrm{d}\mu(y) = f(x), \qquad \text{a.e. } x \in \mathbb{R}^n,$$

where $f \in L^1(\mathbb{R}^n)$, and the limit is taken over any sequence of balls (or cubes) contracting to x. The elegance and simplicity of the argument is remarkable: a geometric property of balls (the covering lemma) implies a quantitative result (a weak-type estimate for the maximal function), and this, in turn, implies qualitative results, such as the fundamental theorem of calculus, a.e. convergence of Poisson integrals, a.e. convergence to the initial data for Cauchy type problems, etc.

Therefore, it has become a paradigm of Fourier analysis inspiring many results and efforts trying to reproduce it in other interesting situations. There are cases where the sets E_α and R_α are different: an example is the spherical maximal function [54], and another is given by taking averages over arcs of curves [41]. But in this chapter we will consider the case where $E_\alpha = R_\alpha$, which, with the extra hypothesis of convexity, is reduced to that of parallelepipeds in $\mathbb{R}^n$ using Fritz John's Lemma 1.1.

Proof of Vitali's Lemma 2.1. Choose B_1 to be an element of $\{B_\alpha\}$ of maximum diameter. Having selected $B_1, \ldots, B_k$, we take B_{k+1} to be one of the $\{B_\alpha\}$ having maximum diameter among those which are disjoint with the preceding. Then each ball B_α is either one of the chosen, or has a non-empty intersection with some B_j so that $\operatorname{diam}(B_j) \ge \operatorname{diam}(B_\alpha)$. In the last case, we have the inclusion $B_\alpha \subset B_j^*$, where B^* denotes the ball with the same center than B and three times its radius. Then

$$\mu\Big(\bigcup_\alpha B_\alpha\Big) \le \mu\Big(\bigcup_j B_j^*\Big) \le 3^n \sum_j \mu(B_j) = 3^n\,\mu\Big(\bigcup_j B_j\Big). \qquad \square$$

An equivalent formulation is the following.

Lemma 2.2. *Given a family of balls* $\{B_\alpha\}$ *in* $\mathbb{R}^n$ *with bounded diameters, there exists a countable subcollection* $\{B_1, B_2, \ldots\}$ *such that*

(i) *the* B_j *are pairwise disjoint,*
(ii) $\mu(\bigcup_\alpha B_\alpha) \le C_n\,\mu(\bigcup_j B_j)$, *where* C_n *is a universal constant* ($C_n \le 3^n$).

Proof of the Hardy–Littlewood inequality (2.2) *in* $\mathbb{R}^n$. Fix $\lambda > 0$ and consider the set $\{x : Mf(x) > \lambda\}$, which is a union of balls B_α such that

$$\frac{1}{\mu(B_\alpha)} \int_{B_\alpha} f(y)\,dy > \lambda.$$

Then Vitali's lemma yields a sequence of pairwise disjoint balls B_j satisfying

$$\frac{1}{\mu(B_j)} \int_{B_j} f(y)\,dy > \lambda$$

and

$$\begin{aligned}\mu\{x : Mf(x) > \lambda\} &= \mu\Big(\bigcup_\alpha B_\alpha\Big) \le C_n \sum_j \mu(B_j) \\ &\le C_n \frac{1}{\lambda} \sum_j \int_{B_j} f(y)\,dy \le C_n \lambda^{-1} \|f\|_j.\end{aligned}$$ □

Closely related is the following:

Lemma 2.3 (Calderón–Zygmund). *Let $f(x) \ge 0$ be an integrable function in $\mathbb{R}^n$, and let $\lambda > 0$. There exist disjoint cubes Q_k such that*

(1) $\lambda < \dfrac{1}{\mu(Q_k)} \displaystyle\int_{Q_k} f(x)\,\mathrm{d}x \le 2^n \lambda$,

(2) $f(x) \le \lambda$ *a.e. for* $x \in \mathbb{R}^n \setminus \bigcup_k Q_k$,

(3) $\mu\Big(\bigcup_k Q_k\Big) \le \dfrac{1}{\lambda} \displaystyle\int_{\mathbb{R}^n} f(x)\,\mathrm{d}x$.

Proof. As was sketched in the introduction, let us divide the space $\mathbb{R}^n$ into a grid of congruent cubes Q so large that $\frac{1}{\mu(Q)} \int_Q f \le \lambda$ for all of them.

Next, we subdivide each such cube into 2^n congruent subcubes, and select those of them for which $\frac{1}{\mu(Q')} \int_{Q'} f > \lambda$, and for such Q' we stop the bisection process. But for the rest, we continue the process until we first arrive at a cube for which the average of f is bigger than λ, at which we stop. The cubes Q' at which we stop are the sequence Q_k of the statement.

By construction, $\frac{1}{\mu(Q_k)} \int_{Q_k} f > \lambda$. Let Q_k^* be the cube containing Q_k which was bisected to produce Q_k. Then $\mu(Q_R^*) = 2^n \mu(Q_R)$, and since we did not stop at Q_k^*, we have $\frac{1}{\mu(Q_k^*)} \int_{Q_k^*} f \le \lambda$, implying that

$$\frac{1}{\mu(Q_k)} \int_{Q_k} f(x)\,dx \le \frac{\mu(Q_k^*)}{\mu(Q_k)} \frac{1}{\mu(Q_k^*)} \int_{Q_k^*} f \le 2^n \lambda.$$

Finally, since $\mu(Q_R) \le \frac{1}{\lambda} \int_{Q_R} f$, and summing on k, we have

$$\mu\Big(\bigcup_k Q_k\Big) \le \frac{1}{\lambda} \sum_k \int_{Q_k} f(x)\,dx \le \frac{1}{\lambda} \int_{\mathbb{R}^n} f(x)\,dx.$$

Since for each $x \notin \bigcup_k Q_k$ there is a sequence $\{C_k\}$ of cubes whose sides converge to zero and such that they satisfy the estimate $\frac{1}{\mu(C_k)} \int_{C_k} f \le \lambda$, it follows from Lebesgue's differentiation theorem that $f(x) \le \lambda$ a.e. in $\mathbb{R}^n \setminus \bigcup_k Q_k$. □

If we apply Lemma 2.3 to a function $f \geq 0$ supported on a square $Q \subset \mathbb{R}^2$, then for each $\lambda > 0$ we find such a sequence of dyadic subsquares. If now we paint all of the same side with identical color (red, blue, yellow), we will obtain a neoplasticist Mondrian painting. It happens that such collection of Mondrian helps to understand the action of singular integral operators on the function f.

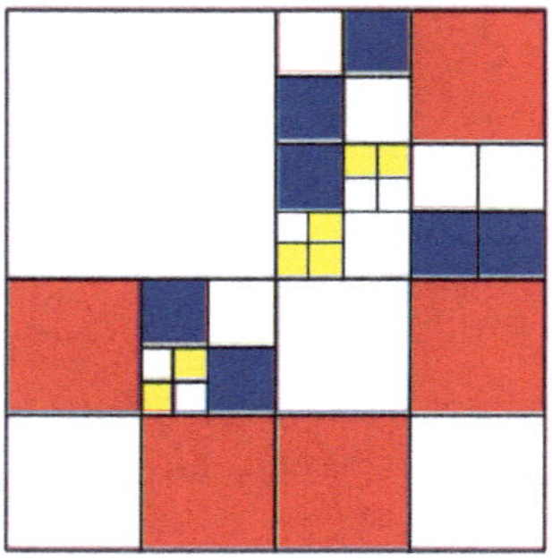

Fig. 2.1: Dyadic Mondrian.

2.2 Vitali's coverings

Let $\mathcal{B}$ be a family of measurable subsets of finite measure.

Definition 2.4. The family $\mathcal{B}$ has the Vitali covering property V_q, $1 \leq q \leq \infty$, if there exists a finite constant C so that for every collection $\{R_i\}_{i\in I} \subset \mathcal{B}$, one can find a subsequence $\{\tilde{R}_k\}_{k\geq 1} \subset \{R_i\}_{i\in I}$ satisfying the following conditions:

(1) $\mu\left(\bigcup_{i\in I} R_i\right) \leq C\mu\left(\bigcup_k \tilde{R}_k\right)$,

(2) $\left\|\sum_k \chi_{\tilde{R}_k}\right\|_q \leq C\mu\left(\bigcup_{i\in I} R_i\right)^{1/q}$.

Here we have used the standard notation χ_A for the indicator function of the set A. To simplify the presentation, and when there is no danger of confusion, we will use the notation $A(x) = \chi_A(x)$ instead.

Theorem 2.5. *The maximal operator M is of weak type (p, p) if and only if $\mathcal{B}$ satisfies the covering property V_q for $1/p + 1/q = 1$, $1 < p < \infty$.*

Proof. Suppose first that $\mathcal{B}$ has property V_q. For $\alpha > 0$, set $F_\alpha = \{x : Mf(x) \geq \alpha\}$. Clearly,

$$F_\alpha = \bigcup_{i\in I} R_i, \qquad R_i \in \mathcal{B},$$

and

$$\frac{1}{\mu(R_i)} \int_{R_i} |f(y)|\, d\mu(y) > \alpha.$$

Let choose $\{\tilde{R}_k\}$ as in the definition of V_q. Then

$$\mu(F_\alpha) = \mu\Big(\bigcup_i R_i\Big) \le C \sum_k \mu(\tilde{R}_k) \le \frac{C}{\alpha} \sum_k \int_{\tilde{R}_k} |f(u)| \, d\mu(u)$$
$$= \frac{C}{\alpha} \int \Big[\sum_k \tilde{R}_k(y)\Big] |f(y)| \, d\mu(y) \le \frac{C}{\alpha} \Big\| \sum_k \tilde{R}_k \Big\|_q \|f\|_p \le \frac{C^2}{\alpha} \mu(F_\alpha)^{1/q} \|f\|_p,$$

which implies, recalling that $1/p + 1/q = 1$,

$$\mu(F_\alpha) \le C^{2p} \frac{\|f\|_p^p}{\alpha^p},$$

i.e., M is of weak -type (p, p).

Conversely, suppose now that M is bounded from L^p to the Lorentz space $L(p, \infty)$ or, equivalently, that M is of weak type (p, p) ([14] contains the properties of Lorentz spaces needed in this proof). Given the family $\{R_i\}$ as in the statement, let us assume the existence of a subsequence $\{\tilde{R}_k\}$ such that $\mu(\bigcup_i R_i) \le C\mu(\bigcup_k \tilde{R}_k)$ and satisfying the following sparseness property:

$$\mu\Big\{\tilde{R}_k \cap \Big(\bigcup_{j<k}\Big)\tilde{R}_j\Big)\Big\} \le \tfrac{1}{2}\,\mu(\tilde{R}_k), \quad \text{for every } k = 1, 2, \ldots \tag{2.3}$$

Then we claim that the sequence $\{\tilde{R}_k\}$ also satisfies $\| \sum_k \tilde{R}_k\|_q \le C(\bigcup_i R_i)^{1/q}$ for some universal constant $C < \infty$. To see this, let

$$E_k = \tilde{R}_k \setminus \bigcup_{j<k} \tilde{R}_j,$$

define the linear operator

$$Tf(x) = \sum_k \Big\{\frac{1}{\mu(\tilde{R}_k)} \int_{\tilde{R}_k} f(y) \, d\mu(y)\Big\} \cdot \chi_{E_k}(x),$$

and note that its adjoint is given by

$$T^* f(x) = \sum_k \Big\{\frac{1}{\mu(\tilde{R}_k)} \int_{E_k} f(y) \, d\mu(y)\Big\} \cdot \chi_{\tilde{R}_k}(x).$$

Observe also that $|Tf(x)| \le Mf(x)$, and that

$$T^*(\chi_{\bigcup_k \tilde{R}_k})(x) \ge \frac{1}{4} \sum_k \chi_{\tilde{R}_k}(x).$$

Since M is (by hypothesis) bounded from L^p to $L(p, \infty)$, the same holds for T with a bound independent of $\{\tilde{R}_k\}$. By duality, the operator T^* must be bounded from the

Lorentz space $L(q,1)$ to L^q, which implies

$$\Big\|\sum_k \chi_{\tilde{R}_k}\Big\|_q \leq C\,\mu\Big(\bigcup_k \tilde{R}_k\Big)^{1/q},$$

where C is a constant depending upon the bounds of M but independent of the family $\{R_i\}$.

To finish, we must show how to select the sequence $\{\tilde{R}_k\}$ having the sparseness property (2.3). First, we can assume without loss of generality that $\{R_i\}$ is a sequence $\{R_1, R_2, \ldots\}$. Then we take $\tilde{R}_1 = R_1$, and assuming that we have chosen $\tilde{R}_1, \ldots, \tilde{R}_k$, we take $\tilde{R}_{k+1}$ to be the first element in the sequence $\{R_i\}$ after $\tilde{R}_k$ with the property that

$$\mu\Big(\tilde{R}_{k+1} \bigcap \Big(\bigcup_{j\leq k} \tilde{R}_j\Big)\Big) \leq \frac{1}{2}\,\mu(\tilde{R}_k),$$

and observe that

$$\bigcup_j R_j \subset \big\{x : M\big(\chi_{\bigcup_k \tilde{R}_k}\big)(x) \geq 1/2\big\}. \qquad \square$$

Remark 2.6. The case $p = 1$, $q = \infty$ cannot be added to the statement of the theorem, because only one of the two implications holds. Namely, property V_∞ implies weak type $(1,1)$, but the converse if false in general. However, the weak-type $(1,1)$ property of the maximal function implies the estimate

$$\int_{\bigcup_k R_k} \exp\Big(\sum_k \chi_{R_k}(x)\Big)\, d\mu(x) \leq C\,\mu\Big(\bigcup_k R_k\Big)$$

for every sequence $\{R_k\}$ satisfying the sparseness property (2.3).

In general, if M is bounded from $L(\log^+ L)^r$ to weak-L^1, then sequences satisfying the spareness property (2.3) must satisfy

$$\int_{\bigcup_k R_k} \exp\Big(\sum_k \chi_{R_k}(x)\Big)^{1/(r+1)} d\mu(x) \leq C\,\mu\Big(\bigcup_k R_k\Big). \tag{2.4}$$

But such a covering lemma is not enough to ensure the boundedness of the maximal function operator from $L(\log^+ L)^r$ to weak-L^1.

The proof of the inequality (2.4) follows easily from the Marcinkiewicz interpolation theorem, keeping track of the growth of the constants in the estimates of the Hardy–Littlewood maximal function in $L^p(\mathbb{R}^n)$, when $p \to 1$.

2.3 The strong maximal function

Let us consider now Lebesgue's measure in $\mathbb{R}^n$ and the family $\mathcal{B}_n$ of parallelepipeds (rectangles) with sides parallel to the coordinate axes. Again we have the associated maximal function M_n, for which the so-called strong maximal theorem can be stated in two equivalent forms:

(a) A quantitative version: M_n is bounded from the Orlicz space $L(\log^+ L)^{n-1}(Q)$ to weak $L^1(Q)$, where Q is a unit cube in $\mathbb{R}^n$: for $\alpha > 0$,

$$\mu\{M_n f(x) > \alpha\} \leq C_n \int_Q \frac{|f(x)|}{\alpha}\left\{1 + \left(\log^+ \frac{|f(x)|}{\alpha}\right)^{n-1}\right\} \mathrm{d}\mu(x)$$

and C_n is some constant depending only upon the dimension n

(b) A qualitative version: the basis $\mathcal{B}_n$ differentiates integrals of functions which are locally in the space $L(\log^+ L)^{n-1}$, that is,

$$\lim_{\substack{x\in R \\ \mathrm{diam}(R)\to 0}} \frac{1}{\mu(R)} \int_R f(y)\, \mathrm{d}\mu(y) = f(x), \quad \text{a.e. } x \in \mathbb{R}^n,$$

for each f which is locally in the space $L(\log^+ L)^{n-1}(\mathbb{R}^n)$.

The strong maximal function is relevant in the study of operators commuting with a multiparameter family of dilation, and here the number of parameter is of capital importance, changing the nature of results.

For example, when $n \geq 2$, M_n does not satisfy the inequality of Hardy–Littlewood. For instance when $n = 2$ and taking

$$f_\varepsilon(x_1, x_2) = \frac{1}{\varepsilon^2} \chi_{\{|x_1|<\varepsilon/2, |x_2|<\varepsilon/2\}},$$

then for $|x_1|, |x_2| > 2\varepsilon$, we get $M_2 f_\varepsilon(x_1, x_2) \sim 1/(|x_1||x_2|)$.

Let $Q_0 = [-1, 1] \times [-1, 1]$. Then

$$\mu\{x \in Q_0 : M_2 f_\varepsilon > \lambda\} \geq \mu\{x \in Q_0 : |x_1 \cdot x_2| < 1/\lambda\} \sim \frac{1}{\lambda}\left|\log \frac{1}{\lambda}\right|, \quad \text{if } \lambda = 1/\varepsilon,$$

and we would like to have an inequality similar to the one satisfied by the Hardy–Littlewood maximal function. It follows that $\|f_\varepsilon\| \geq c \log \frac{1}{\varepsilon}$, and the smallest Orlicz norm for with this holds is the $L(\log^+ L)$ norm.

The results (a) and (b) quoted above are due to Jessen, Marcinkiewicz and Zygmund [57]. The main idea of their proof is to dominate the operator M_n by the composition $M_{x_1} \cdot M_{x_2} \cdots M_{x_n}$, where M_{x_j} denotes the one-dimensional Hardy–Littlewood maximal operator in the direction of the jth-coordinate axis. The elegance and simplicity of that proof are obvious. However, it seems desirable to have a geometric proof based on an adequate covering lemma. The main reason is that for other measures we lack the product structure, and the only way to obtain results for operators similar to the strong maximal function is through a deep understanding of the geometry of rectangles.

Figure 2.2 shows that the Vitali covering lemma cannot hold for parallelepipeds in $\mathbb{R}^n$, $n \geq 2$. Observe that $\mu(R_1) = \mu(R_2) = \cdots = \mu(R_k) = 1$. The sequence of rectangles $\{R_k\}$ satisfies property (2.3), but there is no way of selecting a disjoint subfamily satisfying the requirements of Vitali's covering.

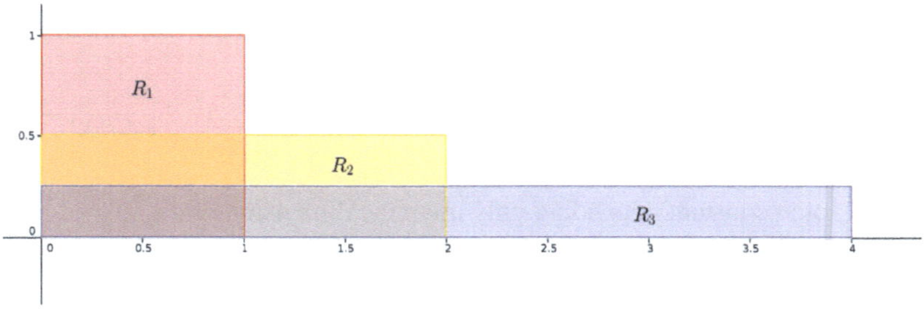

Fig. 2.2: Vitali's covering lemma does not hold for parallelepipeds.

Theorem 2.7. *Let $\{R_k\}$ be a sequence of parallelepipeds in $\mathbb{R}^n$ whose sides are parallel to the coordinate axes and with bounded union. Then there is a subsequence $\{\tilde{R}_j\} \subset \{R_k\}$ so that*

(1) $\mu\Big(\bigcup_k R_k\Big) \leq C_n\, \mu\Big(\bigcup_j \tilde{R}_j\Big)$.

(2) $\Big\| \exp\Big(\sum_j \chi_{\tilde{R}_j}\Big)^{1/(n-1)} - 1 \Big\|_1 \leq C_n\, \mu\Big(\bigcup_j \tilde{R}_j\Big)$, *for a universal constant $C_n < \infty$.*

Before we prove this theorem, let us show that it implies the result of Jessen, Marcinkiewicz and Zygmund mentioned above. Let $\alpha > 0$. For each point $x \in \{M_n f(x) > \alpha\}$, there is a rectangle R_x containing x with

$$\frac{1}{|R_x|} \int_{R_x} |f| > \alpha.$$

Without loss of generality, we assume $\bigcup_x R_x = \bigcup_k R_k$, where the R_k are certain of the R_x. Apply the covering lemma to get $\tilde{R}_k$ with properties (1) and (2) in Theorem 2.7. Then by virtue of (1), we need only show that

$$\mu\Big(\bigcup_k \tilde{R}_k\Big) \leq C_n \int_{\mathbb{R}^n} \frac{|f(x)|}{\alpha} \log^{n-1}\Big(1 + \frac{|f(x)|}{\alpha}\Big)\, dx.$$

Since $\mu(R_k) \leq \frac{1}{\alpha} \int_{R_k} |f|$, and summing, we have

$$\mu\Big(\bigcup_k \tilde{R}_k\Big) \leq \frac{1}{\alpha} \int_{Q_0} |f| \sum_k \chi_{\tilde{R}_k}.$$

The result follows applying conveniently Young's inequality $ab \leq e^a - b + b\log b$, for $a, b > 0$.

Proof of Theorem 2.7. We shall proceed by induction on n. Assume the case $n-1$ and let $R_1, R_2, \ldots$ be ordered such that the x_n side length decreases. For a rectangle R, let R_d denote the rectangle whose center and all the x_i-sidelengths, for $i < n$, are the same as those of R, but whose x_n side length is multiplied by 5. Now we describe the procedure for selecting the $\tilde{R}_k$ from the R_k. Let $\tilde{R}_1 = R_1$. Suppose $\tilde{R}_1, \tilde{R}_2, \ldots, \tilde{R}_k$ have already been chosen. We continue along the list, and each time we consider the rectangle R, we ask whether or not

$$\mu\Big(R \cap \Big(\bigcup_j (\tilde{R}_j)_d\Big)\Big) < \frac{1}{2}\,\mu(R),$$

where the above union is taken over all the $\tilde{R}_j$, $j \leq k$, for which $\tilde{R}_j \cap R \neq \emptyset$. If the answer is no, we move on to consider the next rectangle on the list. If the answer is yes, we make the rectangle $R = \tilde{R}_{k+1}$, and start the process over again.

Now, we prove (1) as follows. If R is an unselected rectangle, then

$$\mu\Big(R \cap \Big[\bigcup_{\substack{\tilde{R}_j \cap R \neq \emptyset \\ \tilde{R}_j \text{ before } R}} (\tilde{R}_j)_d\Big]\Big) \geq \frac{1}{2}\,\mu(R).$$

Let us slice all rectangles with a hyperplane perpendicular to the x_n axis. Then, if the slices are indicated by using S's instead of R's,

$$\mu\Big(S \cap \Big(\bigcup_j (\tilde{S}_j)_d\Big)\Big) \geq \frac{1}{2}\,\mu(S),$$

so that

$$M_{n-1}(\chi_{\bigcup_j (\tilde{R}_j)_d}) > 1/2 \quad \text{on } \textstyle\bigcup_j R_j,$$

where M_{n-1} is acting in the $x_1, x_2, \ldots, x_{n-1}$ coordinates.

By the boundedness of M_{n-1} on, say, L^2, (by induction) we have

$$\mu\Big(\bigcup_j R_j\Big) \leq C\,\mu\Big(\bigcup_k (\tilde{R}_k)_d\Big) \leq C' \sum_k \mu(\tilde{R}_k) \leq C'\,\mu\Big(\bigcup_k \tilde{R}_k\Big).$$

To obtain (2), notice that the $\tilde{R}_j$ satisfy

$$\mu\Big(\tilde{R}_k \cap \Big[\bigcup_{\tilde{R}_j \cap \tilde{R}_k \neq \emptyset,\, j<k} (\tilde{R}_j)_d\Big]\Big) < \frac{1}{2}\,\mu(\tilde{R}_k).$$

Then, we slice again with a hyperplane perpendicular to the x_n axis, to get

$$\mu\Big(\tilde{S}_k \cap \Big(\bigcup_{j<k} \tilde{S}_j\Big)\Big) < \frac{1}{2}\,\mu(\tilde{S}_k),$$

so that, if

$$\widetilde{E}_k = \tilde{S}_k \setminus \bigcup_{j<k} \tilde{S}_j,$$

we have $\mu(\widetilde{E}_k) > \frac{1}{2}\,\mu(\tilde{S}_k)$, and if $\phi \in L(\log L)^{n-1}(S_0)$, we shall show that

$$\int_{S_0} \sum_k \chi_{\tilde{S}_k}\, \phi \le C\, \|\phi\|_{L(\log L)^{n-1}}, \tag{2.5}$$

which will give

$$\int_{S_0} \exp\left[c\Big(\sum_k \chi_{\tilde{S}_k}\Big)^{1/(n-1)}\right] \mathrm{d}x_1 \cdots \mathrm{d}x_{n-1} \le C.$$

Integrating this estimate in x_n finishes things.

To obtain (2.5), we write

$$\begin{aligned}\int_{S_0} \sum_k \chi_{\tilde{S}_k}\, \phi\, \mathrm{d}x_1 \cdots \mathrm{d}x_{n-1} &\le \sum_k \int_{\tilde{S}_k} \phi \le C \sum_k |\widetilde{E}_k| \frac{1}{|\tilde{S}_k|} \int_{\tilde{S}_k} \phi \\ &\le C \int_{S_0} M_{n-1}(\phi)\, \mathrm{d}x_1 \cdot \mathrm{d}x_2 \cdots \mathrm{d}x_{n-1} \le C'\, \|\phi\|_{L(\log L)^{n-1}}. \qquad \square\end{aligned}$$

In our argument above, the slicing was the most important idea. If we try the proof without it, we will not wind up with the estimate we want, on the $\exp(\cdot)^{1/(n-1)}$ norm, but rather on the $\exp(\cdot)^{1/n}$ norm instead. Also the slicing is the mechanism by which we control M_n by the lower parameter operator M_{n-1}, and this enables us to proceed by induction.

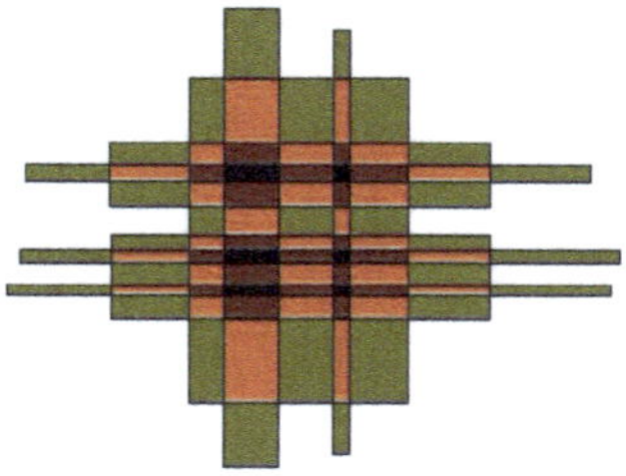

Fig. 2.3: Exponential overlapping.

2.4 A problem of A. Zygmund

Let ϕ be a positive function in $\mathbb{R}^2$, monotonic in each variable separately. We now consider the collection $\mathcal{B}_\phi$ defined as $\mathcal{B}_2 \times \phi$; i.e., the parallelepipeds with sides perpendicular to the axes, and whose dimensions are given by $s \times t \times \phi(s,t)$, for positive real numbers s and t. Note that the strong maximal theorem asserts that since $\mathcal{B}_\phi \subset \mathcal{B}_3$, it differentiates integrals of functions which are locally in $L(\log^+ L)^2(\mathbb{R}^3)$.

However, since the vertical direction of the family $\mathcal{B}_\phi$ of parallelepipeds is constrained to be the image of a monotonic function, then we may hope to obtain some extra regularity, but bare in mind that we cannot obtain any better differentiation properties than those of $\mathcal{B}_2$. Indeed, we are going to prove that the best possible improvement is achieved. Let us state the result in two equivalent forms.

Theorem 2.8. (1) *The collection of parallelepipeds $\mathcal{B}_\phi$ differentiates integrals of functions which are locally in $L(\log^+ L)(\mathbb{R}^3)$, that is,*

$$\lim_{R \Rightarrow x} \frac{1}{\mu(R)} \int_R f(y)\, \mathrm{d}\mu(y) = f(x), \quad a.e.\ x,$$

so long as $f \in L(\log^+ L)(\mathbb{R}^3)$, where μ denotes the Lebesgue measure in $\mathbb{R}^3$.

(2) *The associated maximal function*

$$M_\phi f(x) = \sup_{R \ni x} \frac{1}{\mu(R)} \int_R |f(y)|\, \mathrm{d}\mu(y)$$

satisfies the inequality

$$\mu\{x : M_\phi f(x) > \lambda\} \le C \int \frac{|f(y)|}{\lambda} \left(1 + \log^+ \frac{|f(y)|}{\lambda}\right) \mathrm{d}\mu(y),$$

for some universal constant $C < \infty$.

In order to prove this result, we shall develop, once again, a suitable covering lemma adapted to the family $\mathcal{B}_\phi$. First let us introduce a definition.

Definition 2.9. We say that a family $\mathcal{B}$ of parallelepipeds in $\mathbb{R}^3$ is monotonic if given any two parallelepipeds R_1 and R_2 with respective dimensions $s_1 \times t_1 \times u_1$ and $s_2 \times t_2 \times u_2$, the following implication holds: if $s_1 < s_2$ and $t_1 < t_2$, then $u_1 \le u_2$.

In order to simplify the intuition and the computations, we will restrict to the case where the parallelepipeds of the monotonic family $\mathcal{B}$ are dyadic. The following covering lemma will imply Theorem 2.8.

Lemma 2.10. *Let $\mathcal{B}$ be a monotonic dyadic family of parallelepipeds in $\mathbb{R}^3$. For every $\{R_i\}_{i \in I} \subset \mathcal{B}$ with $\mu(\bigcup_{i \in I} R_i) < \infty$, we can find a finite subsequence $\{\tilde{R}_j\}_{j=1}^M \subset \{R_i\}_{i \in I}$ satisfying both*

(1) $\mu\Big(\bigcup_{i \in I} R_i\Big) \le C \mu\Big(\bigcup_{j=1}^M \tilde{R}_j\Big)$,

(2) $\displaystyle\int_{\bigcup_{j=1}^M \tilde{R}_j} \exp\Big(\sum_{j=1}^M \chi_{\tilde{R}_j}(x)\Big)\, \mathrm{d}\mu(x) \le C \mu\Big(\bigcup_{j=1}^M \tilde{R}_j\Big)$,

where $C < \infty$ is a universal constant.

Proof. First note that since $\mu(\bigcup_i R_i) < \infty$, we can assume, without loss of generality, that $\{R_i\}_{i\in I}$ is a finite sequence $\{R_i\}_{i=1}^N$ such that

$$\mu\Big(R_i \cap \Big(\bigcup_{j\neq i} R_j\Big)\Big) \le \frac{1}{2}\,\mu(R_i), \quad \text{for every } i = 1, \ldots, N.$$

Moreover, let us rearrange the sequence so that the x_3 dimension of the parallelepipeds, $d_3(R_i)$, decreases, that is,

$$d_3(R_k) \ge d_3(R_{k+1}), \quad \text{for all } k.$$

We start our selection process by picking $\tilde{R}_1 = R_1$. Then we choose $\tilde{R}_k$ to be the next element of the sequence satisfying

$$\int_{\tilde{R}_k} \exp\Big(\sum_{j=1}^{k-1} \tilde{R}_j(x)\Big)\,\mathrm{d}\mu(x) \le (1+e^{-1})\,\mu(\tilde{R}_k). \tag{2.6}$$

Note that e^{-1} can be replaced by some other appropriate constant. Next we will show that the sequence $\{\tilde{R}_k\}_{k=1}^M$ satisfies property (2):

$$\begin{aligned}
&\int_{\bigcup_{j=1}^M \tilde{R}_j} \exp\Big(\sum_{j=1}^{M} \chi_{\tilde{R}_j}(x)\Big)\,\mathrm{d}\mu(x)\\
&\quad= \int_{\bigcup_{j=1}^{M-1} \tilde{R}_j\setminus \tilde{R}_M} \exp\Big(\sum_{j=1}^{M-1} \chi_{\tilde{R}_j}(x)\Big)\,\mathrm{d}\mu(x) + \int_{\tilde{R}_M} \exp\Big(\sum_{j=1}^{M} \chi_{\tilde{R}_j}(x)\Big)\,\mathrm{d}\mu(x)\\
&\quad= \int_{\bigcup_{j=1}^{M-1} \tilde{R}_j\setminus \tilde{R}_M} \exp\Big(\sum_{j=1}^{M-1} \chi_{\tilde{R}_j}(x)\Big)\,\mathrm{d}\mu(x) + e\int_{\tilde{R}_M} \exp\Big(\sum_{j=1}^{M-1} \chi_{\tilde{R}_j}(x)\Big)\,\mathrm{d}\mu(x).
\end{aligned}$$

However, the second integral term can be majorized according to our choice (2.6), and we obtain

$$\begin{aligned}
&\int_{\bigcup_{j=1}^M \tilde{R}_j} \exp\Big(\sum_{j=1}^{M} \chi_{\tilde{R}_j}(x)\Big)\,\mathrm{d}\mu(x)\\
&\qquad\le \int_{\bigcup_{j=1}^{M-1} \tilde{R}_j} \exp\Big(\sum_{j=1}^{M-1} \chi_{\tilde{R}_j}(x)\Big)\,\mathrm{d}\mu(x) + (1+e)\,\mu(\tilde{R}_M).
\end{aligned}$$

An iteration of this process yields the following majorization:

$$\int_{\bigcup_{j=1}^M \tilde{R}_j} \exp\Big(\sum_{j=1}^{M} \chi_{\tilde{R}_j}(x)\Big)\,\mathrm{d}\mu(x) \le (1+e)\sum_{k=1}^{M} \mu(\tilde{R}_k) \le 2(1+e)\,\mu\Big(\bigcup_{k=1}^{M} \tilde{R}_k\Big).$$

Our aim is now to show that we have not discarded "too many" parallelepipeds, that is,

$$\mu\Big(\bigcup_{i=1}^{N} R_i\Big) \le C\,\mu\Big(\bigcup_{j=1}^{M} \tilde{R}_j\Big).$$

Suppose R is a discarded parallelepiped. Then, since the family $\{\tilde{R}_j\}_{j=1}^{M}$ has been arranged so that the vertical dimensions $d_3(\tilde{R}_j)$ of the parallelepipeds decrease, it must be that

$$1+e^{-1} < \frac{1}{\mu(R)}\int_R \exp\Big(\sum_{d_3(\tilde{R}_j)\ge d_3(R)} \chi_{\tilde{R}_j}(x)\Big)\,\mathrm{d}\mu(x).$$

Notice that since the family $\mathcal{B}_\phi$ is monotonic, if $d_3(\tilde{R}_j) \ge d_3(R)$, it must be that

$$d_1(\tilde{R}_j) \ge d_1(R) \quad \text{or} \quad d_2(\tilde{R}_j) \ge d_2(R).$$

Thus we can rewrite the above inequality as

$$1+e^{-1} < \frac{1}{\mu(R)}\int_R \exp\Big(\sum_{\substack{d_3(\tilde{R}_j)\ge d_3(R)\\ d_1(\tilde{R}_j)\ge d_1(R)}} \chi_{\tilde{R}_j}(x)\Big)\cdot \exp\Big(\sum_{\substack{d_3(\tilde{R}_j)\ge d_3(R)\\ d_2(\tilde{R}_j)\ge d_2(R)}} \chi_{\tilde{R}_j}(x)\Big)\,\mathrm{d}\mu(x)$$

$$(2.7) \qquad = \frac{1}{\mu(R)}\int_R \exp\Big(\sum_{i=1}^{k} \chi_{\tilde{R}_{\nu_i}}(x)\Big)\cdot \exp\Big(\sum_{j=1}^{l} \chi_{\tilde{R}_{\mu_j}}(x)\Big)\,\mathrm{d}\mu(x),$$

where $\tilde{R}_{\nu_1},\dots,\tilde{R}_{\nu_k}$ are the parallelepipeds in the first sum, and $\tilde{R}_{\mu_1},\dots,\tilde{R}_{\mu_l}$ are those in the second.

With the parallelepiped R fixed, and without loss of generality, we may simplify notation changing the indices so that $\tilde{R}_{\nu_i} = \tilde{R}_i$ for $i = 1,\dots,k$, and $\tilde{R}_{\mu_j} = \tilde{R}_{k+j}$ for $j = 1,\dots,l$. Now let us expand the exponentials inside the integral:

$$\exp\Big(\sum_{i=1}^{k} \chi_{\tilde{R}_{\nu_i}}(x)\Big) = \sum_{n=0}^{\infty}\frac{1}{n!}\Big(\sum_{i=1}^{k} \chi_{\tilde{R}_{\nu_i}}(x)\Big)^n = \sum_{n=0}^{\infty}\frac{1}{n!}\sum_{i_1,\dots,i_n=1}^{k} \chi_{\tilde{R}_{i_1}}(x)\dots\chi_{\tilde{R}_{i_n}}(x),$$

$$\exp\Big(\sum_{j=1}^{l} \chi_{\tilde{R}_{\mu_j}}(x)\Big) = \sum_{m=0}^{\infty}\frac{1}{m!}\Big(\sum_{j=1}^{l} \chi_{\tilde{R}_{\mu_j}}(x)\Big)^m$$

$$= \sum_{m=0}^{\infty}\frac{1}{m!}\sum_{j_1,\dots,j_m=k+1}^{k+l} \chi_{\tilde{R}_{j_1}}(x)\dots\chi_{\tilde{R}_{j_m}}(x).$$

The last summands simply correspond to all possible combinations of the products of up to n terms $\chi_{\tilde{R}_{\nu_i}}$, or up to m terms $\chi_{\tilde{R}_{\mu_j}}$, respectively. Therefore, using this expression in (2.7), we obtain

$$\sum_{n,m=0}^{\infty} \frac{1}{n!\,m!} \sum_{i_1,\dots,i_n=1}^{k} \sum_{j_1,\dots,j_m=k+1}^{k+l} \frac{1}{\mu(R)} \int_R \chi_{\tilde{R}_{i_1}}(x)\dots\chi_{\tilde{R}_{j_m}}(x)\,\mathrm{d}\mu(x)$$
$$= \sum_{n,m=0}^{\infty} \frac{1}{n!\,m!} \sum_{i_1,\dots,i_n=1}^{k} \sum_{j_1,\dots,j_m=k+1}^{k+l} \frac{\mu\{\tilde{R}_{i_1} \cap \dots \cap \tilde{R}_{i_n} \cap \tilde{R}_{j_1} \cap \dots \cap \tilde{R}_{j_m} \cap R\}}{\mu(R)}.$$

It is now clear that in order to show property (1), it is crucial to majorize the overlap of the parallelepipeds $\tilde{R}_{i_1}, \dots, \tilde{R}_{j_m}$ and R. Notice that, if not empty, the intersection must be of the form shown in Figure 2.4. That is, the block corresponding to ε is the intersection of the $\tilde{R}_{i_1}, \dots, \tilde{R}_{i_n}$, and the block corresponding to δ is that of $\tilde{R}_{j_1}, \dots, \tilde{R}_{j_m}$. Note that here we have used the fact that we work with dyadic parallelepipeds: if the intersection is not empty, then the cut has to be "clean", i.e., the parallelepiped that has the longer size of a side slice completely the others on the direction of that side.

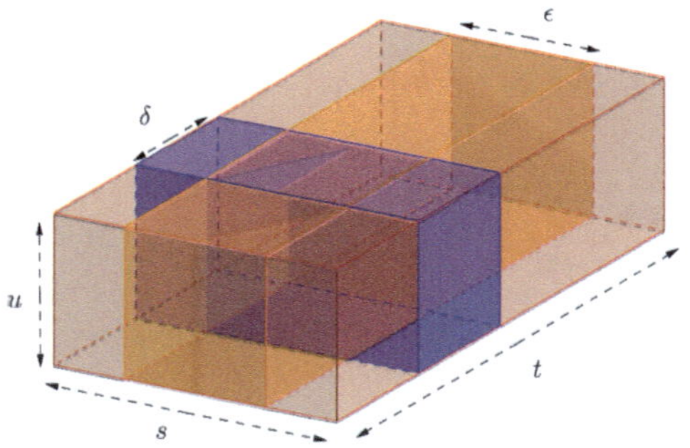

Fig. 2.4: Controlling the intersections.

Thus the intersection $\tilde{R}_{i_1} \cap \dots \cap R_{i_n} \cap \tilde{R}_{j_1} \cap \dots \cap \tilde{R}_{j_m}$ will be a parallelepiped with dimensions $\varepsilon \times \delta \times u$. It then follows that

$$\frac{\mu\{\tilde{R}_{i_1} \cap \dots \cap \tilde{R}_{i_n} \cap \tilde{R}_{j_1} \cap \dots \cap \tilde{R}_{j_m} \cap R\}}{\mu(R)} = \frac{\varepsilon \times \delta}{s \times t},$$

and we have

$$1 + e^{-1} < \sum_{n,m=0}^{\infty} \frac{1}{n!\,m!} \sum_{i_1,\dots,i_n=1}^{k} \sum_{j_1,\dots,j_m=k+1}^{k+l} \frac{\varepsilon \times \delta}{s \times t}. \tag{2.8}$$

Next, let us take a closer look at the intersection of all these parallelepipeds. This time, fix a point inside the intersection $P = (x_0, y_0, z_0)$, and consider the horizontal and vertical planes (see Figure 2.5) intersected with R and the segments

$$I_P^1 = \{(x, y_0, z_0) \in R\} \quad \text{and} \quad I_P^2 = \{(x_0, y, z_0) \in R\}.$$

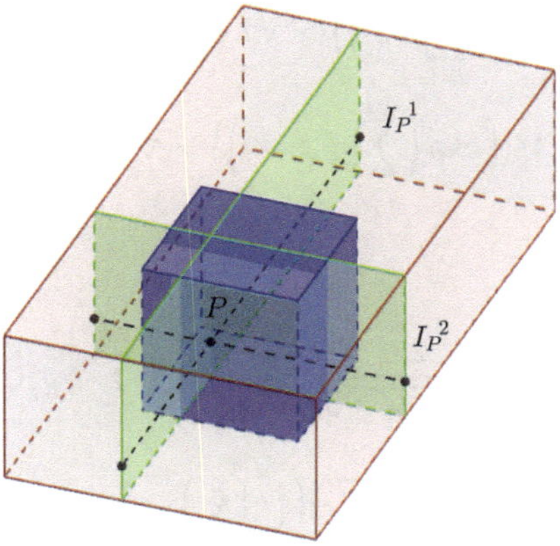

Fig. 2.5: Segments I_P^1 and I_P^2.

By monotonicity, it follows that

$$\frac{|\tilde{R}_{i_1} \cap \cdots \cap \tilde{R}_{i_n} \cap I_P^1|}{|I_P^1|} = \frac{\varepsilon}{s} \quad \text{and} \quad \frac{|\tilde{R}_{j_1} \cap \cdots \cap \tilde{R}_{j_m} \cap I_P^2|}{|I_P^2|} = \frac{\delta}{t}.$$

Here $|\cdot|$ refers to the 1-dimensional Lebesgue measure. Replacing the product of this two quantities in (2.8), we obtain a minorization of directional Hardy–Littlewood maximal functions.

$$\begin{aligned} &1 + e^{-1} \\ &< \sum_{n,m=0}^{\infty} \frac{1}{n!\, m!} \sum_{i_1,\ldots,i_n=1}^{k} \sum_{j_1,\ldots,j_m=k+1}^{k+l} \frac{|\tilde{R}_{i_1} \cap \cdots \cap \tilde{R}_{i_n} \cap I_P^1|}{|I_P^1|} \cdot \frac{|\tilde{R}_{j_1} \cap \cdots \cap \tilde{R}_{j_m} \cap I_P^2|}{|I_P^2|}, \end{aligned}$$

hence

$$\begin{aligned} 1 + e^{-1} &< \left\{ \frac{1}{|I_P^1|} \int_{I_P^1} \exp\Big(\sum_{i=1}^{k} \chi_{\tilde{R}_i}(x) \Big) \,\mathrm{d}x \right\} \cdot \left\{ \frac{1}{|I_P^2|} \int_{I_P^2} \exp\Big(\sum_{j=k+1}^{k+l} \chi_{\tilde{R}_j}(x) \Big) \mathrm{d}x \right\} \\ &< M_1\Big(\exp\Big(\sum_{i=1}^{k} \chi_{\tilde{R}_i}(x) \Big) \Big)(P) \cdot M_2\Big(\exp\Big(\sum_{j=k+1}^{k+l} \chi_{\tilde{R}_j}(x) \Big) \Big)(P), \end{aligned}$$

where M_1 and M_2 denote the extensions to $\mathbb{R}^3$ of the one-dimensional Hardy–Littlewood maximal function in the x_1 and x_2 directions, respectively. We have showed that for any discarded parallelepiped, we have the following inclusion:

$$R \subset \Big\{x : M_1\Big(\exp\Big(\sum_{i=1}^{M}\chi_{\tilde{R}_i}\Big)\Big)\cdot M_2\Big(\exp\Big(\sum_{j=1}^{M}\chi_{\tilde{R}_j}(x)\Big)\Big) > 1+e^{-1}\Big\}$$
$$\subset \Big\{x : M_1\Big(\exp\Big(\sum_{i=1}^{M}\chi_{\tilde{R}_i}\Big)\Big) > \sqrt{1+e^{-1}}\Big\}$$
$$\cup \Big\{x : M_2\Big(\exp\Big(\sum_{j=1}^{M}\chi_{\tilde{R}_j}(x)\Big)\Big) > \sqrt{1+e^{-1}}\Big\}.$$

Finally,

$$\mu\Big(\bigcup_{i\in I}R_i\Big) \le \frac{2}{\sqrt{1+e^{-1}}}\int_{\bigcup_{j=1}^{M}\tilde{R}_j}\exp\Big(\sum_{j=1}^{M}\chi_{\tilde{R}_j}(x)\Big)\,\mathrm{d}\mu(x)$$
$$\le 4\sqrt{1+e^{-1}}\,\mu\Big(\bigcup_{j=1}^{M}\tilde{R}_j\Big). \qquad \square$$

Proof of Theorem 2.8. It is a well-known fact that (1) and (2) are equivalent, and the implication (2) $\Rightarrow$ (1) is the easier of the two. The strategy of the proof is as follows: first we observe that if the family $\mathcal{B}$ satisfies the hypotheses of the covering lemma, and if M denotes the maximal function associated with $\mathcal{B}$, then M satisfies (2). Next, we find such a family $\mathcal{B}$ with the property that $M_\phi f \le CMf$ for some universal constant C.

Let $\mathcal{B}$ satisfy the hypotheses of the covering lemma, and define

$$Mf(x) = \sup_{x\in R^*, R\in\mathcal{B}}\frac{1}{\mu(R)}\int_R |f(y)|\,\mathrm{d}\mu(y),$$

where R^* is the result of dilating R by a factor of 3 with respect to its center.

Claim. *There exists a constant $c < \infty$ such that*

$$\mu(\{Mf(x) > \lambda\}) \le C\int_{\mathbb{R}^3}\frac{|f(x)|}{\lambda}\Big\{1+\log^+\frac{|f(x)|}{\lambda}\Big\}\,\mathrm{d}\mu(x).$$

Proof of the Claim. The set $\{Mf(x) > \lambda\}$ is a union of parallelepipeds $\{R_\alpha\}$ such that

$$\frac{1}{\mu(R_\alpha)}\int_{R_\alpha}|f(y)|\,\mathrm{d}\mu(y) \ge \lambda.$$

We can apply the covering lemma to show the existence of the subfamily $\{R_j\}$ with the prescribed properties. We have

$$\mu\Big(\bigcup_j R_j^*\Big) \le 27\sum_j \mu\Big(\bigcup_j R_j\Big) \le 27\int_{\bigcup_j R_j}\frac{|f(x)|}{\lambda}\sum_j \chi_{R_j}(x)\,\mathrm{d}\mu(x).$$

Next, observe that if u and v are positive real numbers, it follows that $u \cdot v \leq u \log u + e^{v-1}$. Furthermore, for every $\varepsilon > 0$, there exists a constant $C_\varepsilon < \infty$ such that $u \cdot v \leq C_\varepsilon u(1 + \log^+ u) + e^{\varepsilon v - 1}$. In particular,

$$\mu\Big(\bigcup_j R_j^*\Big) \leq C_\varepsilon \int_{\bigcup_j R_j} \frac{|f(x)|}{\lambda} \left\{1 + \log^+ \frac{|f(x)|}{\lambda}\right\} d\mu(x) + e^{-1} \int_{\bigcup_j R_j} \exp\Big(\varepsilon \sum_j \chi_{R_j}(x)\Big) d\mu(x).$$

But

$$\int_{\bigcup_j R_j} \exp\Big(\varepsilon \sum_j \chi_{R_j}(x)\Big) d\mu(x) \leq \Big[\mu\Big(\bigcup_j R_j\Big)\Big]^{1-\varepsilon} \Big(\int_{\bigcup_j R_j} \exp\Big(\sum_j \chi_{R_j}(x)\Big) d\mu(x)\Big)^{\varepsilon} \leq C^\varepsilon \mu\Big(\bigcup_j R_j\Big),$$

where C is the constant appearing in the covering lemma. To finish, we just choose $\varepsilon > 0$ such that $C^\varepsilon e^{-1} \leq 1/2$.

Finally, we must show how to get the family $\mathcal{B}$ from $\mathcal{B}_\phi$. First of all, we reduce the values of s and t so that they are in the set $\{2^n\}_{n\in\mathbb{Z}}$. Then we define $\tilde{\phi}(2^k, 2^l) = 2^m$ if $2^{m-1} < \phi(2^k, 2^l) \leq 2^m$. It is clear that if $\tilde{\mathcal{B}}$ of parallelepipeds obtained in this way, then each element R of $\mathcal{B}_\phi$ is contained in an element $\tilde{R}$ of $\tilde{\mathcal{B}}$ in such a way that $\mu(\tilde{R}) \leq 8\mu(R)$. Next, we consider, in the horizontal plane $z = 0$, the family of dyadic rectangles, and to each one of these rectangles, we attach a parallelepiped having the rectangle as horizontal base and with vertical dimension given by $\tilde{\phi}$. Then we translate, in the vertical direction, each one of these parallelepipeds by integer multiples of its vertical length. The family obtained is $\mathcal{B}$. Clearly, $M_\phi f(x) \leq 8Mf(x)$. □

Theorem 2.8 has important applications to the convergence of Poisson integrals in the Siegel generalized upper-half space. Here $\mathbb{R}^n$ is realized as M_n, the space of $m \times m$ real symmetric matrices, with $n = m(m+1)/2$. There is an associated symmetric space defined as the tube domain

$$T_\Gamma = \{z = x + iy : x \in M_m, y \in \Gamma \subset M_m\},$$

where Γ is the cone of positive definite symmetric matrices. The Poisson kernel is

$$P_y(x) = C_n \left(\frac{\det(y)}{|\det(x + iy)|^2}\right)^{(n+1)/2},$$

and for adequate f defined on $\mathbb{R}^n$, we have the Poisson integral

$$u(x, y) = f * P_y(x),$$

and the question is the convergence of $u(x, y)$ to $f(x)$ when $y \to 0$. There are three types of convergence with their corresponding maximal functions:

(a) *Restricted*:

$$M_{\mathrm{R}} f(x) = \sup_{y \in \Gamma_1} |u(x, y)|,$$

where $\Gamma_1 \subset \Gamma$ is a proper subcone.

(b) *Unrestricted*:

$$M_{\mathrm{U}} f(x) = \sup_{y \in \Delta} |u(x, y)|,$$

where Δ is the family of diagonal matrices in the cone Γ.

(c) *Wild*:

$$M_{\mathrm{W}} f(x) = \sup_{y \in \Gamma} |u(x, y)|.$$

It is well known [54] that $M_R f$ is dominated by the Hardy–Littlewood maximal function, and thus it satisfies a weak-type $(1, 1)$ inequality. And that M_{W} is unbounded on any L^p-space (this phenomenon is related to the Kakeya set of next Chapter 3).

Regarding M_{U}, there is a conjecture:

M_{U} is bounded from $L(\log^+ L)^{m-1}(\mathbb{R}^n)$ to weak $L^1(\mathbb{R}^n)$.

The case $m = 2$ and $n = 3$ of the conjecture is a consequence of Theorem 2.8, because then

$$P_{y_1,y_2}(x_1, x_2, x_3) = (y_1 y_2)^{-3/2} P_{1,1}\left(\frac{x_1}{y_1}, \frac{x_2}{y_2}, \frac{x_3}{\sqrt{y_1 y_2}}\right),$$

and the maximal function M_{U} is majorized by CM_ϕ, where M_ϕ is as in Theorem 2.8, with $\phi(s, t) = \sqrt{st}$.

The extension to higher dimensions of Theorem 2.8, namely taking averages over parallelepipeds of sides parallel to the coordinate axes and dimensions $(s_1, \ldots, s_n, \phi(s_1, \ldots, s_n))$, $n \geq 3$, and ϕ monotonic on each variable separately, is an intriguing open problem:

Does M_ϕ differentiates integrals of functions locally in $L(\log^+ L)^{n-1}(\mathbb{R}^n)$, for $n \geq 3$?

However, the generalized conjecture to basis of parallelepipeds parallel to the coordinate axes and dimensions $(\phi_1(s_1, \ldots, s_n), \ldots, \phi_m(s_1, \ldots, s_n))$, where each function ϕ_j is monotonic non-decreasing on each variable, is false in general, even in the case $m = 3$ and $n = 2$, see [52].

A specially clever presentation of this phenomenon is given in [51].

Theorem 2.11. *There exist functions $\phi_i : \mathbb{Z}^2 \to \mathbb{Z}$, $i = 1, 2, 3$, which are monotonic non-decreasing in each variable and such that the family of parallelepipeds of sides parallel to the coordinate axes and dimensions $(2^{\phi_1(s,t)}, 2^{\phi_2(s_1 t)}, 2^{\phi_3(s_1 t)})$ does not differentiate $L(\log^+ L)(\mathbb{R}^3)$.*

Proof. Let $\psi : \mathbb{Z} \to \mathbb{Z}^3$ be a bijection $\psi = (\psi_1, \psi_2, \psi_3)$. For each $\psi_j : \mathbb{Z} \to \mathbb{Z}$, define

$$\varphi_j(m_1, m_2) = \begin{cases} \max\{\psi_j(m); -m_2 \le m \le m_1\}, & \text{if } m_1 \ge -m_2, \\ \min\{\psi_j(m); m_1 \le m \le -m_2\} & \text{if } m_1 \le -m_2. \end{cases}$$

Then each ϕ_j satisfies

(i) $\phi_j(m, -m) = \psi_j(m)$, for all $m \in \mathbb{Z}$;
(ii) ϕ_j is monotonic non-decreasing in each variable.

Clearly, the family of parallelepipeds obtained in this manner contains all "dyadic" parallelepipeds in $\mathbb{R}^3$ (and all their translates) and, therefore, its associated maximal functions satisfies $Mf \approx M_S f$ (strong maximal function in $\mathbb{R}^3$). □

2.5 Differentiation on lacunary directions

Suppose that in $\mathbb{R}^2$ we consider the class $\mathcal{A}$ of all rectangles of arbitrary side lengths which are oriented in one of the directions $\theta_k = 2^{-k}$, $k = 1, 2, \ldots$, measured from some fixed direction, say the positive x-direction. Define the maximal operator $\mathcal{M}$ by

$$\mathcal{M} f(x) = \sup_{R \in \mathcal{A}, R \ni x} \frac{1}{|R|} \int_R |f(t)| \, dt.$$

Following the ideas of the covering approach outlined above, we prove the following inequality:

$$\mu(\{\mathcal{M} f(x) > \alpha\}) \le \frac{C}{\alpha^2} \|f\|^2_{L^2(\mathbb{R}^2)}.$$

The proof consists of showing that, given a sequence of rectangles $\{R_k\}$ belonging to $\mathcal{A}$, there exists a subfamily $\{\tilde{R}_k\}$ such that

(1) $\mu\Big(\bigcup_k \tilde{R}_k\Big) \gtrsim \mu\Big(\bigcup_k R_k\Big)$,

(2) $\Big\|\sum_k \chi_{\tilde{R}_k}\Big\|_{L^2(\mathbb{R}^2)} \lesssim \mu\Big(\bigcup_k R_k\Big)^{1/2}$.

To prove this, we give a rule for selecting $\tilde{R}_k$, given that we have already selected $\tilde{R}_j$ for $j < k$. Assume that the R_k all have their longest side in a direction in the first quadrant, and that are ordered so that their longer side lengths are decreasing. Then consider the rectangle R following $\tilde{R}_{k-1}$. Consider, in particular,

$$\frac{1}{\mu(R)} \sum_{j<k} \mu(R \cup \tilde{R}_j) = \frac{1}{\mu(R)} \int_R \sum_{j<k} \chi_{\tilde{R}_j} \, dx.$$

If this is less than $1/2$, select R as $\tilde{R}_k$. If not, go to the next rectangle on the list, and apply the same test to it. In this way, we obtain the desired $\tilde{R}_k$. Notice that

$$\Big\|\sum_k \chi_{\tilde{R}_k}\Big\|_2^2 = \int \sum_{j,k} \chi_{\tilde{R}_j}\chi_{\tilde{R}_k} = 2\sum_k \int_{\tilde{R}_k} \sum_{j<k} \chi_{\tilde{R}_j} + \sum_k \mu(\tilde{R}_k)$$
$$\leq 2\sum_k \mu(\tilde{R}_k) \leq C\,\mu\Big(\bigcup_k \tilde{R}_k\Big).$$

This is (2). To show (1), we let R be some rectangle which was unselected. This implies that

$$\frac{1}{\mu(R)} \sum_{\tilde{R}_j \text{ before } R} \mu(R \cap \tilde{R}_j) \geq \frac{1}{2}.$$

Let S be the envelope rectangle to R whose sides are parallel to the coordinate axes. See Figure 2.6.

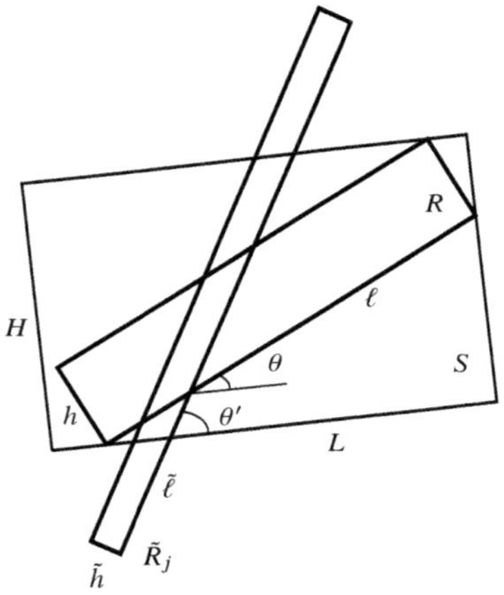

Fig. 2.6: The envelope rectangle.

Then, by dilating S if necessary, we can assume that $\tilde{R}_j$ is centered at the same center as S. Now the point is that

$$\frac{\mu(\tilde{R}_j \cap S)}{\mu(S)} \geq \frac{\mu(\tilde{R}_j \cap R)}{\mu(R)}. \tag{2.9}$$

In fact,

$$\frac{\mu(\tilde{R}_j \cap S)}{\mu(S)} \sim \frac{\widetilde{h}\,H/\theta'}{hL} = \frac{\widetilde{h}}{L\theta'} \quad \text{and} \quad \frac{\mu(\tilde{R}_j \cap R)}{|R|} \sim \frac{\widetilde{h}\,h(\theta'-\theta)}{hl} = \frac{\widetilde{h}}{l(\theta'-\theta)},$$

and our inequality (2.9), taking into account that $l \sim L$, is

$$\frac{1}{\theta'} \geq \frac{1}{(\theta' - \theta)} \quad \text{or} \quad \theta' - \theta \geq c\,\theta',$$

and this, in our case of $\theta_k = 2^{-k}$, is valid with $c = 1/2$. (If $\theta_k = (1-\varepsilon)^k$, then with $c = \varepsilon$.) Then summing over j in (2.9), we have

$$\frac{1}{\mu(S)} \int_S \sum_{\tilde{R}_j \text{ before } R} \chi_{\tilde{R}_j} \geq \frac{1}{\mu(R)} \int_R \sum_{\tilde{R}_j \text{ before } R} \chi_{\tilde{R}_j} \geq \frac{1}{2}\, c.$$

In other words,

$$M_2\Big(\sum_j \chi_{R_j}\Big) > \frac{1}{2}\, c \qquad \text{on } \bigcup_j R_j,$$

and by the boundedness of M_2, we see that

$$\mu\Big(\bigcup_j R_j\Big) \lesssim \Big\|\sum_j \chi_{\tilde{R}_j}\Big\|_2^2 \lesssim \mu\Big(\bigcup_j \tilde{R}_j\Big)$$

by (2.9). We have controlled the operator $\mathscr{M}$ by M_2 here in just exactly the way M_n was previously controlled by M_{n-1}. And while $\mathscr{M}$ is a 3-parameter maximal operator, M_2 is a 2-parameter one.

Finally, we should note that, as before, this covering lemma implies the maximal theorem for $\mathscr{M}$, as claimed. In fact, suppose we have shown that given $\{R_k\}$ there exists a subsequence $\{\tilde{R}_k\}$ so that

(1) $\mu\Big(\bigcup_k \tilde{R}_k\Big) \gtrsim \mu\Big(\bigcup_k R_k\Big)$,

(2) $\Big\|\sum_k \chi_{R_k}\Big\|_{p'} \lesssim \mu\Big(\bigcup_k R_k\Big)^{1/p'}$, where $1/p + 1/p' = 1$.

Then

$$\mu(\{\mathscr{M} f > \alpha\}) \leq \Big(c\,\frac{\|f\|_p}{\alpha}\Big)^p.$$

In fact, we have by definition $\{R_k\}$ so that

$$\{\mathscr{M} f > \alpha\} \subseteq \bigcup_k R_k \quad \text{and} \quad \frac{1}{|R_k|}\int_{R_k} |f| > \alpha \quad \text{for all } k.$$

By the covering lemma, select the class $\{\tilde{R}_k\}$. Then

$$|\tilde{R}_k| \leq \frac{1}{\alpha}\int_{\tilde{R}_k} |f|,$$

and so

$$\mu\Big(\bigcup_k \tilde{R}_k\Big) \leq \frac{1}{\alpha}\int f \sum_k \chi_{\tilde{R}_k} \leq \frac{1}{\alpha}\,\|f\|_p \Big\|\sum_k \chi_{R_k}\Big\|_{p'} \leq \frac{C}{\alpha}\,\|f\|_p\, \mu\Big(\bigcup_k R_k\Big)^{1/p'}$$
$$\leq C'\,\|f\|_p\, \mu\Big(\bigcup_k \tilde{R}_k\Big)^{1/p'},$$

and the estimate on $\mu(\{\mathcal{M} f > a\})$ follows by dividing both sides by $\mu(\bigcup_k \tilde{R}_k)^{1/p'}$.

The argument presented above yields the analogous theorem for a general lacunary set of directions in $\mathbb{R}^2$, namely a set $\{\theta_k\}$ such that $\theta_k \searrow \theta$ and $(\theta_k - \theta_{k+1})/(\theta_{k+1} - \theta_{k+2}) \geq \rho > 1$ for $k = 1, 2, 3, \ldots$

Furthermore, the result is true for general $p > 1$. That is, we have the inequality

$$\mu\{\mathcal{M} f > \alpha\} \leq c_p(\rho)\, \alpha^{-p}\, \|f\|_p^p$$

for $p > 1$, see [50]. But the proof given there does not follows the suprematistic approach of this chapter.

Chapter 3
As a needle turn: Besicovitch or Kakeya sets

The rectangle is not a conscious form. It is the creation of intuitive reason. The face of the new art. The first step to pure creation.

Kazimir S. Malévich

3.1 Understanding the enemy

If we allow the rectangles to have arbitrary directions, then the differentiation theorem turns out to be false, even for indicator functions of measurable sets and the corresponding maximal function can be infinite a.e. for functions on L^p-space, $1 \leq p < \infty$. References [46] and [32] contain proofs of those facts.

There is an interesting remark by J. Littlewood ([69], p. 392): he says that, in order to prove a theorem, the right move is to find the worst enemy of what one want to demonstrate and, then, introduce the adequate hypothesis to force him to change sides. Littlewood gives an excellent example of his principle: suppose that we have an analytic function $F(z)$ on the half strip, $-\pi/2 \leq x \leq \pi/2$, $y \geq 0$, which satisfies $|F(z)| \leq 1$ on the boundary. Is it true that $|F(z)| \leq 1$ inside the strip?

The function $F_0(z) = \exp(e^{-iz} - 1)$ shows that the answer is no. On the other hand, if we add the condition $|F(z)| \leq K \exp[\exp((1-\varepsilon)y)]$ for some $\varepsilon > 0$, then the answer is yes, because that condition kills the enemy and plays an important role in the proof. Curiously enough, Littlewood used this philosophy in his quoted paper to claim, in 1962, that Lusin's conjecture about a.e. convergence of Fourier series should be false, just a few years before the famous proof of Carleson!

Question 3.1. *Find a plane set of minimal area inside which one can move continuously a needle and change the positions of its extremal ends.*

This curious problem, formulated more than a century ago by the Japanese mathematician Soichi Kakeya in [71], has become a central issue of contemporary harmonic analysis and, in a certain sense, introduces an element of contradiction with respect to one of the more established paradigms of mathematical analysis, namely Lebesgue's theory of integration.

A needle will be here an oriented line segment $\overrightarrow{AB}$ of length 1 placed in $\mathbb{R}^2$. The first set coming to our minds where one can turns its position to $\overrightarrow{BA}$ is a circle of radius $1/2$ centered at its midpoint, whose area is $\pi/4$. But this is far from being the

A. Córdoba, *Suprematism in Harmonic Analysis*, Progress in Mathematics 356,
https://doi.org/10.1007/978-3-031-73737-4_3

solution; in particular, there is another plane set whose boundary is a deltoid, where we can also turn the needle and whose area $\pi/8$. Today we know that both sets are irrelevant to answer the question of Kakeya, but that was discovered by the Ukrainian mathematician Abram Besicovitch, one of the creators of geometric measure theory.

Besicovitch was interested in a slightly different problem for which he constructed a closed bounded set in $\mathbb{R}^2$, of measure 0, containing a unit segment on any direction. In particular, the construction implies that there is no such set of minimal area, because for every $\varepsilon > 0$ there is a plane set of area less than ε such that one can move continuously a needle in its interior and revert the position of the extremal points.

Besicovitch published his paper in a not too well-known Russian journal [58] in the turbulent year 1919, without having knowledge of the question asked by Kakeya, as it is stated in the article that he wrote years later in the American Mathematical Monthly, [62]. His motivations, nevertheless, were different and related to the understanding of Fubini's theorem, valid for Lebesgue's integral, but false, in general, for the Riemann method of integration. It is not difficult to find a function f in $\mathbb{R}^2$ which is Riemann integrable (that is, its set of discontinuities has measure 0 in $\mathbb{R}^2$) but so that, for a certain value, the function $x \longrightarrow f(x, y_0)$ is not Riemann integrable because its set of discontinuities has positive measure in the real line.

An example is the function f defined in the unit square Q by

$$f(x,y) = \begin{cases} 1 & \text{if } y \neq y_0, \text{ for certain } 0 < y_0 < 1, \\ 1 & \text{if } y = y_0 \text{ and } x \text{ is irrational}, \\ 0 & \text{if } y = y_0 \text{ and } x \text{ is rational}. \end{cases}$$

Then the double integral $\iint_Q f(x, y)\,dx\,dy = 1$ exists, but $\int_0^1 (\int_0^1 f(x, y)\,dx)\,dy$ does not makes sense because $f(x, y_0)$ is not Riemann integrable.

However, in this example the difficulty disappears if we are clever and rotate the coordinate axes.

Therefore, the following question arises: given a Riemann integrable function in $\mathbb{R}^2$, can we choose an adequate coordinate axes so that the double integral equals the iterated one?

The set constructed by Besicovitch gives a negative answer, but it is also, following Littlewood, the main enemy for the validity of Lebesgue's differentiation theorem when we consider averages over rectangles with arbitrary directions in the plane.

Theorem 3.2. *For every $0 < \varepsilon < 1$, there is a compact plane set K_ε containing a rectangle of dimensions $\varepsilon \times 1$ on any direction, and such that its Lebesgue measure $\mu(K_\varepsilon)$ is bounded above by $C/(1 + |\log \varepsilon|)$, where $C < \infty$ is a universal constant.*

Proof. (a) *Basic construction.* Starting with a triangle T_0 having basis of length b_0 and height h_0, we use the median through the mid-point of its basis to divide it into two subtriangles: $T_0 = T_0^1 \cup T_0^2$.

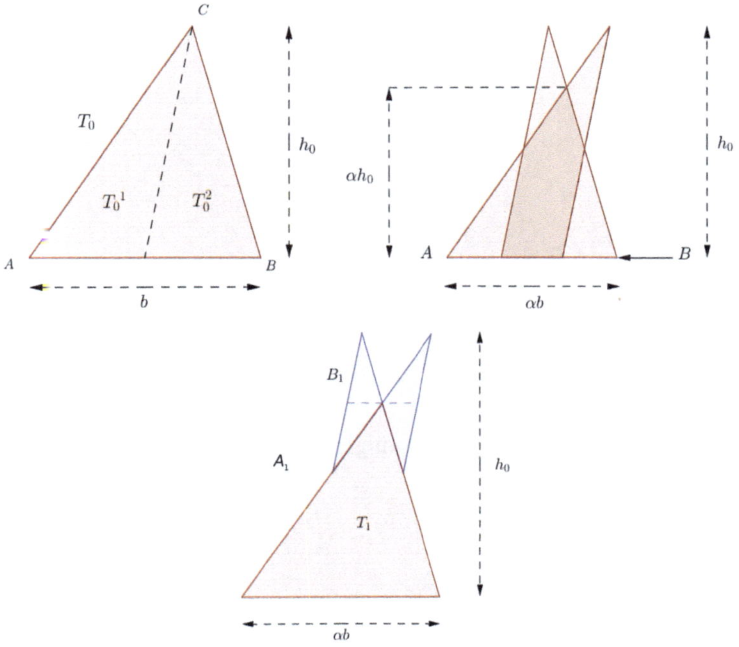

Next, with a factor α, $1/2 < \alpha < 1$, we translate T_0^1 and T_0^2 overlapping their bases and originating a figure A_1 (tree) consisting on a triangle T_1 (the *trunk*) similar to T_0 by the ratio α, and *branches* B_1, which are union of four triangles similar to either T_0^1 or T_0^2 with ratio $1-\alpha$.

An elementary computation yields

$$\mu(A_1) = \mu(T_1) + \mu(B_1) = (\alpha^2 + 2(1-\alpha)^2)\,\mu(T_0) < \mu(T_0)$$

for any choice of α, $1/2 < \alpha < 1$.

(b) The idea is to iterate the basic construction many times. Beginning now with an equilateral triangle T_0 of height h_0, we divide its basis of length b_0 onto 2^n segments of equal length $b_0\,2^{-n}$, and decompose T_0 into 2^n subtriangles T_0^j, $j = 1, \dots, 2^n$

Then with each couple (T_0^{2k-1}, T_0^{2k}), $k = 1, \dots, 2^{n-1}$, and with a parameter α_1, $1/2 < \alpha_1 < 1$, to be fixed later, we construct the tree $A_1^k = T_1^k \cup B_1^k$, and observe that two of those consecutive triangles T_1^k, T_1^{k+1} have always two parallel sides (different of their bases). Therefore, they can all be translated to obtain a new triangle T_1, similar to T_0, whose base and height are, respectively, of lengths $\alpha_1 b_0$ and $\alpha_1 h_0$. We translate also, accordingly, the corresponding branches B_1^k so that each one of the original triangles T_0^k is, after translation, contained in the resulting tree $A_1 = T_1 \cup \{\text{translated branches}\}$, as in the figure above.

We have

$$\mu(A_1) \le \mu(T_1) + \mu(B_1^k) \le \{\alpha_1^2 + 2(1-\alpha_1)^2\}\mu(T_0).$$

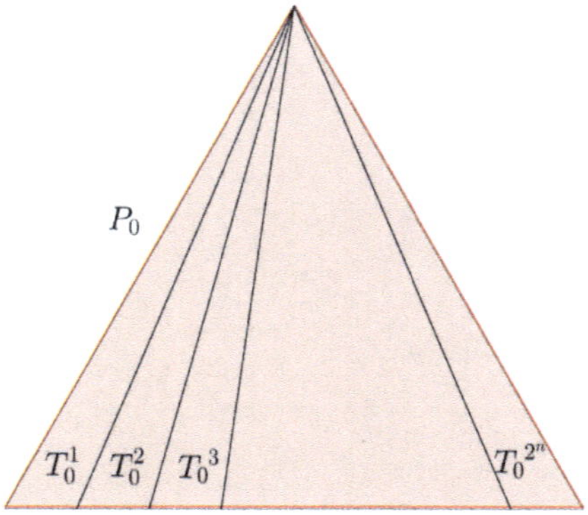

Fig. 3.1: Triangle subdivision.

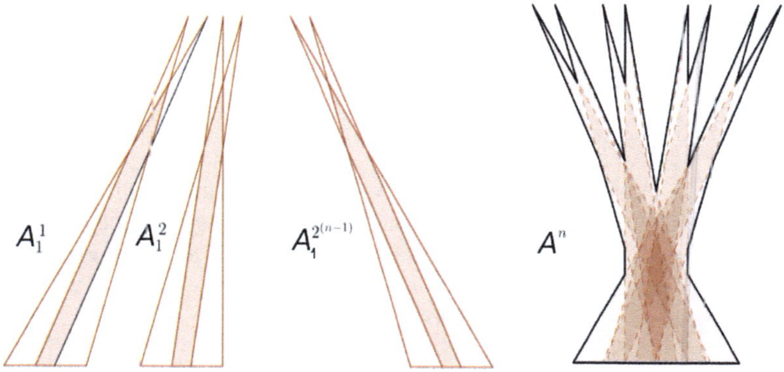

Fig. 3.2: Besicovitch pairings.

Observe now that T_1 is union of 2^{n-1} triangles having equal bases and heights of lengths, respectively, $\alpha_1\, 2^{-n+1}\, b_0$ and $\alpha_1\, h_0$. Therefore, we can repeat the construction with another ratio α_2, $1/2 < \alpha_2 < 1$, to obtain a second tree $A_2 = T_2 \cup \{\text{branches}\}$ satisfying

$$\mu(A_2) \leq \left[\alpha_1^2\alpha_2^2 + 2((1-\alpha_1)^2 + \alpha_1^2(1-\alpha_2)^2)\right] \mu(T_0),$$

and so on.

The iteration of this process n times produces a tree A_n so that

$$\mu(A_n) \leq \{\alpha_1^2\alpha_2^2 \cdots \alpha_n^2 + 2[(1-\alpha_1)^2 + \cdots + \alpha_1^2 \cdots \alpha_{n-1}^2(1-\alpha_n)^2]\}\, \mu(T_0),$$

and having the property that each one of the original triangles T_0^k can be translated inside A_n.

The choice

$$\alpha_k = 1 - \frac{1}{n+2-k} = \frac{n+1-k}{n+2-k}, \qquad k = 1, 2, \ldots, n,$$

yields the estimate

$$\mu(A_n) \le \frac{2}{n+1}\,\mu(T_0).$$

If we begin now with $h_0 = 2$, the union of the set A_n with its rotation by the angles $\pi j/3$, $j = 1, 2, 3, 4, 5$, gives us the set K_ε with an appropriate election of n, namely $2^{-n} \sim \varepsilon$. □

It is clear than inside K_ε one can place a needle on each plane direction. However, adding to K_ε a set of arbitrarily small measure we will be also able to move the needle continuously inside the set. The reason is that K_ε is a union of triangles whose biggest sides can be arranged in pairs having equal directions, and we can always translate a segment between two parallel lines paying with an arbitrary small area, as the following figure shows:

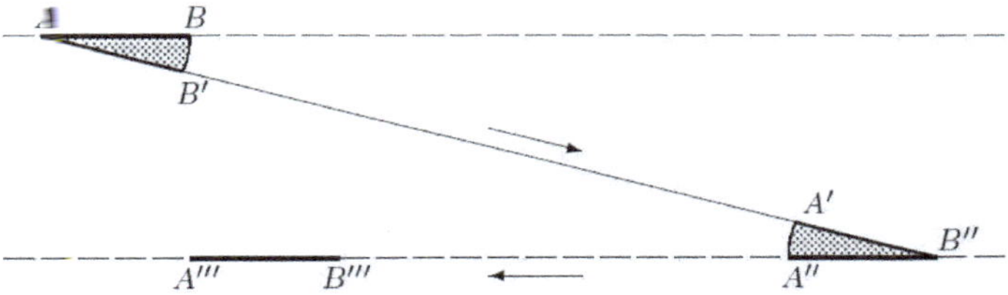

The dashed region can be made of area as small as we want if we move far enough and then take advantage of the fact that a segment moving inside a straight line generates zero area.

Therefore, as was pointed earlier, the original question of Kakeya has the following answer: there is no such set of minimal area, because for every $\varepsilon > 0$ there is a plane region of area less than ε where the needle can be turned.

The Besicovitch construction is optimal, as the following theorem shows.

Theorem 3.3. *There exists an effective constant $C > 0$ so that if a plane set contains a rectangle of dimensions $\varepsilon \times 1$, $0 < \varepsilon < 1$, on each direction, then its area must be bigger than $C/[1 + |\log \varepsilon|]$.*

Proof. Let A be such a set. For each j, $0 \le j \le \varepsilon^{-1}$, let $R_j \subset A$ be a rectangle of dimension $\varepsilon \times 1$ whose direction (i.e., the one of its biggest side) is $\omega_j = \frac{1}{2}\pi\varepsilon j$. Hölder's inequality yields

$$1 \le \sum_j \mu(R_j) = \int \sum_j \chi_{R_j} \le \Big[\mu\Big(\bigcup_j R_j\Big)\Big]^{1/2} \Big[\sum_{j,k} \mu(R_j \cap R_k)\Big]^{1/2}.$$

A straightforward trigonometric computation gives the estimate

$$1 \leq [\mu(A)]^{1/2} \left[\sum_{j,k=1}^{\varepsilon^{-1}} \frac{\varepsilon}{|j-k|+1} \right]^{1/2},$$

implying $\mu(A) \gtrsim (1 + |\log \varepsilon|)^{-1}$. □

Next we shall modify the Besicovitch construction to obtain a limit set K.

Theorem 3.4. *There are compact sets $K \subset \mathbb{R}^2$ of measure $\mu(K) = 0$ containing a unit segment on every plane direction.*

Proof. Starting with an equilateral triangle of side length 2, we divide its basis in 2^{n_0} equal parts, where n_0 is big enough so that when we apply the basic construction of Besicovitch to the resulting adjacent triangles T_0^{2j-1}, T_0^{2j}, $j = 1, \ldots, 2^{n_0-1}$, using a parameter $\alpha = 2/3$, the corresponding branches are located in the region above the line $y = 1$.

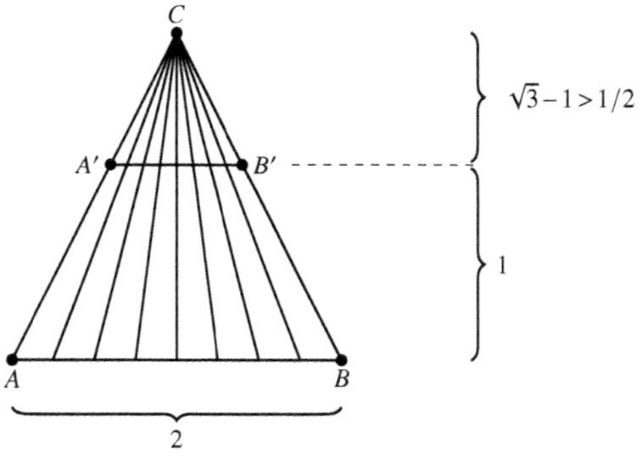

We have freedom to move T_0^{2j-1} in the direction of T_0^{2j} or viceversa, except for the two extremes, that is, we want the segment $AA'BB'$ to be included in the resulting triangles $T_1^1, T_1^2, \ldots, T_1^{2^{n_0-1}}$, and, therefore, contained in the family of romboids

$$R_1^{\nu_1} = T_1^{\nu_1} \cap \{0 \leq y \leq 1\}, \qquad \nu_1 = 1, \ldots, 2^{n_0-1}.$$

Let us observe that the set $\bigcup_{\nu_1} R_1^{\nu_1}$ contains an straight line segment of length 1 on each direction of the original triangle (i.e., the set of directions Γ given by the angle at the vertex C), and satisfy the estimate

$$\mu\Big(\bigcup_{\nu_1} R_1^{\nu_1}\Big) \leq \frac{2}{3}\,\mu(R_0),$$

where R_0 is the romboid whose vertices are $AA'BB'$.

Next we apply the same process to each triangle $T_1^{\nu_1}$ to obtain another family of romboids $\{R_2^{\nu_1,\nu_2}\}$ so that:

(i) $\mu(\bigcup_{\nu_1,\nu_2} R_2^{\nu_1,\nu_2}) \le (\frac{2}{3})^2 \mu(R_0)$.
(ii) A segment of length 1 and direction in the original angle $\widehat{ACB}$ at the vertex C can be placed inside $\bigcup_{\nu_1,\nu_2} R_2^{\nu_1,\nu_2}$.
(iii) $\bigcup_{\nu_1,\nu_2} R_2^{\nu_1,\nu_2}$ contains the lateral boundary segments of all romboids $\{R_1^{\nu_1}\}$.

The iteration of this process yields, for each $j = 1, 2, \ldots$, a family of closed romboids $\{R_j^{\mu_1,\ldots,\nu_j}\}$ and a compact set $K_j = \bigcup_{\nu_1,\ldots,\nu_j} R_j^{\mu_1,\ldots,\nu_j}$ such that:

(iv) $\mu(K_j) \le (\frac{2}{3})^j \mu(R_0)$,
(v) K_j contains a segment of length 1 on each direction of the original set Γ.
(vi) There is a dense set of directions $\{\theta_e\} \subset \Gamma$ such that we can find a straight line segment I_{θ_e}, of direction θ_e, which is contained in all the K_j, $j = 1, 2, \ldots$

Let $K = \bigcap_j K_j$. Then K compact, $\mu(K) = 0$, and for each $\theta \in \Gamma$, there is *a* sequence of those θ_e converging to θ, implying that the corresponding segments I_{θ_e} must converge to a segment I_θ, of direction θ, which, because of compactness, has to be also contained in the set K. □

Definition 3.5. A Besicovitch (or Kakeya) set in the plane is a set of measure 0 containing an straight line segment of fixed length on each direction.

Corollary 3.6. *The Minkowski (box counting) dimension of a Besicovitch set in the plane is* 2.

Proof. Without loss of generality we may assume that K contains segments of length 1 on every direction. For each $\varepsilon > 0$, the set $K_\varepsilon = \{x : \operatorname{dist}(x, K) \le \varepsilon\}$ contains rectangles $\varepsilon \times 1$ on each plane direction. Therefore

$$\mu(K_\varepsilon) \ge \frac{1}{1 + |\log \varepsilon|} \quad \text{and} \quad \lim_{\varepsilon \to 0} \left[2 - \frac{\log \mu(K_\varepsilon)}{\log \varepsilon}\right] = 2.$$ □

3.2 The Kakeya maximal function

With $N \gg 1$ and $\rho > 0$, let us consider the family $\mathcal{B}_N^\rho$ of all plane rectangles of dimensions $\rho \times \rho N$ and arbitrary directions. Given a locally integrable function f in $\mathbb{R}^2$, we have the maximal operator

$$M_N^\rho f(x) = \sup_{x \in R \in \mathcal{B}_N^\rho} \frac{1}{\mu(R)} \int_R |f(y)| \, d\mu(y).$$

Theorem 3.7. *There is a finite constant C (independent of N and ρ) so that the following estimate holds for every $f \in L^2(\mathbb{R}^2)$:*

$$\|M_N^\rho f\|_2 \le C \, [\log(1 + N)]^{1/2} \, \|f\|_2.$$

Remark 3.8. It is enough to prove the case $\rho = 1$, as a convenient dilation argument shows. Since each $R \in \mathcal{B}_N^1$ is contained in a square of side N, it turns out that $M_N^1 f(x) \leq N \cdot M f(x)$, where M denotes the Hardy–Littlewood maximal function. Therefore, the main point of the theorem is not the boundedness of M_N^ρ, but the estimate $(\log N)^{1/2}$ for its norm.

Proof of Theorem 3.7. The family $\mathcal{B}_N^1$ can be divided into two subfamilies, $\mathcal{B}_{N,1}^1 \cup \mathcal{B}_{N,2}^1$, according to the direction θ of it rectangles: $\theta \in [-\pi/4, \pi/4]$ for $\mathcal{B}_{N,1}^1$, and $\theta \in [\pi/4, 3\pi/4]$ for $\mathcal{B}_{N,2}^1$.

Obviously, $\mathcal{B}_{N,1}^1$ and $\mathcal{B}_{N,2}^1$ have associated maximal functions having identical bounds. Thus, it will be enough to consider one of them only.

Let $\mathbb{R}^2 = \bigcup_{\nu \in \mathbb{Z}^2} Q_\nu$ be the standard decomposition of the plane into a grid of disjoints squares of area 1 and sides parallel to the coordinate axes.

Given $f \in L^2(\mathbb{R}^2)$, we choose for each ν a rectangle $R_\nu \in \mathcal{B}_{N,1}^1$ satisfying $Q_\nu \cap R_\nu \neq \emptyset$ and

$$\sup_{x \in Q_\nu} M_{N,1}^1 f(x) \leq 2 \frac{1}{\mu(R_\nu)} \int_{R_\nu} |f(y)| \, \mathrm{d}\mu(y),$$

that is,

$$M_{N,1}^1 f(x) \leq 2 \sum_\nu \left(\frac{1}{N} \int_{R_\nu} |f(y)| \, \mathrm{d}\mu(y) \right) \cdot \chi_{Q_\nu}(x).$$

Our estimate will be then a consequence of the following result.

Lemma 3.9. *There is a finite constant C, independent of N, such that for any choice of rectangles $R_\nu \in \mathcal{B}_{N,1}^1$ satisfying $R_\nu \cap Q_\nu \neq \emptyset$, the operator defined by the formula*

$$T f(x) = \sum_\nu \left(\frac{1}{\mu(R_\nu)} \int_{R_\nu} f(y) \, \mathrm{d}y \right) X_{Q_\nu}(x)$$

satisfies the estimate

$$\|T f\|_2 \leq C \, [1 + \log N]^{1/2} \, \|f\|_2 \qquad \textit{for every } f \in L^2(\mathbb{R}^2).$$

Proof. To prove the lemma, it is convenient to consider the adjoint operator

$$T^* g(x) = \sum_\nu \left(\frac{1}{N} \int_{Q_\nu} g(y) \, \mathrm{d}y \right) \chi_{R_\nu}(x),$$

and note that, because of the dimensions $1 \times N$ of the involved rectangles, it is enough to check the action of T^* on functions supported on a square of dimensions $N \times N$.

Let Q be such a square $N \times N$ having sides parallel to the coordinate axes. Then $Q = E_i \cup \cdots \cup E_N$, where each E_j, $j = \ldots, N$, is a vertical band of thickness 1. We have $g = \sum_1^N g \cdot \chi_{E_j}$ for each $g \in L^2(Q)$, and the lemma will be a consequence of the estimate

$$\|T^*(g\cdot\chi_{Ej})\|_2 \lesssim N^{-1/2}(\log N)^{1/2}\,\|g\cdot\chi_{E_j}\|_2.$$

To see this, let us decompose further $Ej = \bigcup_{i=1}^N Q_i$, where the squares Q_i belong to the initial grid $\{Q_\nu\}$ and the index i indicates position in the vertical direction. Let also R_i be the corresponding assigned rectangle of $\mathcal{B}_{N,1}^1$ in the definition of the operator T. Then

$$\int |T^*(g\cdot\chi_{E_j})(x)|^2\,\mathrm{d}x \lesssim \sum_{i,k=1}^N \frac{1}{N^2}\Big(\int_{Q_i}|g|^2\Big)^{1/2}\Big(\int_{Q_k}|g|^2\Big)^{1/2}\frac{N}{|i-k|+1}$$
$$\lesssim \frac{1}{N}(1+\log N)\,\|g\cdot\chi_{E_j}\|_2^2. \qquad \square$$

Theorem 3.7 admits the following improvement: given $N \gg 1$ denote, by $\mathcal{B}_N$ the class of rectangles in $\mathbb{R}^2$ having eccentricity (ratio of their sides length) equal to N, but otherwise arbitrary directions. Let us consider also the associated maximal function:

$$Mf(x) = \sup_{x\in R\in\mathcal{B}_N}\frac{1}{\mu(R)}\int_R |f(y)|\,\mathrm{d}\mu(y),$$

defined over locally integrable functions f.

Theorem 3.10. *There are finite positive constants C and a so that*

$$\|M_N f\|_s \le C(\log N)^a\,\|f\|_2 \quad \textit{for every function } f\in L^2(\mathbb{R}^2).$$

Proof. First let us observe that, given any rectangle R of eccentricity N, we can find another rectangle $\tilde{R}$ with equal dimensions and direction in the set of angles $\{\pi k N^{-1}\}_{k=1,\ldots,N}$ such that R is contained in the double rectangle $(\tilde{R})^*$. From this fact, it follows that without loss of generality, it suffices to prove the statement above when we only consider rectangles in such restricted set of directions $\{\pi k N^{-1}\}_{k=1,\ldots,N}$.

Using a similar argument, we may also consider that the smaller side has length of the form 2^n, $n\in\mathbb{Z}$.

Next, let us introduce the following notation:

$$T_{2^n}^j f(x) = \sup_{x\in R}\frac{1}{\mu(R)}\int_R |f(y)|\,\mathrm{d}\mu(y),$$

where the sup is now taken over all rectangles having dimensions $2^n\times 2^n N$ and direction $\pi j N^{-1}$; and

$$T_{2^n} = \sup_j T_{2^n}^j, \qquad T^j = \sup_n T_{2^n}^j \qquad \text{and} \qquad T = \sup_j T^j = \sup_n T_{2^n}.$$

Given $\alpha > 0$ we can apply the standard (Vitali) covering lemma to get, for each j, a sequence of rectangles $\{R_l^j\}_{l=1,2,\ldots}$ having direction $\pi j N^{-1}$, pairwise disjoints, and such that

$$E_\alpha^j = \{x : T^j f(x) \geq \alpha\} \subset \bigcup_l (R_l^j)^* \qquad \text{and} \qquad \frac{1}{\mu(R_l^j)} \int_{R_l^j} |f(y)| \, \mathrm{d}\mu(y) \geq \alpha.$$

Then, by the preceding remarks, we know that

$$E_\alpha = \{x : M_N f(x) \geq 4\alpha\} \subset \bigcup_{j=1}^{N} E_\alpha^j.$$

Let n_0 be the biggest integer such that there is a rectangle of dimensions $2^{n_0} \times 2^{n_0} N$ in, at least, one of the sequences above, and consider the collection of all of them. We can select a subfamily F_0 with the following property: no rectangle in F_0 is contained in the double of other, and if a rectangle has been eliminated, then it must be contained in the double of a member of F_0.

By induction, we get for each k a collection $\mathcal{F}_k$ of rectangles having dimensions $2^{n_k} \times 2^{n_k} N$ (with $n_0 > n_1 > \cdots$) in such a way that no rectangle in $\mathcal{F}_k$ is contained in the double of another rectangle in $\mathcal{F}_j$, $j \leq k$, and if R belongs to the primitive N-collections, then either R is in some $\mathcal{F}_k$, or is contained in the double of a member of $\bigcup_k \mathcal{F}_k$.

It follows that

$$E_\alpha \subset \bigcup_{R \in \bigcup_k \mathcal{F}_k} (R^*)^*.$$

Next, with $k = 0, 1, 2, \ldots$, let us define

$$\Delta_k = \left\{R \in \textstyle\bigcup_j \mathcal{F}_j : \dim(R) = 2^n \times 2^n N; n_0 - k \log N \geq n \geq n_0 - (k+1) \log N\right\},$$
$$E_i = \bigcup_{R \in \Delta_i} R, \qquad E_i^{**} = \bigcup_{R \in \Delta_i} (R^*)^*.$$

Observe that the family of sets $\{E_i\}$ is "almost disjoint", i.e., $E_i \cap E_j = \emptyset$ if $|i - j| \geq 2$. This is because if $R_1 \in \Delta_i$, $R_2 \in \Delta_j$, and $i - j \geq 2$, then the biggest side of R_2 is smaller than the smallest side of R_1, and if $R_1 \cap R_2 \neq \emptyset$, then R_2 is contained in R_1^*, in contradiction with our choice of the sequences $\mathcal{F}_k$.

Let $f_i = f/E_i$, $i = 0, 1, \ldots$, and let S_i be the maximal function defined by

$$S_i f(x) = \sup_{x \in R} \frac{1}{\mu(R)} \int_R |f(y)| \, \mathrm{d}y,$$

where the supremum is taken here over rectangles of dimensions $2^n \times 2^n N$, where

$$n_0 + 2 - i \log N \geq n \geq n_0 + 2 - (i+1) \log N.$$

By Theorem 3.7, we know that S_i is bounded on $L^2(\mathbb{R}^2)$, and that its norm is bounded by $C(\log N)^{3/2}$, with C independent of N and i.

Given $x \in E_i^{**}$, there is $R \in \Delta_i$ such that $x \in (R^*)^*$, implying that

$$S_i f_i(x) \geq \frac{1}{16} \frac{1}{\mu(R)} \int_R |f_i(y)| \mathrm{d}\mu(y) \geq \frac{\alpha}{16},$$

that is,

$$E_i^{**} \subset \{x : S_i f_i(x) \geq \alpha/16\}, \qquad \mu(E_i^{**}) \lesssim (\log N)^3 \frac{\|f_i\|_2^2}{\alpha^2},$$

$$\mu(E_\alpha) \leq \sum_i \mu(E_i^{**}) \lesssim (\log N)^3 \frac{1}{\alpha^2} \int f(x)^2 \sum_i \chi_{E_i}(x) \mathrm{d}\mu(x)) \lesssim (\log N)^3 \frac{1}{\alpha^2} \|f\|_2^2,$$

implying that M_N satisfies a weak-L^2 estimate with bound $\lesssim (\log N)^3$.

Since M_N is bounded on L^∞ and satisfies a weak-L^1 inequality with bound N, a straightforward application of the Marcinkiewicz and the Riesz–Thorin interpolation theorems yields the strong L^2-estimate of the theorem. □

3.3 Hausdorff dimension of Kakeya sets

Given a Borel set $E \subset \mathbb{R}^n$ and $\alpha \geq 0$, $\delta > 0$, we define

$$H_\delta^{(E)} := \inf \sum [\mathrm{diam}\,(B_k)]^\alpha,$$

where the infimum is taken over all countable coverings of E by cubes of diameter less than δ. Define also

$$H^\alpha(E) := \lim_{\delta \to 0} H_\delta^\alpha(E).$$

This limit always exists, although can be infinite, and has several basic properties:

(1) H^α is countable additive, invariant under translations and rotations, and satisfies that $H^\alpha(\lambda E) = \lambda^\alpha H^\alpha(E)$ for $\lambda > 0$.
(2) $H^0(E)$ counts the number of points inside E, while $H^n(E) = c_n \mu(E)$ for all Borel sets $E \subset \mathbb{R}^n$, where
$$c_n = \frac{\pi^{n/2}}{2^n\,(n/2)!}.$$
(3) $H^\alpha(E) < \infty$ implies that $H^\beta(E) = 0$ if $\beta > \alpha$. If $\beta < \alpha$ and $H^\alpha(E) > 0$, then $H^\beta(E) = \infty$.

Therefore, given a Borel set $E \subset \mathbb{R}^n$, there is a unique α such that $H^\beta(E) = 0$ if $\beta > \alpha$ and $H^\beta(E) = \infty$ if $\beta < \alpha$.

Definition 3.11 That unique α is called the Hausdorff dimension of E. That is,

$$\dim_H(E) = \sup\{\beta : H^\beta(E) = \infty\} = \inf\{\alpha : H^\alpha(E) = 0\}.$$

If a Borel set $E \subset \mathbb{R}^n$ has positive Lebesgue measure, then $\dim_H(E) = n$.

An interesting observation is that one can substitute cubes by balls, or dyadic cubes, in the definition above without changing the value of $\dim_H(E)$. And that the Minkowski (or box counting dimension) dimension is, in general, bigger than the Hausdorff dimension.

Theorem 3.12. *Every Besicovitch (or Kakeya) set K in $\mathbb{R}^2$ satisfies* $\dim_H(K) = 2$.

Proof. Let $\{Q_\nu\}$ be a covering of K by dyadic cubes of side lengths $\ell(Q_\nu) \leq 2^{-k_0}$:

$$K \subset \bigcup_\nu Q_\nu = \bigcup_{k=k_0}^{\infty} \Big(\bigcup_\nu Q_\nu^k\Big), \qquad \text{where } \ell(Q_\nu^k) = 2^{-k}.$$

Let us introduce the following notation:

$$E^k = \bigcup_\nu Q_\nu^k, \quad N_k = \mathrm{Card}\{Q_\nu^k\} = \text{number of cubes of length } 2^{-k} \text{ in the covering.}$$

It then follows from the definition that we need to show that

$$\sum_{k \geq k_0} N_k\, 2^{-k\alpha} \geq c_\alpha > 0, \qquad \text{for every } \alpha < 2.$$

In fact, it will be convenient to try something stronger, namely:

Claim. *For each $\alpha < 2$, there exists $k \geq k_0$ so that $N_k\, 2^{-k\alpha} \geq c_\alpha > 0$.*

Since K is a Besicovitch set, it contains line segments L_θ of length 1 for each direction $\theta \in S^1$. Therefore,

$$1 = \ell(L_\theta) \leq \sum_{k \geq k_0} \ell(L_\theta \cap E^k).$$

Then for each $\beta > 0$, there is $C_\beta > 0$ and $k \geq k_0$ such that

$$\ell(L_\theta \cap E^k) \geq C_\beta\, 2^{-k\beta}.$$

(This is true because otherwise we would have the existence of $\beta > 0$ such that for all $C_\beta > 0$,

$$1 \leq \sum_{k \geq k_0} \ell(L_\theta \cap E^k) \leq C_\beta \frac{2^{-k_0\beta}}{1 - 2^{-\beta}},$$

which is clearly false for C_β small enough.)

Given $\beta > 0$ and its associated $C_\beta > 0$, let us consider the set

$$\Gamma_k = \{\theta \in S^1 : \ell(L_\theta \cap E^k) \geq C_\beta\, 2^{-k\beta}\}.$$

It follows that

$$1 \leq \mu(S^1) \leq \sum_{k \geq k_0} \mu(\Gamma_k),$$

because each angle $\theta \in S^1$ has, at least, a value $k \geq k_0$ satisfying

$$\ell(L_\theta \cap E^k) \geq C_\beta\, 2^{-k\beta}.$$

It then follows the existence of $k \geq k_0$ so that $\mu(\Gamma_k) \geq C_\beta\, 2^{-k\beta}$ (because of the same reason as before: in the opposite case, we will have the inequality

$$1 \leq c_\beta \sum_{k \geq k_0} 2^{-k\beta} = C_\beta \frac{2^{-k_0\beta}}{1 - 2^{-\beta}},$$

which is false for C_β small enough).

Therefore one can select at least $\frac{1}{2}\beta\, 2^{k(1-\beta)}$ different angles $\{\theta_j\}$ inside Γ_k such that $|\theta_j - \theta_i| \geq 2^{-k}$ if $j \neq i$.

Consider then the corresponding L_{θ_j} segments of those selected directions, and associate to each of them a rectangle R_j of dimensions $2^{-k} \times 1$ in the direction θ_j. We have the following inequalities:

$$\sum_j \mu(R_j) \geq 2^{-k} \frac{1}{2} C_\beta\, 2^{k(1-\beta)} = \frac{1}{2} C_\beta\, 2^{-k\beta}$$

and

$$\sum_j \mu(R_j) = \int_{\bigcup_j R_j} \sum_j \chi_{R_j}(x)\,\mathrm{d}x \leq \mu\Big(\bigcup_j R_j\Big)^{1/2} \Big[\sum_{j,i} \mu(R_i \cap R_j)\Big]^{1/2}$$

$$\lesssim \mu\Big(\bigcup_j R_j\Big)^{1/2} \Big[\sum_{j,i} \frac{-2^k}{|j-i|+1}\Big]^{1/2} \lesssim \mu\Big(\bigcup_j R_j\Big)^{1/2} \left|\log\left(\tfrac{1}{2} C_\beta\, 2^{k(1-\beta)}\right)\right|^{1/2}.$$

That is,

$$\mu\Big(\bigcup_j R_j\Big) \gtrsim \frac{\left[\frac{1}{2} C_\beta\, 2^{-k\beta}\right]^2}{\left|\log(\frac{1}{2} C_\beta\, 2^{k(1-\beta)})\right|} \geq \widetilde{C}_\beta \frac{2^{-2k\beta}}{k}, \tag{3.1}$$

where $\widetilde{C}_\beta > 0$.

Denoting by $\mathcal{M}_k$ the maximal function respect rectangles of dimension $2^{-k} \times 1$ and arbitrary directions, we have

$$\bigcup_j R_j \subset \{x : \mathcal{M}_k(\chi_{E^k})(x) \geq C_\beta\, 2^{-k\beta}\},$$

and the estimates obtained for the Kakeya maximal function yield

$$\mu\Big(\bigcup_j R_j\Big) \leq \frac{C\,(\log 2^k)\,\mu(E^k)}{C_\beta^2\, 2^{-2k\beta}},$$

which, combined with the estimate (3.1) obtained above, produces the inequality

$$\mu(E^k) \gtrsim \widetilde{C}_\beta \, \frac{2^{-2k\beta}}{k} \, \frac{C_\beta^2 \, 2^{-2k\beta}}{C\,(\log 2^k)} \gtrsim \widetilde{\widetilde{C}}_\beta \, \frac{1}{k^2} \, 2^{-4k\beta},$$

for some constant $\widetilde{\widetilde{C}}_\beta > 0$.

On the other hand, since E^k corresponds to the set of dyadic squares in the given covering, we know that $\mu(E^k) = N_k \, 2^{-2k}$. Therefore,

$$N_k \, 2^{-\alpha k} \geq 2^{2k} \, \widetilde{\widetilde{C}}_\beta \, \frac{1}{k^2} \, e^{-(4\beta+\alpha)k} = \widetilde{\widetilde{C}}_\beta \, \frac{1}{k^2} \, 2^{k(2-4\beta-\alpha)} \geq C_\alpha > 0,$$

for some constant $C_\alpha > 0$ so long as $2 - 4\beta - \alpha > 0$, which is true for some $\beta > 0$ if $\alpha < 2$. □

3.4 Further results

3.4.1 Alternative constructions

For the construction of Kakeya sets it is enough to build closed sets of measure zero containing a unit segment on every direction in an open angle. Because then a union of a finite number of its rotations will satisfy the original requirement of containing a segment of straight line on each plane direction.

There is a different construction of those sets due to J.-P. Kahane, see [2, 70].

Let C be the Cantor set of points $x \in [0, 1)$ such that the expansion $x = 0, x_1 x_2 \ldots x_n \ldots$ in base 4 contains only the digits $\{0, 3\}$. That is, we divide the interval $[0, 1)$ into four intervals,

$$[0, 1) = [0, 1/4) \cup [1/4, 1/2) \cup [1/2, 3/4) \cup [3/4, 1),$$

remove the two placed at the center, and repeat the same procedure with the other two, and so on.

Let $C_1 = \frac{1}{2} C = \{\frac{1}{2} x : x \in C\}$, and place copies of C and C_1 at, respectively, the horizontal lines $y = 0$ and $y = 1$:

$$C_1 \subset E_1 = \{(x, y) : 0 \leq x \leq 1/2, y = 1\}, \qquad C \subset E = \{(x, y) : 0 \leq x \leq 1, y = 0\}.$$

Let us define K to be the set of straight line segments joining a point at the top C_1 with other at the bottom C. We have:

K is compact, has measure zero, and contains a segment on each direction determined by joining a point of E_1 with a point of E.

3.4.2 Extensions and variations

Theorem 3.10 admits several variations.

Let $\omega_j = (\cos\theta_j, \sin\theta_j)$, $0 \le \theta_1 < \theta_2 < \cdots < \theta_N \le \pi$ be a set of "equidistributed" directions in the sense that $0 < c \leqslant \theta_{j+1} - \theta_j \leqslant 2c$ and consider the family of rectangles having direction in $\{\omega_j\}$. Then the associated maximal function M verifies the estimate $\|Mf\|_2 \lesssim [1 + \log N]^a \|f\|_2$. for some absolute constant a independent of N and c.

The reference [106] contains a proof of this result which, compared with Section 2.5 for the lacunary case, suggests that, regarding the growth of the L^2-norm of the maximal function, the equidistributed is the worse case. This is a fact that was made precise by N. Katz [72], proving that the logarithmic estimate holds for a general set of N-planar directions.

There is an extension of Theorem 3.10 to higher dimension: fixing a smooth curve $\Gamma\colon [0, 1] \to S^{n-1}$ and $\delta > 0$, let us consider the δ-flat needles having directions in the curve Γ. Then, under mild conditions upon Γ, the logarithmic estimate holds for the maximal functions with constants depending on Γ, but independent of $\delta > 0$. The proof given in [109] is based on an induction argument, and makes use of the Fourier transform.

3.4.3 Partial results in higher dimensions

The construction of Besicovitch can be carried out in every dimension to produce compact sets of measure zero containing a unit segment on each direction of $\mathbb{R}^n$. But the Kakeya maximal function conjecture, in fact the main issue in this area, is quite open as well as the estimates for the Hausdorff dimensions of those Kakeya sets.

With a given δ, $0 < \delta < 1$, let us define a δ-grid of directions in S^{n-1} to be a set $\mathcal{L}_\delta = \{\omega_\alpha\} \subset S^{n-1}$ So that $\mathrm{dist}(\omega_\alpha, \omega_\beta) \ge \delta$, and for a given point $x \in S^{n-1}$, there exists an element $\omega_\alpha \in \mathcal{L}_\delta$ satisfying $\mathrm{dist}(x, \omega_\alpha) \le c\delta$, where $c \ge 1$ is a fixed constant. Clearly, we must have $\mathrm{card}(\mathcal{L}_\delta) \simeq \delta^{-(n-1)}$.

Having such a grid $\mathcal{L}_\delta$, we can consider the family $\mathcal{F}_\delta$ of δ-fat-needles in $\mathbb{R}^n$ (cylinders of height 1 and radius δ, or parallelepipeds of height 1 and basis of side δ) whose directions belong to $\mathcal{L}_\delta$.

Conjecture 3.13 (δ-fat needles conjecture). *Let $\mathcal{A} \subset \mathcal{F}_\delta$ be a family of δ-fat needles having different directions in $\mathcal{F}_\delta$. Then the following estimate holds:*

$$\int_{\mathbb{E}^n} \Big(\sum_{R\in\mathcal{A}} \chi_R \Big)^{n/(n-1)} \le C(n)\, [1 + |\log\delta|]^{a(n)} \sum_{R\in\mathcal{A}} \mu(R),$$

where $C(n)$ and $a(n)$ are independent of δ.

With the same argument used in the two-dimensional case, we get a precise estimate for the Minkowski dimension of every Kakeya set; namely, we have

$$\delta\text{-flat conjecture} \implies \dim K = n.$$

Proof. With $\delta > 0$, we let

$$K_\delta = \{x : \operatorname{dist}(x, K) \le \delta\}.$$

Then, by definition, K_δ contains a δ-fat needle on each direction of $\mathcal{L}_\delta$. Let us call $\mathcal{A}_\delta$ the collections of those needles. We have

$$\begin{aligned} 1 \lesssim \sum_{R\in\mathcal{A}_\delta} \mu(R) &= \int_{K_\delta} \sum_{R\in\mathcal{A}_\delta} \chi_R \le \mu(K_\delta)^{1/n} \int \Big(\sum_{R\in\mathcal{A}_\delta} \chi_R \Big)^{n/(n-1)} \\ &\le \mu(K_\delta)^{1/n} \left(C(n)\, [1 + |\log\delta|^{a(n)}]\right)^{(n-1)/n}, \end{aligned}$$

implying $\dim K = n$. □

In dimension $n \ge 3$, the conjecture above is an open problem. There are, however, some partial results implying lower bound estimates for the Minkowski (or box counting) dimension of Kakeya set.

Definition 3.14. A δ-bush $\mathcal{B} \subset \mathcal{L}_\delta$ is a collection of δ-fat-needles of different directions having a common intersection point.

It is an easy exercise in elementary geometry to prove that if $\mathcal{B} = \{R_j\}$ is a δ-bush, then

$$\int_{\mathbb{R}^n} \Big(\sum_j R_j(x) \Big)^{n/(n-1)} \mathrm{d}x \le C_n \,(1 + |\log\delta|) \sum_j \mu(R_j),$$

with C_n independent of δ.

Suppose now that K is a Kakeya set in $\mathbb{R}^n$ and consider its Minkowski's sausage, i.e.,

$$K_\delta = \{x \in \mathbb{R}^n : \operatorname{dist}(x, K) \le \delta\},$$

which must contain a δ-needle on each direction of the grid $\mathcal{L}_\delta$.

Let $\{R_\alpha\}$ the collection of those δ needles. With a given $M \ge 1$, we can write

$$\{R_\alpha\} = B_0 \cup B_1 \cup \cdots \cup B_r,$$

where the overlapping function of B_0 is bounded by M, and each B_j, $j = 1, \ldots, r$, is a δ-bush whose cardinality is bigger than or equal to M, implying that

$$r \le \frac{\delta^{-(n-1)}}{M}.$$

Then

$$\int \Big(\sum_\alpha R_\alpha\Big)^{n/(n-1)} \lesssim \int \Big(\sum_{R\in B_0} R(x)\Big)^{n/(n-1)} + r^{1/(n-1)} \sum_j \int \Big(\sum_{R\in B_j} R(x)\Big)^{n/(n-1)}$$
$$\lesssim \Big[M^{1/(n-1)} + \frac{\delta^{-1}}{M^{1/(n-1)}}\Big] \cdot \mu\Big(\bigcup_\alpha R_\alpha\Big) \,|\log\delta|.$$

Choosing $M^{1/(n-1)} = \delta^{-1/2}$, we get the estimate

$$\int \Big(\sum_\alpha R_\alpha\Big)^{n/(n-1)} \lesssim \delta^{-1/2} \mu\Big(\bigcup_\alpha R_\alpha\Big).$$

Therefore,

$$1 \lesssim \sum_\alpha \mu(R_\alpha) = \int_K \Big(\sum_\alpha R_\alpha\Big) \leq \mu(K_\delta)^{1/n} \Big(\int \Big(\sum_\alpha R_\alpha\Big)^{n/(n-1)}\Big)^{(n-1)/n}$$
$$\leq |\log\delta|\, \delta^{-(n-1)/(2n)}\, \mu(K_\delta)^{1/n},$$

that is,

$$\mu(K_\delta) \gtrsim \frac{\delta^{(n-1)/2}}{|\log\delta|^n},$$

implying that

$$\liminf_{\delta\to 0} \Big[n - \Big|\frac{\log\mu(K_\delta)}{\log\delta}\Big|\Big] \geq \frac{n+1}{2}.$$

That is, the Minkowski (box-counting) dimension of K_δ is bigger than or equal to $(n+1)/2$.

It is clear that the argument above leave some space for improvement, and the interested reader may consult the recommended bibliography, where in the articles of J. Bourgain, T. Wolff, T. Tao, N. Katz and I. Laba, among others, one can find several better estimates. But they are still far away from the conjectured value n, which would be a consequence of a logarithmic estimate similar to the one known for bushes, for the overlapping function of any family of δ-fat needles having different directions. It happens that some of those estimates are also based on the right control of the overlapping function on the more general configuration called δ-hairbrush.

Definition 3.15. A *δ-hairbrush* $\mathcal{B} \subset \mathcal{F}_\delta$ is a collection of δ-fat needles $\mathcal{B} = \{R_j\}$ so that all of them intersect a fixed one, say R_0, which will be called the *stem* of the brush.

Lemma 3.16. *There are finite constants $C(n)$ and $a(n)$ such that a δ-hairbrush $\mathcal{B} \subset \mathcal{F}_\delta$ in $\mathbb{R}^n$, $n \geq 2$, satisfies*

$$\int_{\mathbb{R}^n} \Big(\sum_{R\in\mathcal{B}} \chi_R\Big)^{n/(n-1)} \leq C(n)\,[1 + |\log\delta|]^{a(n)} \sum_{R\in\mathcal{B}} \mu(R).$$

The main idea of the proof in that the hairbrush condition allows the reduction of the inequality above to the well-known two-dimensional case. Also, elementary geometrical considerations imply the following:

Lemma 3.17. *Given a family $\mathcal{A}$ of δ-fat needles in $\mathbb{R}^n$ such that*

$$\sup_{R\in\mathcal{A}} \left[\operatorname{card}\{S \in \mathcal{A} : R \cap S \neq \emptyset\}\right] = M,$$

we have, for every $R \in \mathcal{A}$,

$$\int_R \Big(\sum_{S\in\mathcal{A}} \chi_S\Big)^{1/(n-1)} \leq C(n)\,[1 + |\log\delta|]^{a(n)}\, M^{(n-2)/(n-1)^2}\,\delta^{n-1}$$

Proposition 3.18. *The combination of both lemmas above yields the estimation $\geq (n+2)/2$ for the Minkowski dimension of any Kakeya set.*

Proof. Let us consider the Minkowski sausage K_δ which contains a δ-fat needle on each direction of the spherical grid $\mathcal{L}_\delta$. If $\{R_\alpha\}$ is the collection of those δ-needles, we have the splitting

$$\{R_\alpha\} = \mathcal{B}_1 \cup \mathcal{B}_2 \cup \cdots \cup \mathcal{B}_m,$$

where in this disjoint union every $\mathcal{B}_j$ is a δ-hairbrush of decreasing cardinality.

With a given $\lambda \geq 1$, to be fixed later, let us consider $r(\lambda)$ so that for $j > r(\lambda)$,

$$\operatorname{card}(\mathcal{B}_j) \leq \lambda\,\delta^{-(n-1)/2},$$

which imposes

$$r(\lambda) \leq \frac{1}{\lambda}\,\delta^{-(n-1)/2}.$$

Then we have the splitting

$$\{R_\alpha\} = \mathcal{B}^* \cup \mathcal{B}^{**}, \qquad \text{where} \quad \mathcal{B}^* = \bigcup_{j=1}^{r(\lambda)} \mathcal{B}_j.$$

Hölder's inequality and Lemma 3.16 yield

$$\int \Big|\sum_{R\in\mathcal{B}^*} \chi_R\Big|^{n/(n-1)} \leq C(n)\,(1 + |\log\delta|)^{a(n)}\,\frac{\delta^{-1/2}}{\lambda^{1/(n-1)}} \sum_{R\in\mathcal{B}^*} \mu(R).$$

On the other hand, Lemma 3.17 gives

$$\begin{aligned}\int \Big|\sum_{R\in\mathcal{B}^{**}} \chi_R\Big|^{n/(n-1)} &= \sum_{R\in\mathcal{B}^{**}} \int_R \Big|\sum_{R\in\mathcal{B}^{**}} \chi_R\Big|^{1/(n-1)} \\ &\leq C(n)[1 + |\log\delta|]^{a(n)}\,\lambda^{\frac{n-2}{(n-1)^2}}\,\delta^{-\frac{n-2}{2(n-1)}} \sum_{R\in\mathcal{B}^{**}} \mu(R)\end{aligned}$$

Choosing now $\lambda = \delta^{-1/2}$, we get the estimate

$$\int \Big|\sum_{\alpha} \chi_{R_\alpha}\Big|^{n/(n-1)} \le C(n)[1 + |\log\delta|]^{a(n)}\, \delta^{-\frac{n-2}{2(n-1)}} \sum_{\alpha} \mu(R_\alpha)$$

and

$$1 \lesssim \sum_{\alpha} \mu(R_\alpha) \lesssim \mu(K_\delta)^{1/n} \left[C(n)\,(1 + |\log\delta|)^{a(n)}\right]^{(n-1)/n} \delta^{\frac{n-2}{2n}},$$

from which we obtain that

$$n - \limsup_{\delta\to 0} \frac{\mu(K_\delta)}{\mu(\delta)} \ge \frac{n+2}{2}. \qquad \square$$

Proof of Lemma 3.16. The first published proof is in [77]. The following is an outlined version from my notes of the seventies.

First, to simplify the geometry, we may consider, without loss of generality, that $R_j \setminus (R_0 \cap R_j)$ is connected for every R_j in the hair brush with stem R_0, and that the axis of each needle R_j in the brush intersects the axis of R_0.

Let us denote by $M(\delta)$ the best constant in the inequality

$$\int \Big|\sum_{R\in\mathcal{B}} \chi_R\Big|^{n/(n-1)} \le M(\delta) \sum_{R\in\mathcal{B}} \mu(R)$$

over all such δ-hairbrushes, and observe that by dilation, the constant is the same if we consider now cylinders of height h and radius δh, for a fixed $h > 0$.

Let $\mathcal{B} = \{R_0, R_1, \ldots, R_\gamma\}$ be a δ-hairbrush (here the height is taken to be 1), having stem R_0, that is, $R_j \cap R_0 \neq \emptyset$ and $R_j \setminus (R_0 \cap R_j)$ is connected. Taking into account that factors of the form $|\log\delta|$ are irrelevant for the dimension estimate, we may assume, without loss of generality, that γ is contained on an interval $2^{-m(n-1)} \le \gamma < 2^{-(m-1)(n-1)}$, so that

$$2^{(m-1)}\delta \le \text{angle}(R_j, R_0) < 2^m\,\delta.$$

Next we split every R_j in the hairbrush into two halves, $R_j^1 \cup R_j^2$, where $R_j^1 \cap R_0 \neq \emptyset$. We have

$$\int \Big(\sum_j \chi_{R_j^2}\Big)^{n/(n-1)} = \sum_j \int_{R_j^2} \Big(\sum_k \chi_{R_k^2}\Big)^{1/(n-1)},$$

and observe that for each j, the family $\{R_k^2 : R_k^2 \cap R_j^2 \neq \emptyset\}$ splits into a finite number (independent of δ) parts, so that on each one of them, $R_k^2 \cap R_j^2 \neq \emptyset$ implies that the direction of R_k has $n-2$ coordinates fixed. Therefore, we can use the L^2-estimate to show that

$$\sum_j \int_{R_j^2} \Big(\sum_k \chi_{R_k^2}\Big)^{1/(n-1)} \lesssim |\log\delta|^{a(n)} \sum_j \mu(R_j^2).$$

Regarding $\sum_k \chi_{R_k^1}$, we have that each R_R^1 is a (2δ)-fat needle, and that $\sum_k \chi_{R_k^1}$ splits into, at most, 2^{n-1} (2δ)-fat hairbrushes $\mathcal{B}_\alpha^*$. Therefore,

$$\int \Big|\sum_k \chi_{R_k^1}\Big|^{n/(n-1)} \lesssim \big[M(2\delta) + |\log\delta|^{a(n)}\big] \sum_{R\in\mathcal{B}} \mu(R),$$

and the recursion formula

$$M(\delta) \le M(2\delta) + |\log\delta|^{a(n)}$$

yields the desired estimate. □

Proof of Lemma 3.17. Without loss of generality, one may assume that $M = 2^{m(n-1)}$, for $m = 0, 1, \ldots, -\log_2\delta$. Furthermore, given such an M, it is obvious that the integral

$$\int_R \Big(\sum_{S\in\mathcal{A}} S\Big)^{1/(n-1)}$$

gets its maximum when all the S needles of $\mathcal{A}$ intersecting R satisfy angle$(R, S) \le 2^m\,\delta$.

Next, we split the set $\{S \in \mathcal{A} : S \cap R \ne \emptyset\}$ into, at most, $-\log_2\delta$ classes $\mathcal{A}_j(R)$ defined by $2^j\delta \le \text{angle}(S, R) < 2^{j+1}\,\delta$, and observe that

$$S \in \mathcal{A}_j(R) \implies \mu(R \cap S) \le \frac{\delta^{n-1}}{2^j}.$$

Let us divide R (along its height) into 2^j needles of length 2^{-j},

$$R = \bigcup_{i=1}^{2^j} R_i,$$

and denote $c_i = \text{card}\{S \in \mathcal{A}_j(R) : S \cap R_i \ne \emptyset\}$. Then we have

$$\int_R \Big(\sum_{S\in\mathcal{A}_j(R)} \chi_S\Big)^{1/(n-1)} \lesssim \Big(\sum_{i=1}^{2^j} c_i\Big)^{1/(n-1)} \frac{\delta^{n-1}}{2^j}$$

$$\lesssim \Big(\sum_i c_i\Big)^{1/(n-1)} 2^{j\frac{n-2}{n-1}} \frac{\delta^{n-1}}{2^j} \lesssim 2^{j\frac{n-2}{n-1}}\,\delta^{n-1},$$

and the result follows adding these estimates on $j \leqslant m$. □

These geometrical or suprematistic methods can be refined, but they seem to be unsatisfactory to get the conjecture. In 1999, J. Bourgain invented a new approach, based on the arithmetic combinatorics of sums and differences of sets which gives some improvements in high dimensions, producing an interesting connection of the Kakeya conjecture with others conjectures in number theory about Dirichlet series.

Chapter 4
Translation invariant operators

An analyst, as a hunter, usually stays at headquarters improving his tools. From time to time, he goes out hunting. But quickly gets back home and continue sharpening his knives.

Koji Ohkitani

4.1 Some general facts

A bounded measurable function $m(\xi)$ is a Fourier multiplier of the space $L^p(\mathbb{R}^n)$, $1 \leq p \leq \infty$, if the formula

$$\widehat{Tf}(\xi) = m(\xi)\,\widehat{f}(\xi),$$

defined for f in the Schwartz space $\mathcal{S}(\mathbb{R}^n)$, has a bounded extension to $L^p(\mathbb{R}^n)$. This means that there exists a finite constant C such that

$$\|Tf\|_p \leq C\,\|f\|_p \qquad \text{for each } f \in \mathcal{S}(\mathbb{R}^n).$$

The minimum value of such C is the norm of the multiplier, and will be denoted either by $|T|_p$ or by $|m|_p$.

These operators are also characterized by the property of being translation invariant, namely, they commute with the translations of the subjacent Euclidean space $\mathbb{R}^n$.

We recall now a few general facts.

Fact 4.1. *If $T \neq 0$ is a translation invariant operator bounded from $L^p(\mathbb{R}^n)$ to $L^q(\mathbb{R}^n)$, then necessarily we must have $q \geq p$.*

Proof. Observe first that for $1 \leq r < \infty$, we have

$$\lim_{h\to\infty} \|\tau_h f + f\|_r = 2^{1/r}\,\|f\|_r,$$

where τ_h denotes translation by the vector h:

$$\tau_h f(x) = f(x-h).$$

Let

$$C_{p,q}(T) = \inf\{C : \|Tf\|_q \leq C\,\|f\|_p, \text{ for all } f \in L^p(\mathbb{R}^n)\}.$$

A. Córdoba, *Suprematism in Harmonic Analysis*, Progress in Mathematics 356,
https://doi.org/10.1007/978-3-031-73737-4_4

Then we have the estimates

$$\|T(\tau_h f + f)\|_q \le C_{p,q}(T)\, \|\tau_h f + f\|_p \quad \text{and} \quad \|T(\tau_h f + f)\|_q = \|\tau_h(Tf) + Tf\|_q.$$

Taking limits as $\|h\| \to \infty$, we get

$$2^{1/q}\, \|Tf\|_q \le C_{p,q}(T)\, 2^{1/p}\, \|f\|_p, \quad \text{i.e.,} \quad \|Tf\|_q \le C_{p,q}(T)\, 2^{1/p-1/q}\, \|f\|_p,$$

which is a contradiction with the definition of $C_{p,q}(T)$ in the case $q < p$. □

Fact 4.2. *If $T\colon L^p(\mathbb{R}^n) \to L^q(\mathbb{R}^n)$, $q \ge p$, commutes with translations, then there exists a tempered distribution $K \in \mathcal{S}'(\mathbb{R}^n)$ such that $T\varphi = K * \varphi$, for every $\varphi \in \mathcal{S}(\mathbb{R}^n)$.*

Proof. Let us observe first that if T commutes with translations, then it also commutes with derivations: $T(D^\alpha \varphi) = D^\alpha(T\varphi)$. Furthermore, a consequence of Sobolev's embedding is that the assumption $D_x^\alpha f \in L^q(\mathbb{R}^n)$, for every $|\alpha| \le n/q + 1$, implies the continuity of f and the estimate

$$\|f\|_\infty \lesssim \sum_{|\alpha| \le n/q+1} \|D_x^\alpha f\|_q.$$

Given $f \in \mathcal{S}(\mathbb{R}^n)$, it happens that $Tf(x) = T(\tau_x f)(0)$. Let us consider the linear mapping defined on $\mathcal{S}(\mathbb{R}^n)$ by the formula

$$\mathcal{L}f :\equiv T(\tau_x f)(0).$$

We have

$$|\mathcal{L}f| = |T(\tau_x f)(0)| \lesssim \sum_{|\alpha| \le n/q+1} \|D^\alpha T f\|_q \lesssim \sum_{|\alpha| \le n/q+1} \|D^\alpha f\|_q.$$

But this estimate implies the existence of a tempered distribution $\widetilde{K} \in \mathcal{S}'(\mathbb{R}^n)$ so that

$$Tf(0) = \langle \widetilde{K}, f \rangle,$$

that is,

$$Tf(x) = T(\tau_x f)(0) = \langle \widetilde{K}, \tau_x \varphi \rangle = K * f(x), \quad \text{where} \quad K(y) = \widetilde{K}(-y),$$

in the distributional sense.

Therefore, if T commutes with translations and it is bounded on some $L^p(\mathbb{R}^n)$, then for each $\varphi \in \mathcal{S}(\mathbb{R}^n)$ we have $\widehat{Tf} = \widehat{K} \cdot \widehat{\varphi} = m \cdot \widehat{\varphi}$, where $m = \widehat{K}$ is the associated multiplier and $K \in \mathcal{S}'(\mathbb{R}^n)$ is the kernel of T.

Let q be the conjugate exponent of p, i.e., $1/p + 1/q = 1$, and assume that $T\varphi = K * \varphi$ is bounded on $L^p(\mathbb{R}^n)$. The identity

$$\int T\varphi(x)\, \psi(x)\, \mathrm{d}x = \int \varphi(x)\, T\psi(x)\, \mathrm{d}x,$$

valid for each pair of Schwartz functions φ and ψ, shows that the boundedness of T on $L^p(\mathbb{R}^n)$ is equivalent to its boundedness on $L^q(\mathbb{R}^n)$. Then Riesz–Thorin interpolation implies that T must be bounded on $L^2(\mathbb{R}^n)$, which together with Plancherel's theorem yields that m has to be a bounded function:

$$\|m\|_\infty = |m|_2 = |T|_2.$$

When the multiplier m is real valued, the operator T coincides with its adjoint T^* but, in general, T^* is given by the complex conjugate multiplier $\bar{m}$. □

Fact 4.3. *If the multiplier T is bounded on $L^1(\mathbb{R}^n)$, or in $L^\infty(\mathbb{R}^n)$, then its kernel $K = \mathrm{d}\mu$ is a measure of finite total variation.*

Proof. The hypothesis implies that the linear map $\mathcal{L}\colon \varphi \to T\varphi(0)$ satisfies $|\mathcal{L}(\varphi)| \lesssim \|\varphi\|_\infty$, and the Riesz representation theorem yields the existence of a unique measure μ of finite total variation so that $T\varphi(0) = \int \varphi(y)\,\mathrm{d}\mu(y)$. Then invariance by translation implies that

$$T\varphi(x) = \int \varphi(y-x)\,\mathrm{d}\mu(y). \qquad \square$$

Fact 4.4 (K. de Leew)**.** *If $m(\xi,\eta)$, $\xi \in \mathbb{R}^j$, $\eta \in \mathbb{R}^k$, is a Fourier multiplier of $L^p(\mathbb{R}^{j+k})$, then for almost every $\xi \in \mathbb{R}^j$, $m(\xi,\cdot)$ is a Fourier multiplier of $L^p(\mathbb{R}^k)$ and, similarly, for almost every $\eta \in \mathbb{R}^k$, $m(\cdot,\eta)$ is a multiplier of $L^p(\mathbb{R}^j)$.*

Proof. Suppose first that $m(\xi,\eta)$ is continuous. Taking functions $f_1, g_1 \in \mathcal{S}(\mathbb{R}^j)$ and $f_2, g_2 \in \mathcal{S}(\mathbb{R}^k)$, we have

$$\Big|\int_{\mathbb{R}^j}\int_{\mathbb{R}^k} m(\xi,\eta)\,\widehat{f_1}(\xi)\,\widehat{f_2}(\eta)\,\widehat{g_1}(\xi)\,\widehat{g_2}(\eta)\,\mathrm{d}\xi\,\mathrm{d}\eta\Big| \le |m|_{L^p(\mathbb{R}^{j+k})}\,\|f_1\|_{L^p(\mathbb{R}^j)}\,\|g_1\|_{L^q(\mathbb{R}^j)}\,\|f_2\|_{L^p(\mathbb{R}^k)}\,\|g_2\|_{L^q(\mathbb{R}^k)},$$

and writing the left-hand side of this inequality in the form

$$\Big|\int_{\mathbb{R}^j}\Big[\int_{\mathbb{R}^k} m(\xi,\eta)\,\widehat{f_2}(\eta)\,\widehat{g_2}(\eta)\,\mathrm{d}\eta\Big]\,\widehat{f_1}(\xi)\,\widehat{g_1}(\xi)\,\mathrm{d}\xi\Big|,$$

we may conclude that

$$M(\xi) = \int_{\mathbb{R}^k} m(\xi,\eta)\,\widehat{f_2}(\eta)\,\widehat{g_2}(\eta)\,\mathrm{d}\eta$$

is a Fourier multiplier of $L^q(\mathbb{R}^k)$ whose norm is controlled by $|m|_p\,\|f_2\|_p\,\|g_2\|_q$, and since every multiplier must satisfy $\|M\|_{L^\infty} = |M|_2 \le |M|_p$, $1 \le p \le \infty$, we get

$$\Big|\int_{\mathbb{R}^j} m(\xi,\eta)\,\widehat{f_2}(\eta)\,\widehat{g_2}(\eta)\,\mathrm{d}\eta\Big| \le |m|_p\,\|f_2\|_p\,\|g_2\|_q,$$

which amounts to show that $|m(\xi,\cdot)|_p \le |m|_p$ for every $\xi \in \mathbb{R}^j$. Similarly, we have that $|m(\cdot,\eta)|_p \le |m|_p$ for every $\eta \in \mathbb{R}^k$.

To remove the restriction about the continuity of the multiplier, we proceed as follows: let $\xi \in \mathbb{R}^j$ be so that (ξ,η) is a Lebesgue point of m for almost every $\eta \in \mathbb{R}^k$, and let us write $m_\varepsilon = m * \phi_\varepsilon$, where $\phi_\varepsilon(x,y) = \varepsilon^{(j+k)}\,\chi_Q(x/\varepsilon, y/\varepsilon)$, and Q is the unit cube in $\mathbb{R}^{j+k}$ centered at the origin.

By the preceding argument, we have

$$\left|\int_{\mathbb{R}^j} m_\varepsilon(\xi,\eta)\,\widehat{f_2}(\eta)\,\widehat{g_2}(\eta)\,\mathrm{d}\eta\right| \le |m_\varepsilon|_p\,\|f_2\|_p\,\|g_2\|_q \le |m|_p\,\|f_2\|_p\,\|g_2\|_q.$$

Then the dominated convergence theorem yields

$$\left|\int_{\mathbb{R}^j} m(\xi,\eta)\,\widehat{f_2}(\eta)\,\widehat{g_2}(\eta)\,\mathrm{d}\eta\right| \le |m|_p\,\|f_2\|_p\,\|g_2\|_q$$

for every $f_2 \in \mathcal{S}(\mathbb{R}^j)$ and $g_2 \in \mathcal{S}(\mathbb{R}^k)$, which allows us to finish the proof. □

The important information contained in these facts is, nevertheless, very unsatisfactory for p different of 1 or 2. Even for these two cases is very hard, in general, to decide when a bounded function is the Fourier transform of a finite measure, or when a tempered distribution has a bounded Fourier transform.

Since the Fourier transform of an integrable function is continuous, we have, at least, a necessary condition for L^1-multiplier: it has to be a continuous function. However, the Hilbert transform in $\mathbb{R}$, defined by

$$Hf(x) = \frac{1}{\pi}\ \text{p.v.}\int \frac{f(x-y)}{y}\,\mathrm{d}y = \frac{1}{\pi}\lim_{\varepsilon\to 0}\int_{|y|\ge\varepsilon} \frac{f(x-y)}{y}\,\mathrm{d}y,$$

or equivalently by

$$\widehat{Hf}(\xi) = -i\,\mathrm{sign}(\xi)\,\widehat{f}(\xi)$$

for every $f \in \mathcal{S}(\mathbb{R})$, is bounded on $L^p(\mathbb{R})$, $1 < p < \infty$, implying that the necessary condition above is only valid for the space L^1.

4.2 Singular integrals (sketch of the Calderón–Zygmund theory)

The Hilbert transform leads us to the next stage of the theory of Fourier multipliers, which contains the important contributions due to Marcinkiewicz, Calderón, Zygmund, Stein, and many others. It is a central paradigm of modern Fourier analysis, and has been discussed at length in several monographs. Here we will consider several of its basic pieces, in order to motivate further developments around the so-called disc multiplier problem.

In the following, we shall consider tempered distributions K having Fourier transforms $m = \widehat{K}$ in $L^\infty(\mathbb{R}^n)$, and such that they coincide outside the origin with a locally integrable function (which will be also denoted K) satisfying

$$\int_{|x|\geq 2|y|>0} |K(x-y)-K(x)|\,\mathrm{d}x \leq c < \infty \quad \text{uniformly on } y>0.$$

The distribution K defines a singular integral operator T by means of the formula

$$Tf(x) = \text{p.v.}\int K(x-y)\,f(y)\,\mathrm{d}y = \lim_{\varepsilon\to 0}\int_{|y|\geq\varepsilon} f(x-y)\,K(y)\,\mathrm{d}y,$$

which is well defined for functions f in the Schwartz class $\mathcal{S}(\mathbb{R}^n)$.

Theorem 4.5. *T has a bounded extension to the space $L^p(\mathbb{R}^n)$, $1<p<\infty$.*

A particular case is given by homogeneous kernels

$$K(x) = \frac{\Omega(x/\|x\|)}{\|x\|^n},$$

where $\int_{S^{n-1}}\Omega(y)\,\mathrm{d}\sigma(y) = 0$ and the Dini–Lipschitz S^{n-1} condition is satisfied, namely,

$$\int_0^1 \frac{\omega(t)}{t}\,\mathrm{d}t < \infty, \qquad \text{where } \omega(t) = \sup_{\substack{|x-y|=t\\|x|=|y|=1}} |\Omega(x)-\Omega(y)|.$$

Let us define

$$T_\varepsilon f(x) = \int_{|y|\geq\varepsilon} \frac{\Omega(y)}{|y|^n}\,f(x-y)\,\mathrm{d}y \qquad \text{and} \qquad T^*f(x) = \sup_{\varepsilon>0}|T_\varepsilon f(x)|.$$

Theorem 4.6. (a) *$\{T_\varepsilon\}_{\varepsilon>0}$ is uniformly bounded on $L^p(\mathbb{R}^n)$ for $1<p<\infty$.*

(b) *T^* is of weak type $(1,1)$, and there exists $\lim_{\varepsilon\to 0}T_\varepsilon f = Tf$ in the L^p-norm when $1<p<\infty$.*

(c) *If $f\in\mathcal{S}(\mathbb{R}^n)$, then $\widehat{Tf}(\xi) = m(\xi)\,\widehat{f}(\xi)$, where m is a homogeneous function of degree 0.*
Given $\xi\in S^{n-1}$, we have the formula

$$m(\xi) = \int_{S^{n-1}}\left[\frac{\pi i}{2}\,\mathrm{sign}(\xi\cdot\eta) + \log\left(\frac{1}{|\xi\cdot\eta|}\right)\right]\Omega(\eta)\,\mathrm{d}\sigma(\eta).$$

The monograph [96] of E. M. Stein contains an excellent exposition of the proof of the theorem above.

Let $m\in L^\infty(\mathbb{R}^n)$ be such that

$$|D^\alpha m(\xi)| \leq C_\alpha\,|\xi|^{-|\alpha|} \qquad \text{for } |\alpha|\leq[n/2]+1,$$

where $[x]$ denotes integer part.

Theorem 4.7. *Under the hypothesis above, m is a bounded Fourier multiplier on $L^p(\mathbb{R}^n)$, for $1 < p < \infty$.*

Proof. First let us consider a smooth partition of unity in the half line $(0, \infty)$, $1 = \sum_{-\infty}^{\infty} \phi_j(t)$, where $\phi_j(t) = \phi(2^{-j}t)$ and the smooth function $\phi \geq 0$ is supported in the interval $(1/2, 5/2)$. Then we decompose the multiplier as follows:

$$m(\xi) = \sum_{j=-\infty}^{\infty} m(\xi)\, \phi_j(|\xi|) = \sum_{j=-\infty}^{\infty} m_j(\xi),$$

and we define

$$K^N(x) = \sum_{j=-N}^{N} \widehat{m_j}(x) = \sum_{j=-N}^{N} K_j(x).$$

The theorem will be a consequence of the following estimate:

$$\int_{|x|\geq 2|y|>0} |K^N(x-y) - K^N(x)|\, \mathrm{d}x \leq C < \infty, \qquad \text{uniformly on } y \text{ and } N.$$

To see this, let us keep in mind the heuristic "ansatz" about the kernel $\widehat{m_j}(x)$, which by the uncertainty principle must be essentially supported in the ball $|x| \leq 2^{-j}$, since its Fourier transform m_j is supported in the annulus $|\xi| \sim 2^j$ and its derivatives decay accordingly, following the hypothesis of the theorem.

More concretely, let us consider, for each $|j| \leq \log N$, the integral

$$\int_{|x|\geq 2|y|>0} |K_j(x-y) - K_j(x)|\, \mathrm{d}x.$$

Case (a). When $|y| \geq 2^{-j}$, one cannot expect cancelations in the difference $K_j(x-y) - K_j(x)$ and, therefore, we are forced to estimate each part separately.

On the one hand, $|K_j(x)| \lesssim 2^{jn}$ when $|x| \leq 2^{-j}$. On the other hand, if $|x| \geq 2^{-j}$,

$$|K_j(x)| \lesssim |x|^{-\theta(n)} \sum_{|\alpha|=\theta(n)} \left| \int e^{-2\pi i x\cdot\xi}\, D_\xi^\alpha m_j(\xi)\, \mathrm{d}\xi \right|,$$

where $\theta(n) = [n/2] + 1$.

Therefore,

$$\begin{aligned}\int_{|x|\geq 2|y|>0} |K_j(x)|\, \mathrm{d}x &\lesssim \sum_{|\alpha|=\theta(n)} \left[\int_{|x|\geq 2|y|} \frac{1}{|y|^{2\theta(n)}}\, \mathrm{d}y \right]^{1/2} \left[\int |D_\xi^\alpha m_j(\xi)|^2\, \mathrm{d}\xi \right]^{1/2} \\ &\lesssim (|y|\, 2^j)^{-\theta(n)+n/2},\end{aligned}$$

and then

$$\sum_{|y|2^j\geq 1} \int_{|x|>2|y|} |K_j(x)|\, \mathrm{d}x \leq C < \infty \qquad \text{uniformly on } y.$$

A similar estimate holds for the integral

$$\sum_{|y|2^j\geq 1}\int_{|x|>2|y|}|K_j(x-y)|\,\mathrm{d}x.$$

Case (b). To finish, we have to estimate the sum of the integrals

$$\int_{|x|\geq 2|y|>0}|K_j(x-y)-K_j(x)|\,\mathrm{d}x$$

when $|y|\leq 2^{-j}$.

Now we expect cancelations to happen, so that the integral of the difference be smaller than each term separately. We use Taylor's expansion:

$$|K_j(x-y)-K_j(x)|$$
$$\lesssim \sum_{1\leq|\alpha|<\theta(n)}\frac{1}{\alpha!}|D^\alpha K_j(x)|\,|y^\alpha| + \sum_{|\alpha|=\theta(n)}\frac{1}{\alpha!}\int_0^1 |D^\alpha K_j(x-ty)|\,\mathrm{d}t\cdot|y|^{\theta(n)}.$$

Let us consider a typical term,

$$|D^\alpha K_j(x)|\,|y^\alpha|,\qquad \text{for } 1\leq\alpha\leq\theta(n)=[n/2]+1,$$

and split again the range of integration into dyadic rings:

$$I_j^s(y)=\int_{2^s|y|\leq|x|\leq 2^{s+1}|y|}|y^\alpha|\,|D^\alpha K_j(x)|\,\mathrm{d}x.$$

Then we have two scenarios:

(1) When $2^s|y|2^j\leq 1$, we write

$$\begin{aligned} I_j^s(y) &\lesssim |y^\alpha|\,(2^s|y|)^{n/2}\Big(\int|\xi|^{2|\alpha|}\,|m_j(\xi)|^2\,\mathrm{d}\xi\Big)^{1/2}\\ &\lesssim |y|^{|\alpha|}\,(2^s|y|)^{n/2}\,2^{j|\alpha|}\,2^{jn/2}=(|y|\,2^j)^{|\alpha|}\,(2^s|y|\,2^j)^{n/2}, \end{aligned}$$

implying the estimate

$$\sum_{\substack{|y|\leq 2^j\\ 2^i|y|2^s\leq 1}} I_j^s(y)<\infty,\qquad \text{uniformly on } y.$$

(2) To finish, we treat the case $1\leq 2^s|y|2^j$. We have

$$D^\alpha K_j(x)=\frac{1}{(2\pi ix)^\beta}\int e^{2\pi ix\cdot\xi}\,D^\beta(\xi^\alpha\,m_j(\xi))\,\mathrm{d}\xi.$$

Therefore,

$$|D^\alpha K_j(x)| \lesssim \sum_{|\beta|=\theta(n)} \frac{1}{|x|^{\theta(n)}} \left| \int e^{2\pi i x\cdot\xi}\, D^\beta(\xi^\alpha\, m_j(\xi))\, \mathrm{d}\xi \right|.$$

Then

$$\begin{aligned} I_s^j(y) &= \int_{2^s|y|\le|x|\le 2^{s+1}|y|} |y|^{|\alpha|}\, |D^\alpha K_j(x)|\, \mathrm{d}x \\ &\lesssim |y|^{|\alpha|} \sum_{|\beta|=\theta(n)} \left[\int_{2^s|y|\le|x|\le 2^{s+1}|y|} \frac{1}{|x|^{2\theta(n)}}\, \mathrm{d}x \right]^{1/2} \left[\int |D^\beta(\xi^\alpha\, m_j(\xi))|^2\, \mathrm{d}\xi \right]^{1/2} \\ &\lesssim (|y|\, 2^j)^{|\alpha|}\, \frac{(2^s|y|2^j)^{n/2}}{(2^s|y|2^j)^{\theta(n)}}, \end{aligned}$$

and we can conclude the proof because we are in the regime $|y|2^j < 1$, $2^s|y|z^j > 1$, and $\theta(n) > n/2$. □

Next let us consider a family $\{\phi_j\}_{j\in\mathbb{Z}}$ of smooth functions satisfying

$$\mathrm{supp}(\phi_j) \subset (2^{j-1}, 2^{j+1}), \qquad \text{and} \qquad |D^\alpha \phi_j(\xi)| \le C_\alpha\, 2^{-j|\alpha|} \quad \text{for } |\alpha| \le \theta(n).$$

With $\{r_j(t)\}$, the set of Rademacher functions in the interval $[0, 1]$, we have the following set of multipliers:

$$m_t(\xi) = \sum r_j(t)\, \phi_j(|\xi|), \quad T_t f(x) = \sum r_j(t)\, T_j f(x) \quad \text{and} \quad \widehat{T_j}(\xi) = \phi_j(|\xi|)\, \widehat{f}(\xi).$$

Theorem 4.7 can be applied to get L^p-bounds, uniformly on $t \in [0, 1]$:

$$\int |T_t f(x)|^p\, \mathrm{d}x \le C_p \int |f(x)|^p\, \mathrm{d}x, \quad 1 < p < \infty.$$

Next we integrate this inequality in the interval $t \in [0, 1]$, and make use of the well-known Khintchine estimates satisfied by the Rademacher functions:

$$\int_0^1 \int |T_t f(x)|^p\, \mathrm{d}x\, \mathrm{d}t = \int \int_0^1 |T_t f(x)|^p\, \mathrm{d}x\, \mathrm{d}t \simeq \int \Big(\sum_j |T_j f(x)|^2 \Big)^{p/2} \mathrm{d}x,$$

for $1 < p < \infty$, implying the estimate:

$$\Big\| \Big(\sum |T_j f|^2 \Big)^{1/2} \Big\|_p \le C_p\, \|f\|_p.$$

Let now assume that $\sum \phi_j(|\xi|) \equiv 1$, and let ψ_j be a smooth function supported on $(2^{j-2}, 2^{j+2})$ so that $\psi_j \equiv 1$ on $\mathrm{supp}(\phi_j)$ and $|D^\alpha \psi_j(\xi)| \le C_\alpha\, |\xi|^{-|\alpha| j}$, for $|\alpha| \le \theta(n)$.

Given f and g in the Schwartz class $\mathcal{S}(\mathbb{R}^n)$, we have

$$\begin{aligned}\int f(x)\,\overline{g(x)}\,\mathrm{d}x &= \int \widehat{f}(\xi)\,\overline{\widehat{g}(\xi)}\,\mathrm{d}\xi = \sum_j \int \phi_j(\xi)\,\widehat{f}(\xi)\,\overline{\psi_j(\xi)\,\widehat{g}(\xi)}\,\mathrm{d}\xi \\ &\le \int \Big(\sum_j |T_j f(x)|^2\Big)^{1/2} \Big(\sum_j |\widetilde{T}_j g(x)|^2\Big)^{1/2} \mathrm{d}x \\ &\le \Big\|\Big(\sum_j |T_j f|^2\Big)^{1/2}\Big\|_p \Big\|\Big(\sum_j |\widetilde{T}_j g|^2\Big)^{1/2}\Big\|_q,\end{aligned}$$

where

$$\widehat{\widetilde{T}_j g}(\xi) = \psi_j(\xi)\,\widehat{g}(\xi).$$

Since we know that

$$\Big\|\Big(\sum |\widetilde{T}_j g|^2\Big)^{1/2}\Big\|_q \le C_q\,\|g\|_q,$$

we can conclude:

Corollary 4.8 (Littlewood–Paley). *Under the hypothesis above, we have the equivalence of norms*

$$\Big\|\Big(\sum_j |T_j f|^2\Big)^{1/2}\Big\|_p \simeq \|f\|_p, \quad \textit{for } 1 < p < \infty.$$

4.3 Weighted inequalities for the Hardy–Littlewood maximal function

In the following, we will consider the boundedness properties of singular integrals and maximal operators with respect to absolutely continuous measures: $\omega(x)\,\mathrm{d}x$, where $\mathrm{d}x$ is, as usual, the Lebesgue measure in $\mathbb{R}^n$.

The history of this subject begins with the 1960 paper [90] of Helson and Szegö, where they proved that the Hilbert transform is bounded in $L^2(\mathrm{d}\mu)$ if and only if

$$\mathrm{d}\mu = \omega(x)\,\mathrm{d}x \quad \text{and} \quad \omega = \exp(u + Hv),$$

where u and v are L^∞-functions such that $\|v\|_\infty < \pi/2$.

Hardy's inequality,

$$\Big[\int_0^\infty \Big|\int_0^x f(y)\,\mathrm{d}y\Big|^p x^{-r-1}\,\mathrm{d}x\Big]^{1/p} \le \frac{p}{r}\Big[\int_0^\infty |yf(y)|^p\,y^{r-1}\,\mathrm{d}y\Big]^{1/p},$$

may also be considered as a precedent of the modern theory. Then we have following theorem, see [89].

Theorem 4.9 (C. Fefferman–Stein). *For each $r > 1$, there exists a finite constant $C_r(n)$ such that*

$$\int_{\mathbb{R}^n} [Mf(x)]^r \,\omega(x)\,\mathrm{d}x \leq C_r(n) \int_{\mathbb{R}^n} |f(x)|^r \,M\omega(x)\,\mathrm{d}x$$

for each pair of locally integrable functions f and w, where M denotes the Hardy–Littlewood maximal function.

Furthermore, we have the following weak-type inequality:

$$\int_{\{Mf>\alpha\}} \omega(x)\,\mathrm{d}x \lesssim \frac{1}{\alpha} \int_{\mathbb{R}^n} |f(x)|\, M\omega(x)\,\mathrm{d}x.$$

Proof. Since $f \longrightarrow Mf$ is bounded from $L^\infty(\omega(x)\,\mathrm{d}x)$ to $L^\infty(\omega(x)\,\mathrm{d}x)$, it suffices to prove the weak-type estimate, because the Marcinkiewicz interpolation theorem will provide the remainder cases.

It will be convenient to fix a decomposition $\{Q_\nu^k\}$ of $\mathbb{R}^n$ into dyadic cubes, with

$$\ell(Q_\nu^k) = 2^k, \qquad \nu \in \mathbb{Z}^n,$$

and observe that given a cube Q, there are 2^n dyadic cubes Q_ν^k such that

$$\ell(Q) \leq \ell(Q_\nu^k) \leq 2\,\ell(Q) \quad \text{and} \quad Q \cap Q_\nu^k \neq \emptyset.$$

The inequality

$$\frac{1}{|Q|}\int_Q |f(y)|\,\mathrm{d}y \geq \alpha$$

implies that at least one of those 2^n dyadic cubes associated to Q must satisfy

$$\frac{1}{|Q_\nu|}\int_{Q_\nu} |f(y)|\,\mathrm{d}y \geq \frac{\alpha}{4^n}.$$

Furthermore, we have the inclusion $Q \subset Q_\nu^*$, where Q_ν^* is a cube concentric with Q_ν and such that $\ell(Q_\nu^*) = 3\ell(Q_\nu)$.

A standard covering argument yields that $\{Mf > \alpha\} \subset \bigcup_\nu Q_\nu^*$, where the Q_ν are disjoint dyadic cubes. Then

$$\begin{aligned}
\int_{\{Mf>\alpha\}} \omega(x)\,\mathrm{d}x &\leq \sum_\nu \int_{Q_\nu^*} \omega \leq \sum_\nu \frac{1}{|Q_\nu^*|}\int_{Q_\nu^*} \omega \cdot 3^n |Q_\nu| \\
&\leq (12)^n \frac{1}{\alpha} \sum_\nu \Big(\frac{1}{|Q_\nu^*|}\int_{Q_\nu^*} \omega\Big) \int_{Q_\nu} |f(y)|\,\mathrm{d}y \\
&\leq (12)^n \frac{1}{\alpha} \sum_\nu \int_{Q_\nu} |f(y)|\, M\omega(y)\,\mathrm{d}y \lesssim \frac{1}{\alpha}\int |f(y)|\, M\omega(y)\,\mathrm{d}y.
\end{aligned}$$

Here, we have used again $|E|$ to denote the Lebesgue measure of the set $E \subset \mathbb{R}^n$. □

Corollary 4.10. *If the locally integrable function ω satisfies $M\omega(x) \le C\omega(x)$, a.e. $x \in \mathbb{R}^n$, then*

$$\int_{\mathbb{R}^n} |Mf(x)|^r \,\omega(x)\,\mathrm{d}x \lesssim \int_{\mathbb{R}^n} |f(x)|^r \,\omega(x)\,\mathrm{d}x$$

is valid for general measurable functions f.

Definition 4.11. The weight w belong to the class A_1 if and only if there exists a finite constant C such that $M\omega(x) \le C\,\omega(x)$ for a.e. $x \in \mathbb{R}^n$. The minimal C in the inequality above is the weight constant.

Proposition 4.12. *For every locally integrable function ω and $r > 1$, the function $[M(\omega^r)(x)]^{1/r}$ is in the class A_1 with constant bounded independently of ω.*

Proof. Let Q be a cube containing the point x, and let Q^* be its expanded version by a factor of 3. Denote by M_{Q^*} the Hardy–Littlewood maximal function restricted to averages over subcubes of Q^*. For every s, $0 < s < 1$, we have the inequality

$$\left(\frac{1}{|Q^*|}\int_{Q^*} [M_{Q^*}f(x)]^s \,\mathrm{d}x\right)^{1/s} \le C_s(n)\,\frac{1}{|Q^*|}\int_{Q^*} |f(x)|\,\mathrm{d}x,$$

where the finite constant $C_s(n)$ depends only upon s and the dimension n.

Define the set

$$E = \left\{y \in Q : [M\omega^r(y)]^{1/r} = [M_{Q^*}\omega^r(y)]^{1/r}\right\},$$

and observe that inside its complement $Q \setminus E$, we have

$$\sup_{y\in Q\setminus E} [M\omega^r(y)]^{1/r} \lesssim \inf_{y\in Q\setminus E} [M\omega^r(y)]^{1/r} \lesssim [M\omega^r(x)]^{1/r}.$$

Therefore,

$$\begin{aligned}\frac{1}{|Q|}\int_Q [M\omega^r(y)]^{1/r}\,\mathrm{d}y &\lesssim \frac{1}{|Q^*|}\int_{Q^*} [M\omega^r(y)]^{1/r}\mathrm{d}y + \frac{1}{|Q|}\int_{Q\setminus E} [M\omega^r(y)]^{1/r}\,\mathrm{d}y \\ &\lesssim \left[\frac{1}{|Q^*|}\int_{Q^*} \omega^r(y)\,\mathrm{d}y\right]^{1/r} + [M\omega^r(x)]^{1/r}.\end{aligned}$$

and taking the supremum in the inequality above over all cubes Q containing the point x, we get

$$M[M\omega^r(x)]^{1/r} \approx [M\omega^r(x)]^{1/r}.$$

□

We have then proven that if $\omega \in A_1$, the Hardy–Littlewood maximal function is of weak type $(1, 1)$ with respect to the measure $\omega(x)\,\mathrm{d}x$. This statement admits a reciprocal:

Theorem 4.13. *If M is of weak type $(1, 1)$ with respect to $\omega(x)\,\mathrm{d}x$, then $\omega \in A_1$.*

Proof. Let $x \in Q_1 \subset Q_2$, where Q_1 and Q_2 are cubes in $\mathbb{R}^n$ and x is a Lebesgue point of ω. With $f = \chi_{Q_1}$, we have

$$Mf(z) \geq \frac{|Q_1|}{|Q_2|} \quad \text{if } z \in Q_2.$$

Therefore,

$$\int_{Q_2} \omega \leq \int_{\{Mf>|Q_1|/|Q_2|\}} \omega \leq C\frac{|Q_2|}{|Q_1|}\int_{Q_1} \omega,$$

that is,

$$\frac{1}{|Q_2|}\int_{Q_2} \omega \leq C\frac{1}{|Q_1|}\int_{Q_1} \omega,$$

and taking limits when $Q_1 \Rightarrow x$, we obtain

$$\frac{1}{|Q_2|}\int_{Q_2} \omega \leq C\,\omega(x),$$

implying that

$$M\omega(x) \leq C\omega(x),$$

i.e., $\omega \in A_1$. □

Definition 4.14. A non-negative w belongs to the class A_p, $1 < p < \infty$, if there exists a finite constant C such that

$$\sup_Q \left[\frac{1}{|Q|}\int_Q \omega\right]\left[\frac{1}{|Q|}\int_Q \omega^{-1/(p-1)}\right]^{p-1} \leq C.$$

Theorem 4.15. *The Hardy–Littlewood maximal function is bounded on $L^p(\omega(x)\,\mathrm{d}x)$ if and only if $\omega \in A_p$.*

Proof. $(\Rightarrow)$ Assume first that M is bounded on $L(\omega\,\mathrm{d}x)$. Given a cube Q, we have

$$M(f\,X_Q)(x) \geq \frac{1}{|Q|}\int_Q |f(y)|\,\mathrm{d}y \cdot \chi_Q(x).$$

Therefore,

$$\left[\frac{1}{|Q|}\int_Q |f(y)|\,\mathrm{d}y\right]^p \int_Q \omega \leq C\int |f(x)|^p\,\omega(x)\,\mathrm{d}x$$

for every f.

Taking

$$f = \omega^{-1/(p-1)},$$

we obtain

$$\left[\frac{1}{|Q|}\int_Q \omega^{-1/(p-1)}\right]^p \int_Q \omega \leq C\int_Q \omega^{-1/(p-1)}(x)\,\mathrm{d}x.$$

(⇐) Let us assume now that $\omega \in A_p$, $1 < p < \infty$, and let q be the conjugate exponent of p, that is, $1/p + 1/q = 1$. For every locally integrable function f, we have

$$\frac{1}{|Q|}\int_Q |f| \lesssim \Big[\frac{1}{|Q|}\int_Q |f|^p\,\omega\Big]^{1/p}\Big[\frac{1}{|Q|}\int_Q \omega^{-q/p}\Big]^{1/q}$$
$$= \Big[\frac{1}{|Q|}\int_Q |f|^p\,\omega\Big]^{1/p}\Big[\frac{1}{|Q|}\int_Q \omega^{-1/(p-1)}\Big]^{(p-1)/p} \lesssim \frac{1}{\int_Q \omega}\Big[\int_Q |f|^p\,\omega\Big]^{1/p},$$

that is,

$$Mf(x) \lesssim [M_\omega f^p(x)]^{1/p}, \qquad \text{where} \qquad M_\omega g(x) = \sup_{Q \ni x} \frac{1}{\int_Q \omega}\int_Q |g(x)|\,\omega(x)\,\mathrm{d}x.$$

The theorem will be a consequence of the following lemmas.

Lemma 4.16. (i) *Assume that $\mu = \omega\,\mathrm{d}x$ is doubling, i.e., there exists a finite constant C such that $\mu(Q^*) \le C\mu(Q)$ for every cube Q. Then $M_\mu f$ is bounded on $L^p(\mu)$.*

(ii) *If we restrict to dyadic cubes in the definition of the maximal function, then the same conclusion of part* (i) *follows without the doubling hypothesis.*

Proof. The proof is identical to the case when μ is the Lebesgue measure in $\mathbb{R}^n$. Observe that this lemma implies that for $\omega \in A_p$, the maximal function is bounded on $L^r(\omega\,\mathrm{d}x)$ for each $r > p$. □

Lemma 4.17. *Given $\omega \in A_p$, $p > 1$, we have*

(i) *$\omega \in A_\infty$, that is, there exists $r > 0$ so that if $E \subset Q$ (E measurable, Q cube), then*

$$\frac{\mu(E)}{\mu(Q)} \le C\Big[\frac{|E|}{|Q|}\Big]^r,$$

where $\mu = \omega\,\mathrm{d}x$, $|E|$ denotes Lebesgue measure, and $C < \infty$ is a universal constant.

(ii) *There exists $\varepsilon > 0$ so that $\omega \in A_{p-\varepsilon}$.*

Lemma 4.18. *Given $\omega \in A_p$, there exist $\delta > 0$ and a finite constant C such that for each cube Q, the following reverse Hölder inequality holds:*

$$\Big[\frac{1}{|Q|}\int_Q \omega^{1+\delta}\Big]^{1/(1+\delta)} \le C\frac{1}{|Q|}\int_Q \omega.$$

Proof that Lemma 4.18 ⇒ *Lemma* 4.17. (i) We have

$$\mu(E) = \int_E \omega \le \Big(\frac{1}{|Q|}\int_Q \omega^{1+\delta}\Big)^{1/(1+\delta)} |E|^{\delta/(1+\delta)}\,|Q|^{1/(1+\delta)}$$
$$\lesssim \frac{1}{|Q|}\int_Q \omega\,|E|^{\delta/(1+\delta)}\,|Q|^{1/(1+\delta)},$$

that is,

$$\frac{\mu(E)}{\mu(Q)} \lesssim \left[\frac{|E|}{|Q|}\right]^{\delta/(1+\delta)}.$$

(ii) Given $\omega \in A_p$, we have that

$$v = \omega^{-1/(p-1)} \in A_q$$

for q such that $1/p + 1/q = 1$. Then Lemma 4.18 yields

$$\left[\frac{1}{|Q|}\int_Q v^{1+\delta}\right]^{1/(1+\sigma)} \lesssim \frac{1}{|Q|}\int_Q v(x)\,\mathrm{d}x$$

for some $\delta > 0$, which implies that $\omega \in A_{p-\varepsilon}$, where $\varepsilon = (p-1)\frac{\delta}{1+\delta}$. This is because

$$\begin{aligned}\left[\frac{1}{|Q|}\int_Q \omega\right]&\left[\frac{1}{|Q|}\int_Q \omega^{-(1+\delta)/(p-1)}\right]^{(p-1)/(1+\delta)}\\ &\lesssim \left[\frac{1}{|Q|}\int_Q v^{-1/(q-1)}\right]^{q-1}\left[\frac{1}{|Q|}\int_Q v^{1+\delta}\right]^{1/(1+\delta)}\\ &\lesssim \left[\frac{1}{|Q|}\int_Q v^{-1/(q-1)}\right]^{q-1}\left[\frac{1}{|Q|}\int_Q v\right] \le C < \infty.\end{aligned}$$

□

It remains to prove Lemma 4.18 and, in order to do it, let us first make the following:

Claim. *Given $w \in A_p$, there exists a constant τ, $0 < \tau < 1$, such that if $E \subset Q$ satisfies $|E| \le \frac{1}{2}|Q|$, then $\mu(E) \le \tau\mu(Q)$.*

Proof of the Claim. For any $x \in Q$, the maximal function M_Q, restricted to subcubes of Q, satisfies

$$M_Q(\chi_{Q\setminus E}) \ge \frac{|Q\setminus E|}{|Q|} \ge \frac{1}{2}.$$

Let $p_1 > p$. We have

$$\left(\frac{1}{2}\right)^{p_1}\int_Q \omega \lesssim \int [M_Q(\chi_{Q\setminus E})(x)]^{p_1}\omega(x)\,\mathrm{d}x \lesssim \int_Q \chi_{Q\setminus E}(x)\,\omega(x)\,\mathrm{d}x.$$

That is,

$$\int_Q \omega \le C\int_{Q\setminus E}\omega$$

for some finite constant C. Therefore,

$$\int_E \omega = \int_Q \omega - \int_{Q\setminus E}\omega \le \left(1-\frac{1}{C}\right)\int_Q \omega = \tau\int_Q \omega,$$

where $\tau = 1 - 1/C$.

□

End of the proof of Lemma 4.18. Using the dilation invariance of the estimate and after multiplication of ω by an appropriate real number, we may assume, without lost of generality, that $|Q| = 1$ and $\int_Q \omega = 1$.

Using lemma 4.16, for each $\lambda_k = 2^{(n+1)k}$ we find a family $\{Q_j^k\}$ of dyadic subcubes of $Q \subset \mathbb{R}^n$ such that

$$2^{(n+1)k} \le \omega_{Q_j^k} = \frac{1}{|Q_j^k|} \int_{Q^k} \omega \le 2^{(n+1)k+n} \qquad \text{for each } j,$$

and

$$\omega(x) \le 2^{(n+1)k} \qquad \text{a.e. in } D_k^c = Q \setminus \bigcup_j Q_j^k.$$

It is then easy to see that each cube in the family $\{Q_j^{k+1}\}$ is a subcube of some cube Q_i^k of the previous generation, and that

$$|Q_i^k \cap D_{k+1}| < \frac{1}{2} |Q_i^k|$$

holds for each cube of the k-generation, implying that

$$|D_k \cap D_{k+1}| \le \frac{1}{2} |D_k|.$$

Therefore, by the claim above, we have the estimate:

$$\int_{D_R} \omega \le \tau \int_{D_{k-1}} \omega \le \tau^k \int_{D_0} \omega, \qquad \text{with } 0 < \tau < 1,$$

and

$$\begin{aligned} \int_Q \omega^{1+\delta} &= \int_{Q \setminus D_0} \omega^{1+\delta} + \sum_{k=0}^{\infty} \Big[\int_{D_R \setminus D_{R-1}} \omega(x)\,\mathrm{d}x \Big] 2^{(n+1)(k+1)\delta} \\ &\le 2^{\delta} + \sum_{k=0}^{\infty} \tau^k \,(2^{(n+1)\delta})^k\, 2^{(n+1)\delta}, \end{aligned}$$

which converges for some value $\delta > 0$. □

4.4 Weighted inequalities for singular integrals

It is convenient now to introduce the "sharp function operator", defined over locally integrable functions f by the following formula:

$$f^{\#}(x) = \sup_{x \in Q} \frac{1}{|Q|} \int_Q |f(y) - f_Q|\,\mathrm{d}y,$$

where for each cube Q,

$$f_Q = \frac{1}{|Q|} \int_Q f(y)\, dy.$$

Theorem 4.19. *Given $\omega \in A_\infty$ and $f \in \bigcup_{1<p<\infty} L^p(\mathbb{R}^n)$, we have the inequality*

$$\int_{\mathbb{R}^n} (Mf(x))^p\, \omega(x)\, dx \le C(p,\omega) \int_{\mathbb{R}^n} (f^\sharp(x))^p\, \omega(x)\, dx,$$

where $C(p,\omega) < \infty$ and M is the Hardy–Littlewood maximal operator.

Proof. It will be based on the following estimate:

$$\mu(\{MF(x) > \alpha, F^\sharp(x) < \gamma\alpha\}) \le C\gamma^\delta\, \mu(\{MF(x) > \alpha/2^{n+1}\}), \tag{4.1}$$

valid for every $\gamma, \alpha > 0$, where $C < \infty$ is a finite constant, independent of γ, α and F, and $\delta > 0$ is given by the A_∞ property of ω:

$$\frac{\mu(E)}{\mu(Q)} \le \left[\frac{|E|}{|Q|}\right]^\delta,$$

and, as usual, M denotes the Hardy–Littlewood maximal function.

Let us show first that (4.1) implies the theorem:

$$\begin{aligned}
\int_{\mathbb{R}^n} [Mf(x)]^p \omega(x)\, dx &= p \int_0^\infty \alpha^{p-1}\, \mu(\{Mf(x) > \alpha\})\, d\alpha \\
&\le p \int_0^\infty \alpha^{P-1} \mu(\{Mf(x) > \alpha, f^\sharp(x) < \gamma\alpha\})\, d\alpha \\
&\quad + p \int_0^\infty \alpha^{p-1} \mu(\{f^\#(x) > \gamma\alpha\})\, d\alpha \\
&\le pC \int_0^\infty \alpha^{p-1} \mu(\{Mf > \alpha/2^{n+1}\})\, d\alpha \cdot \gamma^\delta + p \int_0^\infty \alpha^{p-1} \mu(\{f^\# > \gamma\alpha\})\, d\alpha \\
&\le pC\gamma^\delta 2^{(n+1)(p-1)}\, 2^{n+1} \int_0^\infty \alpha^{p-1}\, \mu(\{Mf > \alpha\})\, d\alpha \\
&\quad + p \left(\frac{1}{\gamma}\right)^p \int_0^\infty \alpha^{p-1}\, \mu(\{f^\sharp > \alpha\})\, d\alpha \\
&\le C\gamma^\delta 2^{(n+1)p} \int_{\mathbb{R}^n} [Mf(x)]^p\, \omega(x)\, dx + \frac{1}{\gamma^p} \int_{\mathbb{R}^n} [f^\sharp(x)]^p\, \omega(x)\, dx,
\end{aligned}$$

and we take γ small enough so that $C\gamma^\delta 2^{(n+1)p} \le 1/2$ to get the theorem.

To finish, we shall prove (4.1).

As before, let $\{Q_j^0\}$ be the family of disjoint dyadic cubes of $\mathbb{R}^n$ such that

$$\frac{\alpha}{2^{n+1}} \le f_{Q_j^0} < 2^n \frac{\alpha}{2^{n+1}} = \frac{\alpha}{2},$$

and let us consider also the class $\{Q_k^1\}$ of dyadic cubes corresponding to the stopping time

$$\alpha \le f_{Q_k^1} < 2^n \alpha.$$

Since we know that each Q_k^1 is a subcube of some Q_j^0, it will be enough to show that, for each j, the following inequality holds:

$$\mu(\{x \in Q_j^0 : Mf(x) > \alpha, f^\sharp(x) < \gamma\alpha\}) \le c\,\gamma^\delta\,\mu(Q_j^0).$$

Furthermore, using condition A_∞, it suffices to prove the following estimate for the Lebesgue measure:

$$|\{x \in Q_j^0 : Mf(x) > \alpha, f^\sharp(x) < \gamma\alpha\}| \le \gamma\,|Q_j^0|.$$

To see this, let us consider, for every Q_j^0, the members of the next generation contained inside. We have two possibilities:

(1) If

$$\frac{1}{|Q_j^0|}\int_{Q_j} |f - f_{Q^0}| \ge \gamma\alpha,$$

then the set $\{x \in Q_j^0 : Mf(x) > \alpha, f^\sharp(x) < \gamma\alpha\}$ is empty, and there is nothing to be proved.

(2) Otherwise, we have

$$\begin{aligned}\gamma\alpha &> \frac{1}{|Q_j^0|}\int_{Q_j^0} |f - f_{Q_j^0}| \ge \frac{1}{|Q_j^0|}\sum_{Q_k^1\subset Q_j^0}\int_{Q_k^1}|f - f_{Q_j^0}| \\ &\ge \frac{1}{|Q_j^0|}\sum_{Q_k^1\subset Q_j^0}|Q_k^1|\,|f_{Q_k^1} - f_{Q_j^0}| \ge \frac{\alpha}{2}\sum_{Q_k^1\subset Q_j^0}\frac{|Q_k^1|}{|Q_j^0|},\end{aligned}$$

which implies the searched estimate because

$$\bigcup_k Q_k^1 \supset \{x \in Q_j^0 : Mf(x) > \alpha\}.$$ □

Let $Tf = (\text{p.v. } K) * f$ be a singular integral operator satisfying

$$\widehat{K}(\xi) = m(\xi) \in L^\infty(\mathbb{R}^n),$$
$$|K(x)| \le |x|^{-n} \quad \text{and} \quad |\nabla K(x)| \le |x|^{-n-1}, \quad \text{for } |x| > 0.$$

Theorem 4.20. *We have*

$$Tf^\sharp(x) \le C_r\,[M(|f|^r)(x)]^{1/r}$$

for every $r > 1$, where the finite constant C_r is independent of f.

Proof. Given a cube Q and $x \in Q$, let us consider the splitting of f given by its restriction to Q^* and $\mathbb{R}^n \setminus Q^*$:

$$f = f|_{Q^*} + f|_{\mathbb{R}^n \setminus Q^*} = f^{(1)} + f^{(2)}.$$

Then

$$Tf = Tf^{(1)} + Tf^{(2)},$$

and

$$\begin{aligned}\frac{1}{|Q|}\int_Q |Tf - (Tf)_Q| &\le \frac{1}{|Q|}\int_Q |Tf^{(1)} - (Tf^{(1)})_Q| + \frac{1}{|Q|}\int_Q |Tf^{(2)} - (Tf^{(2)})_Q| \\ &= \mathrm{I} + \mathrm{II}.\end{aligned}$$

We have

$$\begin{aligned}\mathrm{I} &\le \frac{2}{|Q|}\int_Q |Tf^{(1)}| \le 2\left[\frac{1}{|Q|}\int_Q |Tf^{(1)}|^r\right]^{1/r} \le 2\,C_r(n)\left[\frac{1}{|Q|}\int |f^{(1)}|^r\right]^{1/r} \\ &\le 2\,C_r(n)\left[\frac{1}{|Q|}\int_{Q^*} |f|^r\right]^{1/r} \le 2C_r(n)\,(Mf^r)^{1/r},\end{aligned}$$

where $C_r(n)$ is independent of f and Q.

For the other term, we write

$$\mathrm{II} = \frac{1}{|Q|}\int_Q \left|\int_Q (Tf^{(2)}(y) - Tf^{(2)}(z))\,\mathrm{d}z\right|\mathrm{d}y.$$

Given $y, z \in Q$, we have

$$\begin{aligned}|Tf^{(2)}(y) - Tf^{(2)}(z)| &= \left|\int [K(y-u) - K(z-u)]\, f^{(2)}(u)\,\mathrm{d}u\right| \\ &\le \sum_\nu \int_{Q^*_{3^{\nu+1}} \setminus Q^*_{3^\nu}} C\,|\nabla K(t-u)|\,|y-z|\,|f^{(2)}(u)|\,\mathrm{d}u\end{aligned}$$

where Q^*_N denotes the dilation of Q by a factor N, t is the point in the line segment limited by y and z prescribed by the mean value theorem, and $C < \infty$ is an absolute constant.

Therefore, for $y, z \in Q$, we have the estimate

$$\begin{aligned}|Tf^{(2)}(y) - Tf^{(2)}(z)| &\le C\sum_{\nu=0}^{\infty} \operatorname{diam}(Q)\,\frac{1}{|Q_{3^*_{\nu+1}}|}\int_{Q^*_{3^{\nu+1}}} |f^{(2)}|\,\frac{1}{2^\nu \operatorname{diam}(Q)} \\ &\lesssim Mf(x) \lesssim (Mf^r(x))^{1/r}.\end{aligned}$$

□

Corollary 4.21. *Under the hypothesis above, T is a bounded operator on $L^p(\omega\,\mathrm{d}x)$ for every weight $\omega \in A_p$, $1 < p < \infty$.*

Proof. We have

$$\int |Tf(x)|^p\, \omega(x)\,\mathrm{d}x \lesssim \int |Tf^{\sharp}(x)|^p\, \omega(x)\,\mathrm{d}x \le C_{p,r} \int (Mf^r(x))^{p/r}\, \omega(x)\,\mathrm{d}x.$$

Since $\omega \in A_{p-\varepsilon}$ for some $\varepsilon > 0$, one can choose $r > 1$ so that $p/r = 1 - \varepsilon$, and therefore

$$\int |Tf(x)|^p\, \omega(x)\,\mathrm{d}x \le \widetilde{C}_p(\omega) \int |f(x)|^p\, \omega(x)\,\mathrm{d}x,$$

where $\widetilde{C}_p(\omega)$ depends only upon p and the bounds of w as an A_p weight. □

Let

$$T_j f = K_j * f$$

be a set of singular integral operators having uniform bounds, i.e., $\|\widehat{K}_j\|_\infty \le C$, $|K_j(x)| \le C\,|x|^{-n}$ and $|\nabla K_j(x)| \le C\,|x|^{-n-1}$.

Theorem 4.22. *The vector valued operator* $\{f_j\} \xrightarrow{T} \{T_j f_j\}$ *is bounded on* $L^p(\ell^r)$, *for* $1 < p < \infty$, $1 < r < \infty$.

Proof. A duality argument shows that the boundedness of the operator on $L^p(\ell^r)$ is equivalent to the boundedness on $L^q(\ell^s)$ for $1/p + 1/q = 1$ and $1/r + 1/s = 1$. Therefore, it will suffice to prove the case $p \ge r$.

The case $p = r$ follows from the classical Calderón–Zygmund theory. If $p > r$, we have

$$\left[\int_{\mathbb{R}^n} \Big[\sum_j |T_j f_j(x)|^r\Big]^{p/r} \mathrm{d}x\right]^{r/p} = \sup_{\|g\|_q = 1} \int_{\mathbb{R}^n} \sum_j |T_j f_j(x)|^r\, g(x)\,\mathrm{d}x,$$

where $r/p + 1/q = 1$, and

$$\begin{aligned}\sum_j \int_{\mathbb{R}^n} |T_j f_j(x)|^r\, g(x)\,\mathrm{d}x &\le \sum_j \int_{\mathbb{R}^n} |T_j f_j(x)|^r\, (Mg^s(x))^{1/s}\,\mathrm{d}x\\ &\le C_{r,s} \int_{\mathbb{R}^n} \Big[\sum_j |f_j(x)|^r\Big] (Mg^s(x))^{1/s}\,\mathrm{d}x\\ &\le C_{r,s} \Big\|\Big(\sum_j |f_j|^r\Big)^{1/r}\Big\|_p \Big[\int_{\mathbb{R}^n} (Mg^s(x))^{q/s}\Big]^{1/q},\end{aligned}$$

and choosing $s > 1$ so that $q/s > 1$, we have

$$\|(Mg^s)^{1/s}\|_q \lesssim \|g\|_q \lesssim 1.$$

□

Chapter 5
The multiplier problem for the ball

Finally, when the number of sides becomes so numerous, and the sides themselves so small, that the figure cannot be distinguished from a circle, he is included in the Circular or Priestly order, and this is the higher class of all.

Edwin Abbot (Flatland)

5.1 Fefferman's counterexample

An important and very natural example of Fourier multiplier is given by the indicator function of a ball centered at the origin and radius R. It satisfies the following formula:

$$S_R f(x) = \int_{|\xi|\le R} \exp(2\pi i x\cdot\xi)\,\widehat{f}(\xi)\,\mathrm{d}\xi.$$

A simple dilation argument shows that the boundedness of the family of operators S_R can be reduced to the boundedness of one of them, for example S_1.

One would expect to recover f as a limit of $S_R f$ when R goes to infinity, as it happens in the one-dimensional setting, where S_1 is closely related to the Hilbert transform.

In $\mathbb{R}^n$, $n \geq 2$, the kernel K of S_1 can be explicitly computed in terms of Bessel functions:

$$K(x) = \int_{|\xi\le 1} \exp(2\pi i x\cdot\xi)\,\mathrm{d}\xi = C_n\, J_{n/2}(2\pi|x|)\,|x|^{-n/2},$$

where J_ν denotes the Bessel function of the first kind and order ν.

However, the decay $K(x) = O(|x|^{-(n+1)/2})$ is too far from the critical $|x|^{-n}$ needed in the theory of singular integrals, and the action of S_1 over functions in the Schwartz class whose Fourier transform is identically equal to 1 on the unit ball shows that it cannot be bounded on $L^p(\mathbb{R}^n)$ if $p < 2n/(n+1)$ and also, by duality, if $p > 2n/(n-1)$, $n \geq 2$.

It was considered for a long time that inside that critical range, $2n/(n+1) < p < 2n/(n-1)$, the answer to the boundedness of S_1 should be positive. But in his celebrated paper [115], C. Fefferman disproved that conjecture: it was like a bomb destroying an established program. However, it also helped to visualize a whole set of new interesting questions whose answers are needed to obtain a deep understanding of n-dimensional Fourier analysis.

A. Córdoba, *Suprematism in Harmonic Analysis*, Progress in Mathematics 356,
https://doi.org/10.1007/978-3-031-73737-4_5

After the work of many artists, we can state now that the project is quite well developed and understood in dimension 2, but it is much more difficult for $n > 2$, where it remains today as a quite unexplored terra incognita.

Theorem 5.1. *In $L^p(\mathbb{R}^n)$, $n \geq 2$, the multiplier S_1 is bounded if and only if $p = 2$.*

Remark 5.2. It is sufficient to prove the theorem above in dimension two, because boundedness of S_1 in dimension n implies boundedness in dimension $n-1$ by the result (Fact 4.4) of Karel de Leew mentioned in Chapter 4.

The proof will follow by contradiction with the help of the following lemma, due to Y. Meyer.

Lemma 5.3. *Assume the boundedness of S_1 in the space $L^p(\mathbb{R}^2)$, $1 < p < \infty$. Given a sequence $\{v_j\}$ of unit vectors in $\mathbb{R}^2$, let us define the Fourier multipliers operators $\{T_j\}$ by the formula*

$$\widehat{T_j f}(\xi) = A_j(\xi)\,\widehat{f}(\xi),$$

where each A_j is the indicator function of the half-plane $\{x \in \mathbb{R}^2 : x \cdot v_j \geq 0\}$.

Then for any sequence of functions $\{f_j\}$ on the Schwartz class, we have the estimate

$$\Big\|\Big(\sum_j |T_j f_j|^2\Big)^{1/2}\Big\|_p \lesssim \Big\|\Big(\sum_j |f_j|^2\Big)^{1/2}\Big\|_p.$$

Proof. It is enough to obtain the inequality above in the particular case of a finite set $\{f_j\}$ such that their Fourier transforms $\{\widehat{f_j}\}$ have all of them compact support, so long as the obtained estimate keeps independent of those assumptions.

Let

$$\widehat{T_j^r f}(\xi) = D_j^r(\xi)\,\widehat{f}(\xi),$$

where D_j^r denotes the indicator function of the disc (also denoted by D_j^r) of radius r centered at the point rv_j. It then follows that

$$T_j f_j(x) = \lim_{r\to\infty} T_j^r f_j(x) \quad \text{uniformly},$$

and that

$$\Big\|\Big(\sum |T_j f_j|^2\Big)^{1/2}\Big\|_p \leq \limsup_{r\to\infty}\Big\|\Big(\sum |T_j^r f|^2\Big)^{1/2}\Big\|_p.$$

Therefore, it suffices to prove the estimate

$$\Big\|\Big(\sum_j |T_j^r f_j|^2\Big)^{1/2}\Big\|_p \leq C_p \Big\|\Big(\sum_j |f_j|^2\Big)^{1/2}\Big\|_p, \tag{5.1}$$

with C_p finite and independent of r, as well as of the assumptions above.

Next we observe that a straightforward dilation argument allows the reduction to prove only the case $r = 1$. Since

$$T_j^r f_j(x) = \exp(2\pi i v_j \cdot x)\, S_1(\exp(-2\pi i v_j\, -) f(-))(x),$$

the left-hand side of (5.1) is equal to

$$\Big\| \Big(\sum_j |S_1(\exp(-2\pi i v_j \cdot -) f_j(-))(x)|^2 \Big)^{1/2} \Big\|_p,$$

and the boundedness of S_1 in $L^p(\mathbb{R}^2)$ implies, by Khintchine's inequality, its boundedness in $L^p(\ell_2)(\mathbb{R}^2)$. This finishes the proof. □

Remark 5.4 The Hilbert transform in the direction $v \in S^{n-1}$ is by definition the operator

$$H_v f(x) = \lim_{\varepsilon \to 0} \frac{1}{\pi} \int_{|t| \geq \varepsilon} \frac{f(x - tv)}{t}\, dt. \tag{5.2}$$

With the notation of Lemma 5.3, it satisfies the identities

$$\widehat{H_{v_j} f}(\xi) = -i\, \mathrm{sign}(v_j \cdot \xi)\, \widehat{f}(\xi) \quad \text{and} \quad H_{v_j} f(x) = i\,[f(x) - 2T_j f(x)].$$

Thus, an estimate analogous to that of Lemma 5.3 holds if we substitute T_j by the Hilbert transform H_{v_j}.

To finish the proof of Theorem 5.1, let us observe that the construction of the Besicovitch set of Chapter 3 produces, for every $N > 1$, a collection $\{R_j\}_{j=1,\dots,N}$ of plane rectangles of dimensions $(1/N) \times 1$ and respective directions $v_j \in \mathbb{S}^{n-1}$, such that

(i) $\mu(\bigcup_{j=1}^N R_j) \leq C/\log(1 + N)$;
(ii) if $\tilde{R}_j$ is a $3v_j$ translation of R_j, then the family $\{\tilde{R}_j\}$ is pairwise disjoint.

Denoting by H_j the Hilbert transform in the direction v_j, it happens that

$$|H_j R_j(x)| \geq \frac{1}{4\pi} \quad \text{when } x \in \tilde{R}_j.$$

Therefore, the assumption that S_1 is bounded on $L^p(\mathbb{R}^2)$, $1 \leq p < 2$, implies

$$\begin{aligned} 1 = \mu\Big(\bigcup \tilde{R}_j\Big) &= \int \Big(\sum_j |\tilde{R}_j(x)|^2 \Big)^{p/2} d\mu(x) \\ &\leq \Big(\frac{1}{4\pi}\Big)^p \int \Big(\sum_j |H_j R_j(x)|^2 \Big)^{p/2} d\mu(x) \lesssim \int \Big| \sum_j R_j(x) \Big|^{p/2} d\mu(x) \\ &\leq \Big(\int \sum_j R_j(x)\, d\mu(x) \Big)^{p/2} \mu\Big(\bigcup_j R_j\Big)^{(2-p)/2} \lesssim \Big[\frac{1}{\log(1 + N)} \Big]^{(2-p)/2}, \end{aligned}$$

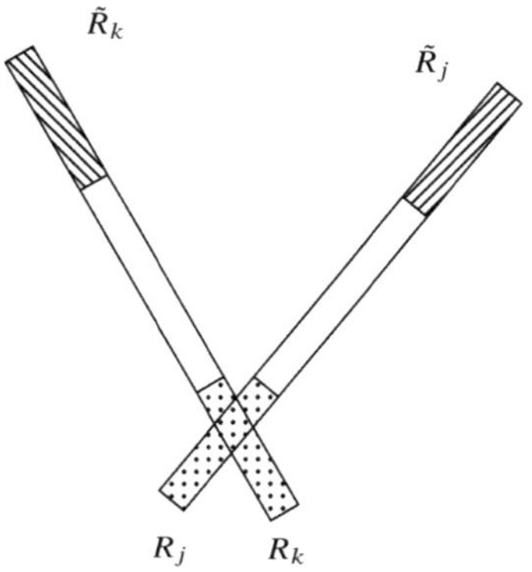

Fig. 5.1: Rectangles and dilated rectangles.

which is clearly false for N big enough. This implies that S_1 is bounded on $L^p(\mathbb{R}^n)$, $n \geq 2$, if and only if $p = 2$. □

5.2 Bochner–Riesz operators: Spherical summation

The Bochner–Riesz operators, defined for $\lambda > 0$ by

$$\widehat{T^\lambda f}(\xi) = (1 - |\xi|^2)_+^\lambda \widehat{f}(\xi),$$

are smoother versions of the ball multiplier. Their kernels can be also explicitly evaluated in terms of Bessel functions:

$$K_\lambda(x) = \pi^{-\lambda}\Gamma(1 + \lambda)\,|x|^{-n/2-\lambda} J_{n/2+\lambda}(2\pi|x|),$$

although this fact will be of no importance in the subsequent discussion of their L^p-boundedness.

If $\lambda > (n-1)/2$ (critical exponent), then K_λ is integrable and T^λ is bounded on every $L^p(\mathbb{R}^n)$, for $1 \leq p \leq \infty$.

But when $0 < \lambda < (n-1)/2$, the decay of K_λ is far away from the Calderón–Zygmund theory of singular integrals and, in fact, an obvious counterexample ($\widehat{f}$ with bounded support and $\widehat{f}(\xi) \equiv 1$ when $|\xi| \leq 1$) shows that T^λ cannot be bounded outside the range

$$\frac{2n}{n+1+2\lambda} < p < \frac{2n}{n-1-2\lambda}.$$

But inside that range, the estimate holds for radial functions $f(|x|)$, and also for products of the type $f(|x|)P_k(x)$, where P_k is a spherical harmonic of degree k.

Based on that evidence, the following conjecture was formulated:

Conjecture 5.5. *The operator T^λ is bounded on $L^p(\mathbb{R}^n)$ so long as*

$$\frac{2n}{n+1+2\lambda} < p < \frac{2n}{n-1-2\lambda} \quad \textit{and} \quad 0 < \lambda < \frac{n-1}{2}.$$

Besides the case $n = 1$, the conjecture has been proved only in dimension two, being the main purpose of this section to present the ideas involved in its proof.

Theorem 5.6. *For $0 < \lambda < 1/2$, the operator T^λ is bounded on $L^p(\mathbb{R}^2)$ when*

$$\frac{4}{3+2\lambda} < p < \frac{4}{1-2\lambda}.$$

The multiplier $m_\lambda(\xi) = (1-|\xi|^2)_+^\lambda$ is rather complicated, and a good strategy for understanding its properties is to find out the basic blocks of the Calderón–Zygmund theory lying inside m_λ.

Since m_λ is radial, and basically constant on thin annuli, it seems reasonable to introduce the following decomposition:

$$m_\lambda(\xi) = \sum_{k=0}^{\infty} (1-|\xi|^2)_+^\lambda \, \phi_k(|\xi|),$$

where each ϕ_k, for $k \geq 1$, is a smooth function supported in the interval $[1 - 2^{-k}, 1 - 2^{-k+1}]$ such that $|D^\alpha \phi_k| \leq C_\alpha 2^{k\alpha}$, with C_α independent of k; satisfying $\sum_{k=1}^\infty \phi_k(x) \equiv 1$ on the interval $1/2 \leq x \leq 1$. Let us define $\phi_0 = 1 - \sum_{k=1}^\infty \phi_k$.

Observe that

$$\phi_j(|\xi|)(1-|\xi|^2)_+^\lambda \simeq 2^{-j\lambda} \phi_j(|\xi|),$$

so our problem is reduced to get good estimates for the growth, as $j \to \infty$, of the norm of the multiplier given by $\phi_j(|\xi|)$. More precisely, the proof of Theorem 5.6 will follow easily if we can show that the operator

$$\widehat{T_k f}(\xi) = \phi_k(|\xi|)\, \widehat{f}(\xi)$$

satisfies the estimate

$$\|T_k f\|_4 \lesssim k^{1/4} \|f\|_4 \quad \text{for every } f \in \mathcal{S}(\mathbb{R}^2).$$

This is because interpolation with the estimate $\|T^k f\|_\infty \lesssim 2^{k/2} \|f\|_\infty$ yields

$$\|T_k f\|_p \lesssim k^{t/4} 2^{k(1-t)/2} \|f\|_p.$$

When $1/p = t/4$ and $2 \leq p < 4/(1-2\lambda)$, then $1 - t < 2\lambda$, and the series $\sum_{j=1}^\infty 2^{-j\lambda} j^{1/4} 2^{j(1-t)/2}$ is convergent.

Remark 5.7. The series above is convergent if the power $j^{1/4}$ is changed to j^α so long as $1 - t < 2\lambda$. That is, any power k^α in the estimate of the norm of T_k in $L^4(\mathbb{R}^2)$ suffices to conclude the validity of Theorem 5.6, allowing certain flexibility to extend the theorem to more general contexts.

Theorem 5.6 will then be a mere corollary of the following:

Proposition 5.8. *Let*

$$\widehat{S_\delta f}(\xi) = \phi(\delta^{-1}(1 - |\xi|))\, \widehat{f}(\xi),$$

where $0 < \delta < 1/2$, *and where* ϕ *is a smooth function supported on the interval* $(-1, 1)$. *There exists a finite constant* C, *independent of* δ, *such that*

$$\|S_\delta f\|_4 \le C\, |\log \delta|^{1/4}\, \|f\|_4$$

for every f *in the Schwartz class* $\mathcal{S}(\mathbb{R}^2)$.

Proof. There is no loss of generality assuming that S_δ is applied to functions f whose Fourier transform $\widehat{f}$ is supported in the angular region $\{0 \le \arg(\xi) \le \pi/3\}$.

Let us consider a smooth partition of unity ψ_j, $j = 1, 2, \ldots, [\delta^{-1/2}]$, in the unit circle, such that

$$\psi_j(\theta) = \psi\big(\delta^{-1/2}\,(\theta - 2\pi j\, \delta^{1/2})\big),$$

where ψ is a smooth function supported in the unit interval (without loss of generality, we may assume that $\delta^{-1/2}$ is an integer).

We have

$$\widehat{S_\delta f}(|\xi|e^{i\theta}) = \sum_{j=1}^{\delta^{-1/2}} \phi(\delta^{-1}(1 - |\xi|))\, \psi_j(\theta)\, \widehat{f}(\xi) = \sum_j m_j(\xi)\, \widehat{f}(\xi) = \sum_j \widehat{T_j f}(\xi)$$

and

$$\begin{aligned}\int \Big|\sum_j T_j f\Big|^4 \mathrm{d}x &= \int \Big|\sum_{j,k} T_j f(x)\, T_k f(x)\Big|^2 \mathrm{d}x = \int \Big|\sum_{j,k} \widehat{T_j f} * \widehat{T_k f}(\xi)\Big|^2 \mathrm{d}\xi \\ &\lesssim \sum_{j,k} \int |\widehat{T_j f} * \widehat{T_k f}(\xi)|^2\, \mathrm{d}\xi = \int \Big(\sum_j |T_j f(x)|^2\Big)^2 \mathrm{d}x.\end{aligned}$$

where the crucial point in the inequalities above is that the family of sets $\operatorname{supp}(\widehat{T_j f}) + \operatorname{supp}(\widehat{T_k f})$ has a finite overlapping independently of δ.

Next, we make use of the pointwise estimate

$$\Big(\sum_j |T_j f(x)|^2\Big)^{1/2} \le \Big(\sum_j T_{2j} f(x)|^2\Big)^{1/2} + \Big(\sum_j |T_{2j+1} f(x)|^2\Big)^{1/2},$$

and observe that both terms in the right-hand side have identical norms.

Regarding the first term, the supports of $m_{2j}(\xi)$ are basically perpendicular to the horizontal axis, and we can find a configuration of horizontal bands E_j, uniformly spaced in the vertical direction, such that

$$\operatorname{supp}(m_{2j}) \subset E_{2j}.$$

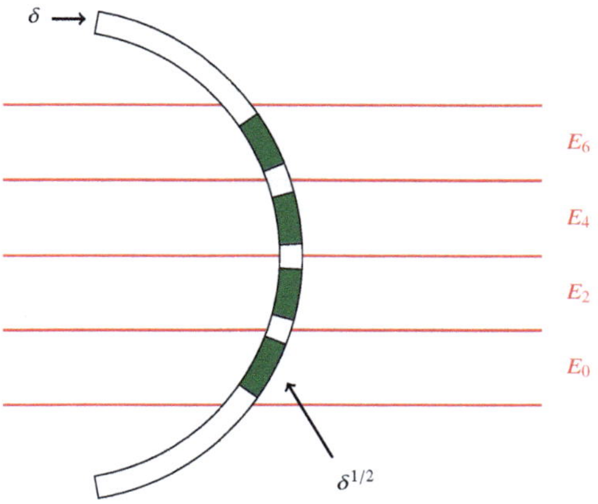

Fig. 5.2: G-function.

Therefore, $m_{2j}(\xi) = m_{2j}(\xi)E_{2j}(\xi)$. Defining

$$\widehat{P_{2j}f}(\xi) = E_{2j}(\xi)\widehat{f}(\xi),$$

we have

$$\Big(\sum_j |T_{2j}f(x)|^2\Big)^{1/2} = \Big(\sum_j T_{2j}P_{2j}f(x)|^2\Big)^{1/2}.$$

An elementary integration by parts argument yields that the kernel K_0 of T_0 satisfies the following: for every pair of integers $\alpha, \beta \geq 0$, there exists a finite constant $C_{\alpha,\beta}$, independent of δ, such that

$$|K_0(x, y)| \leq C_{\alpha,\beta}\, \delta^{3/2}\, |\delta x|^{-\alpha}\, |\delta^{1/2} y|^{-\beta},$$

where (x, y) are the standard cartesian coordinates in $\mathbb{R}^2$.

Therefore, the operator T_0 is pointwise majorized by the positive operator whose kernel is given by

$$C \sum_{n=0}^{\infty} 2^{-n} \frac{1}{\mu(R_n^0)} \chi_{R_n^0}(x, y),$$

where

$$R_n^0 = \{(x, y) : |x| \leq 2^n \delta^{-1}, |y| \leq 2^n \delta^{-1/2}\},$$

for a suitable constant C independent of δ. Using an appropriate rotation, we obtain a similar majorization for every piece T_j.

Due to the decay 2^{-n}, it is enough to show that, for each n, the L^4-norm of the function

$$\Big(\sum_j \Big|\frac{1}{\mu(R_n^{2j})} \chi_{R_n^{2j}} * P_{2j}f(x)\Big|^2\Big)^{1/2}$$

is bounded by $C|\log \delta|^{1/4}\, \|f\|_4$.

To see this, we take $\omega \geq 0$ in the space $L^2(\mathbb{R}^2)$ and consider the integrals

$$\sum_j \int \left|\frac{1}{\mu(R_n^{2j})} \chi_{R_n^{2j}} * P_{2j} f(x)\right|^2 \omega(x)\,dx \lesssim \sum_j \int |P_{2j} f(y)|^2 \frac{1}{\mu(R_n^{2j})} * \omega(y)\,dy$$
$$\lesssim \sum_j \int |P_{2j} f(y)|^2 M\omega(y)\,dy \leq \left\|\Big(\sum_j |P_{2j} f|^2\Big)^{1/2}\right\|_4^2 \|M\omega\|_2,$$

where M is the Kakeya maximal function of Chapter 3 corresponding to rectangles of eccentricity $N = \delta^{-1/2}$. Therefore, we have

$$\left\|\Big(\sum_1 |T_{2j} f|^2\Big)^{1/2}\right\|_4^2 \leq \sup_{\|\omega\|_2=1} \int \sum_j |T_{2j} f(x)|^2 \omega(x)\,dx$$
$$\lesssim |\log \delta|^{1/2} \left\|\Big(\sum_j |P_{2j} f|^2\Big)^{1/2}\right\|_4^2,$$

and the proposition will follows from the next lemma:

Lemma 5.9. *In $\mathbb{R}^n$, let us consider a lattice $\mathcal{L} = \{Q_\nu\}_{\nu\in\mathbb{Z}^n}$. That is, the parallelepipeds Q_ν are congruent with pairwise disjoint interiors and are such that $\mathbb{R}^n = \bigcup_\nu Q_\nu$.*

Define, for each $\nu \in \mathbb{Z}^n$, the operator

$$\widehat{P_\nu f}(\xi) = Q_\nu(\xi)\,\widehat{f}(\xi)$$

and the square function

$$Gf(x) = \Big(\sum_\nu |P_\nu f(x)|^2\Big)^{1/2}.$$

Then for each $s > 1$, there exists a finite constant C_s such that, for every locally integrable function $\omega \geq 0$ and for every $f \in \mathcal{S}(\mathbb{R}^n)$, we have

$$\int_{\mathbb{R}^n} [Gf(x)]^2 \omega(x)\,dx \leq C_s \int_{\mathbb{R}^n} |f(x)|^2 A_s\omega(x)\,dx,$$

where $A_s\omega(x) = [M\omega^s(x)]^{1/s}$, and M denotes the strong maximal function in $\mathbb{R}^n$.

Proof. Using a convenient dilation, we may assume, without loss of generality, that $\mathcal{L}$ is the unit lattice, i.e., that Q_ν is a unit cube centered at the point $\nu \in \mathbb{Z}^n$.

Let ψ be a smooth function with compact support so that $\psi \equiv 1$ inside Q_0. For each $y \in Q_0$, let us consider the Fourier multiplier given by

$$m_y(z) = \sum_\nu \exp(2\pi i\nu \cdot y)\,\psi(z - \nu).$$

The kernel $\mu_y = \widehat{m}_y$ is a measure of finite total variation, uniformly in $y \in Q_0$. More precisely,

$$\mu_y = \sum_\nu \widehat{\psi}(y+\nu)\,\delta_{y+\nu},$$

where δ_x denotes the Dirac delta function. Therefore,

$$|\mu_y * f(x)|^2 \lesssim \sum_\nu (1+|\nu|)^{2n}\,|\widehat{\psi}(y+\nu)|^2\,|f(x-y-\nu)|^2,$$

and using the fact that $\widehat{\psi}$ is rapidly decreasing, for N big enough, we have

$$\int_{\mathbb{R}^n} |\mu_y * f(x)|^2\,\omega(x)\,\mathrm{d}x \le C_N \sum_\nu (1+|\nu|)^{2n-N} \int_{\mathbb{R}^n} |f(x)|^2\,\omega(x+y+\nu)\,\mathrm{d}x.$$

Choosing $N \geq 4n$, we get

$$\begin{aligned}
&\int_{Q_0}\int_{\mathbb{R}^n} \mu_y * f(x)|^2\,\omega(x)\,\mathrm{d}x\,\mathrm{d}y \\
&\qquad\le C_N \sum_\nu (1+|\nu|)^{2n-N} \int_{\mathbb{R}^n} |f(x)|^2 \Big(\int_{Q_0} \omega(x+y+\nu)\,\mathrm{d}y\Big)\,\mathrm{d}x \\
&\qquad\lesssim \int_{\mathbb{R}^n} |f(x)|^2 M\omega(x)\,\mathrm{d}x,
\end{aligned}$$

where M is the Hardy–Littlewood maximal function.

But, on the other hand, if $\widehat{T_\nu f}(\xi) = \psi(\xi-\nu)\,\widehat{f}(\xi)$, we have

$$\mu_y * f(x) = \sum_\nu \exp(2\pi i\nu\cdot y)\,T_\nu f(x)$$

and

$$\int_{Q_0}\int_{\mathbb{R}^n} |\mu_y * f(x)|^2\,\omega(x)\,\mathrm{d}x\,\mathrm{d}y = \sum_\nu \int_{\mathbb{R}^n} |T_\nu f(x)|^2\,\omega(x)\,\mathrm{d}x,$$

which yields the searched estimate:

$$\begin{aligned}
\sum_\nu \int_{\mathbb{R}^n} |P_\nu f(x)|^2\,\omega(x)\,\mathrm{d}x &= \sum_\nu \int_{\mathbb{R}^n} |P_\nu T_\nu f(x)|^2\,\omega(x)\,\mathrm{d}x \\
&\lesssim C_s \sum_\nu \int_{\mathbb{R}^n} |T_\nu f(x)|^2\,A_s\omega(x)\,\mathrm{d}x \lesssim C_s \int_{\mathbb{R}^n} |f(x)|^2\,A_s\omega(x)\,\mathrm{d}x. \qquad \square
\end{aligned}$$

Remark 5.10. Lemma 5.9 admits a stronger (non-weighted) version on which the lattice is substituted by a more general family of pairwise disjoint parallelepipeds having sides parallel to the coordinate axis. The proof in the one-dimensional case was obtained by J. L. Rubio de Francia [123], and by J.-L. Journé [119] in arbitrary dimension.

5.3 Vector valued inequalities

Let us now consider dilations T_R^λ of the Bochner–Riesz operators:

$$\widehat{T_R^\lambda f}(\xi) = (1 - |\xi|^2/R^2)_+^\lambda \widehat{f}(\xi),$$

and let $\{R_j\}$ be any sequence of positive numbers. Then we have a vector valued version of Theorem 5.6.

Theorem 5.11. *In $\mathbb{R}^2$, given $\lambda > 0$ and $4/(3+2\lambda) < p < 4/(1-2\lambda)$, there exists a finite constant $C_{p,\lambda}$ such that*

$$\Big\|\Big(\sum_j |T_{R_j}^\lambda f_j|^2\Big)^{1/2}\Big\|_p \leq C_{p,\lambda} \Big\|\Big(\sum_j |f_j|^2\Big)^{1/2}\Big\|_p.$$

Proof. As in proof of Theorem 5.6, we may reduce the estimate to the case of a multiplier supported on thin annuli. Suppose that $\phi\colon \mathbb{R} \to \mathbb{R}$ is a smooth function supported in $[-1,+1]$, and define

$$\widehat{S_j^\delta f}(\xi) = \phi(\delta^{-1}(1 - |\xi|/R_j)\, \widehat{f}(\xi),$$
$$\widehat{T_j^n f}(\xi) = \psi_n(\arg(\xi))\, \widehat{S_j^\delta f}(\xi) = m_{j,n}(\xi)\, \widehat{f}(\xi),$$

where $\{\psi_n\}$ is the smooth angular partition of unity introduced in Proposition 5.8.

Also without loss of generality, we may assume that

$$\operatorname{supp}(\widehat{f_j}) \subset \{0 \leq \arg(\xi) \leq \pi/3\}, \qquad \text{for every } j.$$

Let us recall that ψ_n is supported in $|\theta - 2\pi n/N| \leq 2\pi/N$, with $N = \delta^{-1/2}$ (which we may consider to be an integer without loss of generality). This means that $m_{j,n}(\xi)$ is supported on a "rectangle" of dimensions $R_j\delta \times R_j\delta^{1/2}$, and with direction $2\pi n/N$.

There are three steps in the proof:

(a) An argument similar to that used in Theorem 5.6 allows us to reduce the proof to the inequality

$$\Big\|\Big(\sum_j |S_j^\delta f_j|^2\Big)^{1/2}\Big\|_4 \leq C\,|\log\delta|^\alpha \Big\|\Big(\sum_j |f_j|^2\Big)^{1/2}\Big\|_4.$$

(b) With adequate decompositions of the multipliers and several geometric arguments, we prove

$$\Big\|\Big(\sum_j |S_j^\delta f_j|^2\Big)^{1/2}\Big\|_4 \leq C\,|\log\delta| \Big\|\Big(\sum_{j,n} |T_j^n f_j|^2\Big)^{1/2}\Big\|_4.$$

(c) An estimate of the kernels of T_j^n, together with Lemma 5.9 and Theorem 3.7, yields

$$\Big\|\Big(\sum_{j,n}|T_j^n f_j|^2\Big)^{1/2}\Big\|_4 \leq C\,|\log\delta|^\beta \Big\|\Big(\sum_j |f_j|^2\Big)^{1/2}\Big\|_4.$$

Part (a) is left to the reader.

Regarding part (b), we begin with the selection of just one dyadic interval $2^k < R \leq 2^{k+1}$ out of each $|\log_2\delta|$ correlative intervals, and we allow in the left-hand side of (b) only those indices j for which R_j lays in a selected interval. Also we take only one T_j^n for each four correlative indices n, and only those supported in the angular sector $\{0 \leq \arg(\xi) \leq \pi/3\}$. All these operations will contribute with a factor $C|\log\delta$ to the inequality (b).

The left-hand side of (b) is less than the fourth root of twice

$$\sum_{R_j\leq R_k}\int\Big|\Big(\sum_n T_j^n f_j\Big)\Big(\sum_m T_k^m f_k\Big)\Big|^2.$$

We can distinguish two kind of pairs (j,k): either $R_j \leq R_k \leq 2R_j$, or $R_j \leq \delta R_k$. Let us denote $\sum_{\rm I}$ and $\sum_{\rm II}$ the two corresponding halves of the sum above. We have

$$\sum_{\rm I} = \sum_{\rm I}\int\Big|\sum_{n,m}(\widehat{T_j^n f_j}) * (\widehat{T_k^m f_k})\Big|^2 \leq 4\sum_{\rm I}\int\Big|\sum_{n\leq m}\widehat{T_j^n f_j} * \widehat{T_k^m f_k}\Big|^2,$$

and an easy geometric argument shows that, for fixed j and k, the supports of $\widehat{T_j^n f_j} * \widehat{T_k^m f_k}$ are disjoint for different pairs $n \leq m$. Therefore,

$$\sum_{\rm I} \leq 4\int\sum_{\rm I}\sum_{n\leq m}|\widehat{T_j^n f_j} * \widehat{T_k^m f_k}|^2 \leq 4A,$$

where

$$A = \Big\|\Big(\sum_{j,n}|T_j^n f_j|^2\Big)^{1/2}\Big\|_4^4.$$

Next, let us observe that when (j,k) are in case $\sum_{\rm II}$, then $R_j \leq \delta R_k$, and the supports of the convolutions

$$\widehat{T_j^{n_1} f_j} * \widehat{T_k^{m_1} f_k} \quad\text{and}\quad \widehat{T_j^{n_2} f_j} * \widehat{T_k^{m_2} f_k}$$

are disjoint if $m_1 \neq m_2$. Then

$$\sum_{\rm II} = \sum_{\rm II}\int\Big|\sum_m\Big(\sum_n T_j^n f_j\Big)T_k^m f_k\Big|^2$$

$$\leq \Big[\int\Big(\sum_j\Big|\sum_n T_j^n f_j\Big|^2\Big)^{1/2}\Big]^{1/2}\cdot\Big[\int\Big(\sum_{k,m}|T_k^m f_k|^2\Big)^2\Big]^{1/2} \leq \sqrt{2}\Big[\sum_{\rm I}+\sum_{\rm II}\Big]^{1/2}A^{1/2},$$

and this implies (b).

To prove part (c), we first observe that for each fixed j we can choose a grid of rectangles as in Lemma 5.9 so that each of the multipliers $m_{j,n}(\xi)$ is supported in one of the rectangles in the grid, which will be denoted Q_j^n.

If $\widehat{P_j^n f} = Q_j^n \cdot \widehat{f}$ is the corresponding Fourier multiplier, we have the identity

$$T_j f_j = T_j^n P_j^n f_j.$$

Furthermore, an integration by parts argument (as in Theorem 5.6) shows that the kernel of T_j^n is majorized by

$$C \sum_{\nu=0}^{\infty} 2^{-\nu} \frac{1}{|R_{\nu,j}^n|} R_{\nu,j}^n(x),$$

where $R_{\nu,j}^n$ is a rectangle having dimensions $2^\nu \delta^{-1} \times 2^\nu \delta^{-1/2}$ and direction angle $2\pi\delta^{1/2} n$, where C is independent of n, j and $\delta > 0$.

Thus, in order to estimate A, we need to estimate, uniformly in ν, the L^4-norm of

$$\Big[\sum_{j,n} \Big(\frac{1}{|R_{\nu,j}^n|} R_{\nu,j}^n * |P_j^n f_j\,|\Big)^2\Big]^{1/2}.$$

If $w \geq 0$ is in $L^2(\mathbb{R}^2)$, we have

$$\begin{aligned}
\sum_{j,n} \int \Big[\frac{1}{|R_{\nu,j}^n|} R_{\nu,j}^n * |P_j^n f_j|\Big]^2 \omega(x)\,\mathrm{d}x &\leq \sum_{j,n} \int |P_j^n f_j(y)|^2 \Big[\frac{1}{|R_{\nu,j}^n|} R_{\nu,j}^n * \omega(y)\Big]\,\mathrm{d}y \\
&\leq \sum_{j,n} \int |P_j^n f_j(y)|^2\, M\omega(y)\,\mathrm{d}y \leq C_s \sum_j \int |f_j(y)|^2 A_s(M\omega)(y)\,\mathrm{d}y \\
&\leq \widetilde{C}_s \Big\|\Big(\sum_j |f_j|^2\Big)^{1/2}\Big\|_4^2 \|M\omega\|_2 \leq C\,|\log\delta|^{1/2}\, \|\omega\|_2 \Big\|\Big(\sum_j |f_j|^2\Big)^{1/2}\Big\|_4^2,
\end{aligned}$$

where we have used again Lemma 5.9 and the estimates for the Kakeya maximal function of Chapter 3. □

5.4 An angular square function

In $\mathbb{R}^2$. let us consider the angles $\omega_j = 2\pi j/N$, for $j = 0, 1, \ldots, N-1$, the regions

$$\Delta_j = \{\xi : \omega_j \leq \arg(\xi) \leq \omega_{j+1}\},$$

the Fourier multipliers

$$\widehat{S_j f}(\xi) = \Delta_j(\xi)\widehat{f}(\xi)$$

and the associated square function

$$G(f)(x) = \left(\sum |S_j f(x)|^2\right)^{1/2}$$

Theorem 5.12. *There exist constants a and C, independent of N, such that*

$$\|G(f)\|_4 \le C\,[\log N]^a\,\|f\|_4$$

for every f in the Schwartz class $\mathcal{S}(\mathbb{R}^2)$.

Let $\{\psi_j\}$ be angular cutoff functions like in Theorem 5.11 so that $\psi_j \equiv 1$ on Δ_j and $\mathrm{supp}(\psi_j) \subset \Delta_{j-1} \cup \Delta_j \cup \Delta_{j+1}$, and write

$$\widehat{\tilde{S}_j f}(\xi) = \psi_j(\xi)\,\widehat{f}(\xi) \quad \text{and} \quad \widetilde{G}f(x) = \left(\sum |\tilde{S}_j f(x)|^2\right)^{1/2}.$$

Then with the help of the Kakeya maximal function estimates of Chapter 3, Theorem 5.12 will be a consequence of the following.

Proposition 5.13. *There exist finite constants a and C, independent of N, such that:*

$$\|\widetilde{G}(f)\|_4 \le C\,[\log N]^a\,\|f\|_4.$$

Let $\{\phi_n\}$ be smooth functions such that $\mathrm{supp}(\phi_n) \subset (2^{n-1}, 2^{n+1})$ and satisfying $|D_x^\alpha \phi_n| \le C_\alpha 2^{-n\alpha}$, and define

$$\widehat{\tilde{S}_{j,n} f}(\xi) = \phi_n(|\xi|)\,\psi_j(\xi)\,\widehat{f}(\xi) \quad \text{and} \quad \widetilde{G}_n f(x) = \left(\sum_j |\tilde{S}_{j,n} f(x)|^2\right)^{1/2}.$$

Then, in a first step, we shall prove the estimate

$$\|\widetilde{G}_n f\|_4 \le C\,[\log N]^a\,\|f\|_4$$

with constants C and a independent of n and N. But a simple dilation argument shows that it is enough to prove that estimate in the case $n = 1$.

In a second step, we will put together all those estimates to get the proof of the proposition. It will be also convenient to observe that, without loss of generality, we may assume that the functions f have their Fourier transforms supported in the region

$$\{\xi : 0 \le \arg(\xi) \le \pi/4\}.$$

We have to compute

$$\|\widetilde{G}_1 f\|_4^4 = \sum_{j,k} \int_{\mathbb{R}^2} \left|\tilde{S}_j f(x)\,\tilde{S}_k f(x)\right|^2 dx = \sum_{|j-k|\le 4} (\cdots) + \sum_{|j-k|>4} (\cdots) = \mathrm{I} + \mathrm{II}.$$

Next we decompose further the support of $\phi_1(|\xi|)\psi_j(\xi)$ into $N^{1/2}$ parts Δ_j^α, $\alpha = 1, \dots, N^{1/2}$, where

$$\Delta_j^\alpha = \left\{\xi \in \sup(\phi_1 \cdot \psi_j) : \alpha N^{-1/2} \leq |\xi| - 1 \leq (\alpha+1)N^{-1/2}\right\}$$

(again, without loss of generality, we may assume that $N^{1/2}$ is an integer). It happens that when $|j-k| \geq 4$, the overlapping of the sets $\Delta_j^\alpha + \Delta_k^\beta$ is bounded, uniformly on N.

Therefore, using once more Plancherel's theorem, we get

$$\mathrm{II} \leq \sum_{|j-k|\geq 4} \sum_{\alpha,\beta} \int_{\mathbb{R}^2} |\tilde{S}_j^\alpha f(x)\, \tilde{S}_j^\beta f(x)|^2 \, dx \leq \left\| \Big(\sum_{j,\alpha} |\tilde{S}_j^\alpha f|^2 \Big)^{1/2} \right\|_4,$$

where

$$\widehat{\tilde{S}_j^\alpha f}(\xi) = \phi_1(|\xi|)\, \psi_j(\xi)\, \eta_\alpha(|\xi|)\, \widehat{f}(\xi)$$

and η_α is an adapted cut-off family of smooth functions, i.e., $\sum \eta_\alpha(|\xi|) \equiv 1$ on the support of $\phi_1\, \psi_j$, and also

$$\operatorname{supp} \eta_\alpha \subset (1 + \alpha N^{-1/2}, 1 + (\alpha+2)N^{-1/2}) \quad \text{and} \quad |D_x^m \eta_\alpha| \leq C_m N^{m/2}.$$

Observe that

$$\begin{aligned}
\mathrm{I} &\leq \int \Big(\sum |\tilde{S}_j f|^4\Big) \leq \int \Big(\sum_j |\tilde{S}_j f|^2\Big) \cdot \Big(\sup_j |\tilde{S}_j f(x)|^2\Big)\, dx \\
&\leq \Big(\int \sum_{j,k} |\tilde{S}_j f(x)\, \tilde{S}_k f(x)|^2 \, dx\Big)^{1/2} \Big[\int |M_N f(x)|^4\Big]^{1/2} \\
&\leq \frac{1}{2}\,\varepsilon\,(\mathrm{I}+\mathrm{II}) + \frac{1}{2}\frac{1}{\varepsilon}\, C\,(\log N)^a\, \|f\|_4^4
\end{aligned}$$

for every $\varepsilon > 0$. Therefore, choosing ε conveniently, it will be enough to prove the estimate

$$\mathrm{II} \lesssim [\log N]^a\, \|f\|_4^4.$$

To do that, let $\omega \geq 0$ be such that $\|\omega\|_2 = 1$, and consider the integral

$$\begin{aligned}
\sum_{j,\alpha} \int |\tilde{S}_j^\alpha f(x)|^2\, \omega(x)\, dx &= \sum_{\alpha=1}^{N^{1/2}} \int \Big|\sum_j \tilde{S}_j^\alpha f(x)\Big|^2 \omega(x)\, dx \\
&= \sum_{\alpha=1}^{N^{1/2}} \sum_{\tau=0}^{N^{1/2}} \int \Big| \sum_{j=\tau N^{1/2}}^{(\tau+1)N^{1/2}} \tilde{S}_j^\alpha f(x)\Big|^2 \omega(x)\, dx \\
&= \sum_{\alpha=1}^{N^{1/2}} \sum_{\tau=0}^{N^{1/2}} \int \Big| \sum_{j=\tau N^{1/2}}^{(\tau+1)N^{1/2}} \tilde{S}_j^\alpha E_j^\alpha f(x)\Big|^2 \omega(x)\, dx,
\end{aligned}$$

where the E_j^α are disjoint rectangles, parallel to the bisector of Δ_j, with dimensions of the order of $N^{-1/2}\times N^{-1}$, and such that $E_j^\alpha \supset \operatorname{supp}(\phi_1\, \psi_j\, \eta_\alpha)$ and the operator E_j^α is defined by the formula

$$\widehat{E_j^\alpha f}(\xi) = E_j^\alpha(\xi)\, \widehat{f}(\xi),$$

where we have used again our convention of denoting with the same symbol E_j^α the set, its indicator function, and the associated Fourier multiplier.

We are now in position to apply Lemma 5.9:

$$\begin{aligned}\sum_{j,\alpha} \int \left|\tilde{S}_j^\alpha f(x)\right|^2 \omega(x)\, dx &= \sum_{j,\alpha} \int \left|\tilde{S}_j^\alpha E_j^\alpha f(x)\right|^2 \omega(x)\, dx \\ &\lesssim \sum_{\alpha=1}^{N/2} \sum_{\tau=0}^{N^{1/2}} \int |Q_\alpha^\tau f(x)|^2\, M_N\omega(x)\, dx,\end{aligned}$$

because as in Theorem 5.11, the kernel of $\tilde{S}_j^\alpha$ is majorized by a sum

$$C \sum_{\nu=0}^{\infty} 2^{-\nu} \frac{1}{|R_{j,n}^\alpha|} R_{j,n}^\alpha(x),$$

where the $R_{j,n}^\alpha$ are rectangles of eccentricity $N^{1/2}$ in the j-direction.

Here Q_α^τ, for each $\alpha, \tau = 1, \ldots, N^{1/2}$, is now a square of side length $\sim N^{-1/2}$, and the family $\{Q_\alpha^\tau\}$ has a universally bounded overlapping.

Then another application of lemma 5.9, together with the known estimates for the Kakeya maximal function, yield the desired inequality.

To finish, divide into residue classes $\operatorname{mod}[\log_2 N]$ the indices of the "dyadic" cut-off functions ϕ_n, divide also the index j of directions into classes $\operatorname{mod}(4)$, and apply the same method of Theorem 5.11. □

5.5 Maximal Bochner–Riesz: pointwise convergence

For $\lambda > 0$ the maximal function

$$T_*^\lambda f(x) = \sup_{R>0} |T_R^\lambda f(x)|, \quad \text{where} \quad \widehat{T_R^\lambda f}(\xi) = (1 - |\xi|^2/R^2)_+^\lambda\, \widehat{f}(\xi),$$

dominates the a.e. convergence of $T_R^\lambda f(x)$ to $f(x)$ when R goes to infinity.

Using the standard decomposition of the unit ball into coronas of thickness δ near the boundary ($\delta = 2^{-k}, k = 1, 2, \ldots$), the L^p-boundedness properties of T_*^λ are reduced to get good estimates for

$$S_*^\delta f(x) = \sup_{R>0} |S_R^\delta f(x)|,$$

where

$$\widehat{S_R^\delta f}(\xi) = \phi(|\xi|, R)\, \widehat{f}(\xi)$$

and ϕ is any smooth function satisfying

$$\begin{aligned}
&\sup \phi(-,t) \subset (t - t\,\delta, t + t\,\delta), \quad |D_r^\alpha \phi(r,t)| \leq C_\alpha (t\delta)^{-\alpha}, \\
&D_t \phi(r,t) = \psi(r,t)(t\delta)^{-1}, \\
&\qquad \text{with } \operatorname{supp} \psi(r,t) \subset (t - t\delta, t + t\delta) \text{ and } |D_r^\alpha \psi(r,t)| \leq C_\alpha (t\delta)^{-\alpha}.
\end{aligned}$$

An example is

$$\phi(|\xi|, t) = \varphi\Big(\frac{|\xi| - t}{\delta t}\Big),$$

where φ is a smooth function supported in $(-1, +1)$.

Theorem 5.14. *S_*^δ is bounded on $L^2(\mathbb{R}^n)$.*

Corollary 5.15. $\lim_{R\to\infty} T_R^\lambda f(x) = f(x)$ *almost everywhere for every $f \in L^2(\mathbb{R}^n)$, when $\lambda > 0$.*

Proof of Theorem 5.14. We have

$$(S_R^\delta f(x))^2 = \int_0^R \frac{d}{dt}(S_t^\delta f(x))^2\, \mathrm{d}t = 2\int_0^R S_t^\delta f(x) \cdot \frac{d}{dt}(S_t^\delta f(x))^2\, \mathrm{d}t,$$

implying the estimate

$$\begin{aligned}
S_*^\delta f(x) &\leq \delta^{-1/2} \Big(\int_0^\infty |S_t^\delta f(x)|^2 \frac{\mathrm{d}t}{t} \Big)^{1/2} + \delta^{1/2} \Big(\int_0^\infty \Big| \frac{\mathrm{d}}{\mathrm{d}t} S_t^\delta f(x) \Big|^2 t\, \mathrm{d}t \Big)^{1/2} \\
&= \delta^{-1/2}\, G_1^\delta f(x) + \delta^{1/2}\, G_2^\delta f(x)
\end{aligned}$$

and

$$\begin{aligned}
\int_{\mathbb{R}^n} |G_1^\delta f(x)|^2\, \mathrm{d}x &= \int_{\mathbb{R}^n} \Big(\int_0^\infty |\phi(|\xi|, t)|^2 \frac{\mathrm{d}t}{t} \Big) |\widehat{f}(\xi)|^2\, \mathrm{d}\xi \leq C\,\delta\, \|f\|_2^2, \\
\int_{\mathbb{R}^n} |G_2^\delta f(x)|^2\, \mathrm{d}x &= \int_{\mathbb{R}^n} \Big(\int_0^\infty |\psi(|\xi|, t)|^2 \frac{\mathrm{d}t}{t} \Big) |\widehat{f}(\xi)|^2\, \mathrm{d}\xi\, \delta^{-1} \leq C\,\delta^{-1}\, \|f\|_2^2. \quad \square
\end{aligned}$$

In dimension $n = 2$, Theorem 5.14 admits several improvements.

First, let us consider $\{R_j\}$ to be lacunary, i.e., $R_{j+1}/R_j > \rho > 1$ for every j, and consider the associated maximal operator

$$M_*^\lambda = \sup_j |T_{R_j}^\lambda f(x)|.$$

Then we have:

Theorem 5.16. *For $0 < \lambda < 1/2$, the maximal operator M_*^λ is bounded in $L^2(\mathbb{R}^2)$ if and only if $4/(3 + 2\lambda) < p < 4/(1 - 2\lambda)$.*

Proof. Using the notation introduced in Theorem 5.11 and similar arguments, it will be sufficient to show the estimate

$$\Big\| \Big(\sum_j |S_j^\delta f|^2 \Big)^{1/2} \Big\|_4 \leq C \, |\log \delta|^a \, \|f\|_4 \tag{5.3}$$

for some finite constants a and C, independent of $\delta > 0$, where

$$\widehat{S_j^\delta f}(\xi) = \phi(\delta^{-1} (1 - |\xi|/R_j) \, \widehat{f}(\xi).$$

Observe that, without loss of generality, it is enough to proof (5.3) for functions f whose Fourier transforms $\widehat{f}$ are supported in the angle $\{\xi : 0 \leq \arg(\xi) \leq \pi/3\}$.

Since every dyadic interval $[2^l, 2^{l+1}]$ Contains, at most, $1 + |\log \delta|$ elements of the lacunary set $\{R_j\}$ when $1 < \rho \leq 2$, we can also assume, again with no loss of generality, that $\rho \geq 2$, so that each dyadic interval contains, at most, only one R_j.

Then (5.3) follows easily with the help of the inequalities developed in the proof of Theorem 5.11. The details are left to the reader. □

Theorem 5.17. *For* $0 < \lambda < 1/2$, *the maximal operator*

$$T_*^\lambda f(x) = \sup_{R>0} |T_R^\lambda f(x)|$$

is bounded $L^p(\mathbb{R}^2)$ *if* $2 \leq p < 4/(1 - 2\lambda)$.

Taking into account the estimate

$$\sup_{R>0} |S_R^\delta f(x)| \lesssim \delta^{-1/2} \, Mf(x),$$

where M denotes the Hardy–Littlewood maximal function and $S_R^\delta f$ is given by

$$\widehat{S_R^\delta f}(\xi) = \phi(|\xi|, R) \, \widehat{f}(\xi),$$

as was defined before, the proof of Theorem 5.17 follows by standard interpolation from the estimate for the square function

$$Gf(x) = \Big(\int_0^\infty |S_t^\delta f(x)|^2 \, \frac{dt}{t} \Big)^{1/2}$$

given in the following result.

Lemma 5.18. *There exist finite constants a and C such that*

$$\|Gf\|_4 \leq C \, |\log \delta|^a \, \delta^{1/2} \, \|f\|_4, \quad \textit{for any } f \in \mathcal{S}(\mathbb{R}^2).$$

Proof. There is no loss of generality assuming that $\operatorname{supp}(\widehat{f}) \subset \{\xi : 0 \leq \arg(\xi) \leq \pi/3\}$ and that $\delta^{-1/2} = N^{1/2}$ is an integer.

Let $\{\psi_j\}_{j=1}^N$ be a smooth partition of unity so that ψ_j is a rotation of ψ_0 and homogeneous of degree 0. Furthermore, let us impose that $\psi_j \equiv 0$ outside $\Delta_{j-1} \cup \Delta_j \cup \Delta_{j+1}$, where Δ_j is the angular region

$$\{\xi : j\delta^{1/2} \leq \arg(\xi) \leq (j+1)\delta^{1/2}\}.$$

That is,

$$\sum_j \psi_j(e^{2\pi i\theta}) \equiv 1 \quad \text{and} \quad |D_\theta^\alpha \psi_j| \leq C_\alpha\, \delta^{-\alpha/2}$$

for suitable constants C_α independent of δ.

Let also $\{\Theta_k\}$ be a smooth "dyadic" partition of unity in the interval $[1, \infty)$ such that $\operatorname{supp}(\Theta_k) \subset (2^{k-1}, 2^{k+2})$ and $|D^\alpha \Theta_k| \leq C_\alpha\, 2^{-k\alpha}$ for every k, α and appropriate constants C_α.

Furthermore, for each $j = 1, \ldots, N^{1/2}$ and k, let us introduce smooth cut-off functions $\varphi_\sigma^{j,k}$, $\sigma = 1, \ldots, N$, so that

$$\sum_k \sum_\sigma \varphi_\sigma^{j,k}(\xi) = 1 \quad \text{when } \xi \in \operatorname{supp}(\psi_j)$$

and $\operatorname{supp}(\varphi_\sigma^{j,k}) \subset (R_\sigma^{j,k})^*$, where $(^*)$ denotes here the quadruple of a rectangle $R_\sigma^{j,k}$ contained in $\Delta_j \cap \{2^{k-1} < |\xi| < 2^{k+2}\}$ of dimensions $2^{k-1}\delta \times 2^{k-1}\delta^{1/2}$ and having its two sides, respectively, in the direction of the bisector of Δ_j and its perpendicular (see the Figure 5.3 below).

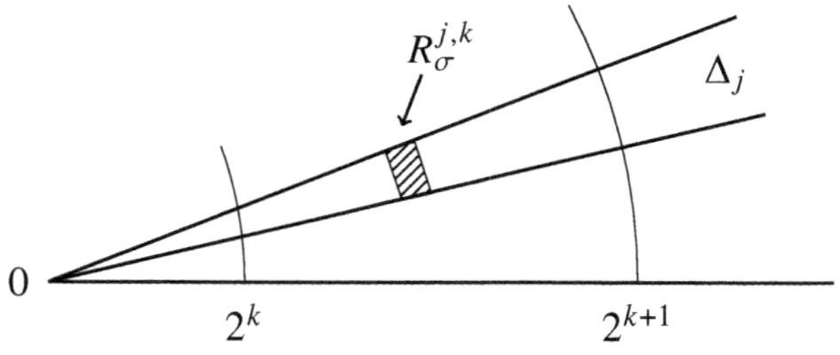

Fig. 5.3: The rectangles $R_\sigma^{j,k}$.

Furthermore, let us assume that the functions $\varphi_\sigma^{j,k}$ are well adapted to $R_\sigma^{j,k}$; more precisely,

$$|D_v^\alpha \varphi_\sigma^{j,k}| \leq \begin{cases} C_\alpha\, (\delta\, 2^k)^{-\alpha}, & \text{if } v \text{ is a bisector direction,} \\ C_\alpha\, (\delta^{1/2}\, 2^k)^{-\alpha} & \text{if } v \text{ is orthogonal to bisector.} \end{cases}$$

Then

$$\widehat{S_t^\delta f}(\xi) = \sum_{k,j,\sigma} \widehat{Q_{k,j,\sigma}^t f}(\xi),$$

where

$$\widehat{Q^t_{k,j,\sigma} f}(\xi) = \Theta_k(\xi)\, \phi(|\xi|,t)\, \psi_j(\xi)\, \varphi^{j,k}_\sigma(\xi)\, \widehat{f}(\xi).$$

Next, we simplify the geometry of the supports dividing the k-index into residue classes mod$[|\log_2(\delta)|]$, so that two rectangles $R^{j_1,k_1}_{\sigma_1}$ and $R^{j_2,k_2}_{\sigma_2}$ corresponding to different values $k_1 > k_2$ in the same residue class I, satisfy that the small dimension of $R^{j_1,k_1}_{\sigma_1}$ is bigger than 2^{k_2}, that is, $2^{k_1}\delta > 2^{k_2}$.

We may also assume that the other indices j and σ are taken, respectively, in residue classes J and Σ mod 8.

Then, except for an allowed factor $|\log\delta|$, the lemma will follow from the estimate

$$\|G_{I,J,\Sigma}(f)\|_4 \lesssim |\log\delta|^a\, \|f\|_4$$

for any choice of residue classes I, J and Σ, where

$$G_{I,J,\Sigma}(f)(x) = \Big(\int_0^\infty \Big|\sum_{k\in I}\sum_{j\in J}\sum_{\sigma\in\Sigma} Q^t_{k,j,\sigma} f(x)\Big|^2 \frac{dt}{t}\Big)^{1/2}.$$

To simplify notation, in the following we shall drop the symbol I, J, Σ, and assume that the indices of summation belong to those fixed residue classes.

We have

$$\|G(f)\|_4^4 = \int_0^\infty\int_0^\infty \frac{dt}{t}\frac{du}{u}\Big(\int_{\mathbb{R}^2}\Big|\sum_{k_1,j_1,\sigma_1} Q^t_{k_1,j_1,\sigma_1} f(x) \sum_{k_2,j_2,\sigma_2} Q^u_{k_2,j_2,\sigma_2} f(x)\Big|^2 dx\Big),$$

and applying Plancherel's theorem to the inner integral, we get

$$\|G(f)\|_4^4 = \int_0^\infty\int_0^\infty \frac{dt}{t}\frac{du}{u}\Big(\int_{\mathbb{R}^2}\Big|\sum_{k_1,j_1,\sigma_1}\sum_{k_2,j_2,\sigma_2} \Omega^{k_2,j_2,\sigma_2}_{k_1,j_1,\sigma_1}(\xi)\Big|^2 d\xi\Big),$$

where $\Omega^{k_2,j_2,\sigma_2}_{k_1,j_1,\sigma_1}$ is the convolution of

$$\Theta_{k_1}(\xi)\, \phi(|\xi|,t)\, \psi_{j_1}(\xi)\, \varphi^{k_1,j_1}_{\sigma_1}(\xi) \quad \text{with} \quad \Theta_{k_2}(\xi)\, \phi(|\xi|,u)\, \psi_{j_2}(\xi)\, \varphi^{k_2 j_2}_{\sigma_2}(\xi).$$

Taking now into consideration the conditions imposed upon the supports of Θ_k, ψ_j and $\varphi^{j,k}_\sigma$, we get

$$\|G(f)\|_4^4 \lesssim \int_{\mathbb{R}^2}\Big(\int_0^\infty \sum_{k,j,\sigma} |Q^t_{k,j,\sigma} f(x)|^2 \frac{dt}{t}\Big)^2 dx,$$

that is, it suffices to estimate the L^4-norm of

$$\Big[\int_0^\infty \sum_{k,j,\sigma} |Q^t_{k,j,\sigma} f(x)|^2 \frac{dt}{t}\Big]^{1/2}$$

To do this, let us observe first that for each set of indices (k, j, σ) and any function $g \in \mathcal{S}(\mathbb{R}^2)$, we have the pointwise inequality

$$\int_0^\infty |Q^t_{k,j,\sigma}(g)(x)|^2 \,\frac{\mathrm{d}t}{t} \lesssim \delta \sup_t |Q^t_{k,j,\sigma}(g)(x)|^2.$$

This is because

$$\begin{aligned}\{t : Q^t_{k,j,\sigma} g(x) \neq 0\} &\subset \left\{t : \operatorname{supp}(\widehat{Q^t_{k,j,\sigma} g}) \text{ is not empty}\right\}\\ &\subset \left\{t : \exists \xi, \Theta_k(\xi)\, \phi(|\xi|, t)\, \varphi^{k,j}_\sigma(\xi)\, \psi_j(\xi) \neq 0\right\} = E_{k,j,\sigma},\end{aligned}$$

and from the definitions of Θ_k, ϕ, ψ_j and $\varphi^{k,j}_\sigma$, it follows easily that

$$\int_0^\infty E_{k,j,\sigma}(t)\,\frac{\mathrm{d}t}{t} \lesssim \delta.$$

An integration by parts argument yields estimates for the kernel (Fourier transform of $\Theta_k\, \phi(\xi)\, \phi(|\xi|, t)\, \varphi^{k,j}_\sigma(\xi) \psi_j(\xi)$) in terms of averages over rectangles of eccentricity $\delta^{-1/2} = N^{1/2}$, implying the estimate

$$\sup_t |Q^t_{k,j,\sigma}\, g(x)| \lesssim M_{\delta^{-1/2}}(\widetilde{Q}_{k,j,\sigma}\, g)(x),$$

where $M_{\delta^{-1/2}}$ denotes the Kakeya maximal function with rectangles of eccentricity $\delta^{-1/2}$,

$$\widehat{\widetilde{Q}_{k,j,\sigma}\, g}(\xi) = \widetilde{\Theta}_k(\xi)\, \widetilde{\varphi}^{k,j}_\sigma(\xi)\, \widehat{f}(\xi),$$

and $\widetilde{\Theta}_k$ and $\widetilde{\varphi}^{k,j}_\sigma$ are adequate cut-off satisfying, respectively, the same estimates than Θ_k and $\varphi^{k,j}_\sigma$, and being identically equal to 1 on their supports.

Then for every $\omega \in L^2(\mathbb{R}^2)$, we get

$$\begin{aligned}\int_0^\infty \int_{\mathbb{R}^2} |Gf(x)|^2\, \omega(x)\,\mathrm{d}x\,\frac{\mathrm{d}t}{t} &\lesssim \delta \sum_{j,k,\sigma} \int_{\mathbb{R}^2} |\widetilde{Q}_{j,k\sigma} f(x)|^2\, M_{\delta^{-1/2}}\omega(x)\,\mathrm{d}x\\ &\le C_s\, \delta \sum_{\tau=1}^{N^{1/2}} \sum_{k,j} \int_{\mathbb{R}^2} \left|\widetilde{\widetilde{Q}}{}^\tau_{k,j} f(x)\right|^2 \left[M_1 (M_{\delta^{-1/2}}\omega)^s(x)\right]^{1/s} \mathrm{d}x\end{aligned}$$

for every $s > 1$, where

$$\widetilde{\widetilde{Q}}{}^\tau_{k,j}(\xi) = \sum_{\sigma=\tau}^{\tau+N^{1/2}} Q_{k,j,\sigma}(\xi).$$

Therefore $\widetilde{\widetilde{Q}}{}^\tau_{k,j}$, for each $1 \le \tau \le N^{1/2}$, is a cut-off function adapted to a square of side-length $2^k\delta^{1/2}$, and so that the family of sets $\{\operatorname{supp}(\widetilde{\widetilde{Q}}{}^\tau_{k,j})\}$ has a finite overlapping independently of δ.

Another application of Lemma 5.9 to the inequality above, together with the known estimates for the Kakeya maximal function, allows us to finish the proof. □

5.6 Restriction theorems

Statement in $\mathbb{R}^n$, $n > 1$, there are exponents $1 \leq p, q \leq \infty$ and a constant $C(p, q; n) < \infty$ so that for each f in the Schwartz class, the following inequality holds:

$$\Big(\int_{S^{n-1}} |\widehat{f}(\xi)|^q \, d\sigma(\xi)\Big)^{1/q} \leq C(p, q; n)\Big(\int_{\mathbb{R}^n} |f(x)|^p \, dx\Big)^{1/p}.$$

A trivial case is $p = 1$, $q = \infty$, but there are others much more interesting.

Once we agree in the possibility of such statement, it is not difficult to find out the values of p and q which may be allowed.

Let φ be a standard C^∞-bump function such that $\varphi(x) = 1$ in the unit ball $\{|x| \leq 1\}$ and $\varphi(x) = 0$ outside $\{|x| \leq 2\}$.

With $\delta > 0$, let us consider

$$\varphi_\delta(x) = \varphi\Big(\frac{x_1 - 1}{\delta}, \frac{x_2}{\delta^{1/2}}, \dots, \frac{x_n}{\delta^{1/2}}\Big).$$

Then

$$\widehat{\varphi}_\delta(\xi) = e^{-2\pi \xi_1} \, \delta^{(n+1)/2} \, \widehat{\varphi}(\delta\xi_1, \delta^{1/2}\xi_2, \dots, \delta^{1/2}\xi_n).$$

Let $\psi_\delta = \widehat{\varphi}_\delta$. We have

$$\Big(\int_{S^{n-1}} |\widehat{\psi}_\delta(\xi)|^q \, d\sigma(\xi)\Big)^{1/q} \simeq (\delta^{(n-1)/2})^{1/q}$$

$$\Big(\int_{\mathbb{R}^n} |\psi_\delta(x)|^p \, dx\Big)^{1/p} \simeq \delta^{\frac{n+1}{2}[1-1/p]},$$

implying ($\delta \to 0$) that $1/q \geq \frac{n+1}{n-1}[1 - 1/p]$.

Examining the adjoint of the restriction operator, we see that restriction implies that given $g \in L^{q'}(S^{n-1})$, we must have

$$\Big(\int_{\mathbb{R}^n} |\widehat{g \, d\sigma}(x)|^{p'} \, dx\Big)^{1/p'} \lesssim \Big(\int_{S^{n-1}} |g(\xi)|^{q'} \, d\sigma(\xi)\Big)^{1/q'},$$

where $1/p + 1/p'1 = 1$ and $1/q + 1/q' = 1$.

Therefore choosing $g \equiv 1$ and taking into account the formula

$$\widehat{d\sigma}(x) = c_n \frac{J_{(n-2)/2}(2\pi|x|)}{|x|^{(n-2)/2}},$$

we get the relation $p'(n-1)/2 > n$, i.e., $1 \leq p < 2n/(n+1)$.

The restriction conjecture claims that the statement above is true under the conditions

$$1 \le p < \frac{2n}{n+1} \qquad \text{and} \qquad \frac{1}{q} \ge \frac{n+1}{n-1}\left[1 - \frac{1}{p}\right].$$

This is a remarkable fact for several reasons. First, the Fourier transform is defined almost everywhere, and here is considered its restriction to a set of measure zero. Second, it is a linear object (the Fourier transform) and S^{n-1} is highly non-linear. In fact, this is crucial because restriction to hyperplanes would implies restrictions of one-dimensional Fourier transforms to points, and that amounts to defining $\widehat{f}(0) = \int f$ for non-integrable f. The rôle of curvature is here essential, and it makes sense to extends these results to other surfaces so long as they have non-zero curvatures.

The particular case $p = 2$, $q = 2(n+1)/(n+3)$ is well known, and its proof was obtained by P. Tomas and E. Stein using complex interpolation methods. The general case in dimension $n = 2$ is due to C. Fefferman and E. Stein, but for $n \ge 3$ is *terra incognita* and, so far, there are only partial results. In the following, we will present a "suprematistic" proof of the 2-dimensional restriction theorem, see [108].

Theorem 5.19. *If $f \in L^p(\mathbb{R}^2)$, $1 \le p < 4/3$, then $\widehat{f}$ exists almost everywhere on $|\xi| = 1$, and we have*

$$\left(\int_{|\xi|=1} |\widehat{f}(\xi)|^q \, d\sigma(\xi)\right)^{1/q} \le A_{p,q}\, \|f\|_p,$$

where $q = p'/3 = p/(3(p-1))$.

Proof. It is enough to show that

$$\lim_{\delta\to 0}\left(\delta^{-1}\int_{1-\delta/2\le|\xi|\le 1+\delta/2} |\widehat{f}(\xi)|^q \, d\xi\right)^{1/q} \le A_{p,q}\, \|f\|_p$$

under the same conditions as those of the theorem.

By duality, it is enough to prove the following estimate:

$$\delta^{-1}\Big\|\sum_j a_j\, \widehat{\varphi}_j\Big\|_{p'} \le C\Big(\sum_j |a_j|^{q'}\, \delta^{1/2}\Big)^{1/q'},$$

where p' and q' are conjugate exponents to p and q, respectively, the a_j's are complex numbers, and the φ_j's are characteristic functions of the "blocks" ($\delta^{1/2}$ in the tangential direction, δ in the normal direction) of the canonical partition of the annulus $1 - \delta/2 \le |\xi| \le 1 + \delta/2$. Observe that $p' > 4$. Therefore,

$$\left(\int \Big|\sum_{j,k=1}^{\delta^{-1/2}} a_j\, a_k\, \widehat{\varphi}_j\, \widehat{\varphi}_k\Big|^{p'/2}\right)^{2/p'\cdot 1/2} \le \left(\int \Big|\sum_{j,k=1}^{\delta^{-1/2}} a_j\, a_k\, \varphi_j * \varphi_k\Big|^{r}\right)^{1/r\cdot 1/2},$$

where $1/r + 2/p = 1$.

Since the supports of $\{\varphi_j * \varphi_k\}$ are "almost disjoint", and $\mu\{\operatorname{supp}(\varphi_j * \varphi_k)\} \le \delta^{3/2}\,(|j-k|+1)$, it follows that

$$\|\varphi_j * \varphi_k\|_\infty \le \frac{\delta^{3/2}}{|j-k|+1}.$$

Hence we have

$$\Big\|\sum_j a_j\,\widehat{\varphi}_j\Big\|_{p'} \le C\,\delta^{\frac{3}{4}(1+1/r)}\Big(\sum_{j,k=1}^{\delta^{-1/2}} \frac{|a_j|^r\,|a_k|^r}{(1+|j-k|)^{r-1}}\Big)^{1/2\cdot 1/r}$$

A fractional integration argument yields

$$\begin{aligned}\Big\|\sum_j a_j\,\widehat{\varphi}_j\Big\|_{p'} &\le C\,\delta^{\frac{3}{4}(1+1/r)}\Big[\Big(\sum_j |a_j|^{2r/(3-r)}\Big)^{3-r}\Big]^{1/(2r)}\\ &\le C\,\delta\Big(\sum_j |a_j|^{q'}\,\delta^{1/2}\Big)^{1/q'}.\end{aligned}$$

□

Observe that the only property used of the unit circle is its positive curvature. Therefore, we have in fact, proved a more general theorem. The "suprematistic" proof given above has been extended to the case of a cone Γ in $\mathbb{R}^3$ with measure $\mathrm{d}\sigma(\xi) = r^{-1}d\mu(\xi)$, where $d\mu$ is the Lebesgue measure on Γ and r is distance to the cone vertex, see [97].

Theorem 5.20. *We have the a priori inequality*

$$\Big(\int_\Gamma |\widehat{f}(\xi)|^q\,\mathrm{d}\sigma(\xi)\Big)^{1/q} \le C_{p,q}\,\|f\|_p \tag{5.4}$$

for every $f \in \mathcal{S}(\mathbb{R}^3)$ and $1 \le p < 4/3$, $1/q = 3(1-1/p)$.

Proof. Due to the fact that the Fourier transform is well behaved with respect to translations and rotations, we may assume, without loss of generality, that the cone has its vertex at the origin and that it is generated by the half-line $z = x$, $z > 0$, around the z-axis. We then have the cartesian equation

$$z^2 = x^2 + y^2, \quad z > 0,$$

and we can parameterize it by

$$z = r, \quad x = r\cos\theta, \quad y = r\sin\theta,$$

and with this parametrization $\mathrm{d}\sigma$ is $\mathrm{d}\sigma = \mathrm{d}r\,\mathrm{d}\theta$.

It is sufficient to consider the first quadrant $0 \le \theta \le \pi/2$, since the symmetry of the cone will allow us to treat the other cases equally, and the result will only be affected by a different constant. We divide the cone into dyadic pieces Γ_n, $n \in \mathbb{Z}$, where Γ_n is the part of the cone comprised between the planes of heights 2^n and 2^{n+1}.

Given that $\delta > 0$, we expand Γ_0 in the radial direction by a thickness δr, in such a way that we obtain a "fat" cone Γ_0^*, and at a height z with $1 \le z \le 2$, we obtain the ring $0 \le \theta \le \pi/2$, $z(1 - \delta/2) \le r \le z(1 + \delta/2)$.

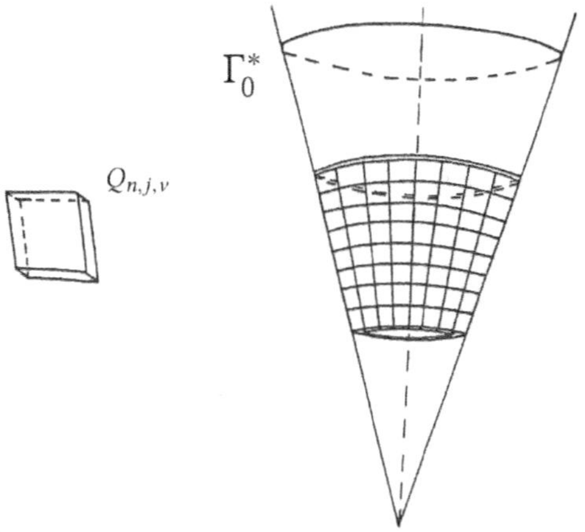

Fig. 5.4: The tiles $Q_{n,j,\nu}$.

We now divide Γ_0^* into $\delta^{-1/2} \times \delta^{-1/2}$ tiles $Q_{0,j,\nu}$, where $Q_{0,j,\nu}$ is the tile (or plate) of dimensions $\delta \times \delta^{1/2} \times \delta^{1/2}$ formed by the points of Γ_0^* at a height $1 + j\delta^{1/2} \le z \le 1 + (j+1)\delta^{1/2}$ and with an angular direction $\delta^{1/2}\nu \le \theta \le \delta^{1/2}(\nu + 1)$.

Once we obtain the decomposition of Γ_0^*, we obtain Γ_n^* and, by homogeneity, a similar partition of Γ_n^* into tiles. In this way, Γ_n^* will be divided into $\delta^{-1/2} \times \delta^{-1/2}$ pieces $Q_{n,j,\nu}$, where $Q_{n,j,\nu}$ is the tile of dimensions $2^n\delta \times 2^n\delta^{1/2} \times 2^n\delta^{1/2}$ at a height $2^n + 2^n j\delta^{1/2} \le z \le 2^n + 2^n(j+1)\delta^{1/2}$ and an angular direction $\delta^{1/2}\nu \le \theta \le \delta^{1/2}(\nu+1)$. For purposes of calculations, we may assume that $Q_{n,j,\nu}$ are rectangular tiles adapted to the surface. See Figure 5.4.

We set

$$\Gamma^* = \bigcup_n \Gamma_n^*,$$

and we call $\varphi_{n,j,\nu}$ the characteristic function of $Q_{n,j,\nu}$.

To prove inequality (5.20), in desingularising the measure, it is sufficient to consider

$$\lim_{\delta \to 0} \left(\frac{1}{\delta} \int_{\Gamma^*} |\widehat{f}(\xi)|^q \frac{d\xi}{|\xi|^2} \right)^{1/q} \le C \, \|f\|_p,$$

and to prove this inequality by duality, it is enough to take functions

$$g(\xi) = \sum_{n,j,\nu} a_{n,j,\nu} \, \varphi_{n,j,\nu}(\xi)$$

that are constant in each tile, so that for each sufficiently small $\delta > 0$,

$$\Big\| \sum_{n,j,\nu} \frac{a_{n,j,\nu}}{2^{2n/q}} \widehat{\varphi}_{n,j,\nu} \Big\|_{p'} \le C\, \delta^{1/q} \Big\| \sum_{n,j,\nu} a_{n,j,\nu}\, \varphi_{n,j,\nu} \Big\|_{q'}, \tag{5.5}$$

where p' and q' are the dual exponents for p and q, respectively.

Because only the limiting case $\delta \to 0$ is of interest, we consider any fixed portion $2^{-M} \le z \le 2^{M}$ of the cone, and we shall prove (5.5) for M fixed with bounds independent of M once δ is sufficiently small.

We take then $a_{n,j,\nu} = 0$ for $|n| > M$. We consider also δ such that

$$|n - m| \le |\log_2 \delta|/2,$$

because this geometrically means, if for example $m \le n$, that the diameter of Γ_m^* is smaller than each tile in Γ_n^*. This is necessary for the computations that follow.

Since $p' > 4$, we can apply the Hausdorff–Young theorem to the left-hand side of (5.5), with $1/s + 2/p' = 1$; then

$$\begin{aligned}
&\Big(\int \Big| \sum_{n,j,\nu} \frac{a_{n,j,\nu}}{2^{2n/q}} \widehat{\varphi}_{n,j,\nu}(\xi) \Big|^{p'} \mathrm{d}\xi \Big)^{1/p'} \qquad (5.6)\\
&= \Big(\int \Big| \sum_{n,m} \sum_{j,k} \sum_{\nu,\mu} \frac{a_{n,j,\nu}\, a_{m,k,\mu}}{2^{2n/q}\, 2^{2m/q}} \widehat{\varphi}_{n,j,\nu}(\xi) \cdot \widehat{\varphi}_{m,k,\mu}(\xi) \Big|^{p'/2} \mathrm{d}\xi \Big)^{2/p' \cdot 1/2}\\
&\le \Big(\int \Big| \sum_{n,m} \sum_{j,k} \sum_{\nu,\mu} \frac{a_{n,j,\nu}\, a_{m,k,\mu}}{2^{2n/q}\, 2^{2m/q}} \varphi_{n,j,\nu} * \varphi_{m,k,\mu}(x) \Big|^{s} \mathrm{d}x \Big)^{1/2s}\\
&\le C \Big(\int \Big| \sum_{m \le n} \sum_{j,k} \sum_{\nu,\mu} \frac{a_{n,j,\nu}\, a_{m,k,\mu}}{2^{2n/q}\, 2^{2m/q}} \varphi_{n,j,\nu} * \varphi_{m,k,\mu}(x) \Big|^{s} \mathrm{d}x \Big)^{1/2s}\\
&\le C \Big[\sum_{r = n-m \ge 0} \Big(\int \Big| \sum_{n} \sum_{j,k} \sum_{\nu,\mu} \frac{a_{n,j,\nu}\, a_{m,k,\mu}}{2^{2n/q}\, 2^{2m/q}} \varphi_{n,j,\nu} * \varphi_{m,k,\mu}(x) \Big|^{s} \mathrm{d}x \Big)^{1/s} \Big]^{1/2}\\
&= \Big[\sum_{r \ge 0} A_r \Big]^{1/2},
\end{aligned}$$

where, for fixed $r \ge 0$, A_r is the above expression with parentheses in which, once n is taken, m is $m = n - r$.

Note also that $r \le |\log_2 \delta|/2$ because we are in the fixed portion $2^{-M} \le z \le 2^{M}$ of the cone.

The proof follows by carefully analyzing the different terms inside the integrand. To do this, we need to estimate the size of the supports and how they overlap.

The role of the z-coordinate appears in the vertical overlapping, while the horizontal one is used to study the convolution of a measure $f(\theta)\, \mathrm{d}\theta$ on circles of distinct radii living in $\mathbb{R}^2$.

To estimate A_r, as we shall see in the calculations below, if m is close to n, the overlapping is small and the size is large. But if m moves away from n, the overlapping increases and the size decreases. It is marvelous to observe that one exactly compensates the other.

To measure the vertical overlapping, if $m \leq n$, we consider in Γ_m^* the $2^m/(\delta^{1/2}2^n)$ rings $\Gamma_{m,l}^*$, where $\Gamma_{m,l}^*$ is the ring of height $\delta^{1/2}2^n$ obtained by the union $\bigcup_{k,\mu} Q_{k,m,\mu}$, with μ varying over all angular directions and k such that

$$\delta^{1/2}\,2^n\,l \leq \delta^{1/2}\,2^m\,k \leq \delta^{1/2}\,2^n(l+1).$$

Hence

$$\left[\Gamma_{m,l}^* + \mathrm{supp}(\varphi_{n,j,\nu})\right] \cap \left[\Gamma_{m,l'} + \mathrm{supp}(\varphi_{n,j',\nu'})\right] \neq \emptyset$$

for $l+j = l'+j'$, while this intersection has a finite overlapping independent of δ if $l+j \neq l'+j'$.

Thereby, fixing n, m, a point in $\Gamma_{ml}^* + \mathrm{supp}(\varphi_{n,j,\nu})$ belongs at most to $2^m/(\delta^{1/2}2^n)$ different sums of this kind.

Since

$$\mathrm{supp}\left(\varphi_{n,j,\nu} * \sum_\mu \varphi_{m,k,\mu}\right) \cap \mathrm{supp}\left(\varphi_{n,j,\nu} * \sum_\mu \varphi_{m,k',\mu}\right) \neq \emptyset$$

for all k, k' with $\mathrm{supp}(\sum_\mu \varphi_{m,k,\mu}) \subset \Gamma_{m,l}^*$ and $\mathrm{supp}(\sum_\mu \varphi_{m,k',\mu}) \subset \Gamma_{m,l}^*$, and there are $2^n/2^m$ of these indices, it results that a point in $\mathrm{supp}(\varphi_{n,j,\nu} * \sum_\mu \varphi_{m,k,\mu})$ belongs at most to

$$\left(\frac{2^m}{\delta^{1/2}\,2^n}\right)\cdot\left(\frac{2^n}{2^m}\right) = \delta^{-1/2}$$

different supports of this class.

This provides us with the vertical overlapping. We therefore have

$$\begin{aligned}
\sum_r A_r &= \sum_{\substack{0\leq r\leq \frac{d}{2}|\log_2\delta|,\\ r=n-m}} \left\{\int \left|\sum_n \sum_{j,k} \sum_{\nu,\mu} \frac{a_{n,j,\nu}\,a_{m,k,\mu}}{2^{2n/q}\,2^{2m/q}}\,\varphi_{n,j,\nu} * \varphi_{m,k,\mu}(x)\right|^s \mathrm{d}x\right\}^{1/s} \\
\text{(5.7)}\qquad &\leq C \sum_{\substack{n-m\leq|\log_2\delta|\\ m\leq n}} \left\{\sum_{n\leq n}\sum_{j,k} \frac{\delta^{-(s-1)/2}}{2^{2ns/q}\,2^{2ms/q}} \left|\left(\sum_\nu a_{n,j,\nu}\,\varphi_{n,j,\nu}\right)\right.\right. \\
&\qquad \left.\left. * \left(\sum_\mu a_{m,k,\mu}\,\varphi_{m,k,\mu}(x)\right)\right|^s \mathrm{d}x\right\}^{1/s}.
\end{aligned}$$

Let us now go on to estimate the horizontal overlapping. We have to bear in mind that if two supports $Q_{n,j,\nu}$ and $Q_{m,k,\mu}$ have small angular separation $|\nu-\mu|$, then the size of the sum $Q_{n,j,\nu} + Q_{m,k,\mu}$ is going to be as large as $Q_{n,j,\nu}$, and there will be considerable overlapping with other supports. The contrary happens if there is a large angular separation.

Once we have fixed the heights j and k, in order to study the situation, it is sufficient to observe how it works in a section $S_{j,k}$ perpendicular to the cone axis at a height

$$2^n + 2^m + j\delta^{1/2}2^n + k\delta^{1/2}\,2^m,$$

because $\operatorname{supp}(\varphi_{n,j,\nu} * \sum_\mu \varphi_{m,k,\mu})$ is found at this height, and the final sum $Q_{n,j,\nu} + Q_{m,k,\mu}$ would be obtained by moving its section with $S_{j,k}$ parallel to the vertical direction of $Q_{n,j,\nu}$.

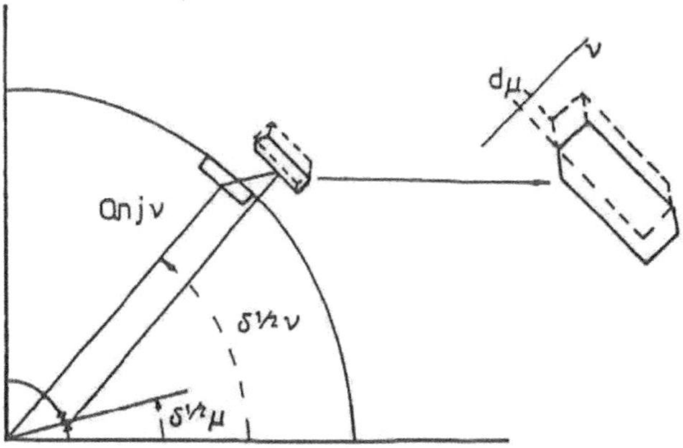

Fig. 5.5: Sumset of tiles.

In the $S_{j,k}$ plane, $\operatorname{supp}(\varphi_{n,j,\nu} * \varphi_{m,k,\mu})$ is centered at the point

$$2^n\,(1+\delta^{1/2}j)\,e^{\pi\nu i/2} + 2^m\,(1+\delta^{1/2}k)\,e^{\pi\mu i/2},$$

and is essentially like a block of length $2^n\delta^{1/2}$ and width

$$\delta^{1/2}\,2^n + \delta^{1/2}\,2^m\sin((|\nu-\mu|+1)\delta^{1/2}).$$

Also, $\operatorname{supp}(\varphi_{n,j,\nu} * \varphi_{m,k,\mu})$ is translated a distance

$$\delta^{1/2}\,2^m\sin(|\nu-\mu|\delta^{1/2}) \simeq \delta\,2^m\,|\nu-\mu|$$

from $\operatorname{supp}(\varphi_{n,j,\nu} * \varphi_{m,k,\mu+1})$, as is observed if it is projected in the ν direction.

To measure the number of sums $\operatorname{supp}(\varphi_{n,j,\nu} * \varphi_{m,k,\mu})$ that overlap, we will distinguish the following two cases:

Case (1). $|\nu-\mu|+1 \le 2^{n-m}$.

In this case, a point belongs to no more than N different supports, with N such that

$$\delta\,2^m\,|\nu-\mu| + \delta\,2^m\,(|\nu-\mu|-1) + \cdots + \delta\,2^m\,(|\nu-\mu|-N) \le \delta\,2^n,$$

that is, $N \le 2^{n-m}/(|\nu-\mu|+1)$.

Moreover,

$$|\operatorname{supp}(\varphi_{n,j,\nu} * \varphi_{m,k,\mu})| \simeq |\operatorname{supp}(\varphi_{n,j,\nu})| = \delta^2\, 2^{3n},$$

and

$$\|\varphi_{n,j,\nu} * \varphi_{m,k,\mu}\|_\infty \simeq |\operatorname{supp}(\varphi_{m,k,\mu})| = \delta^2\, 2^{3m}.$$

Hence

$$\int |\varphi_{n,j,\nu} * \varphi_{m,k,\mu}(x)|^s\, \mathrm{d}x \le C\delta^{2s+2}\, 2^{3ms}\, 2^{3n}.$$

Case (ii). $|\nu-\mu|+1 \ge 2^{n-m}$.
Then $\operatorname{supp}(\varphi_{n,j,\nu} * \varphi_{m,k,\mu})$ has a finite overlapping when μ varies, and

$$\begin{aligned}|\operatorname{supp}(\varphi_{n,j,\nu} * \varphi_{m,k,\mu})| &\le C\delta^2\, 2^{2n}\, 2^m (|\nu-\mu|+1),\\ \|\varphi_{n,j,\nu} * \varphi_{m,k,\mu}\|_\infty &\le C\delta^2\, 2^n\, 2^{2m}/(|\nu-\mu|+1).\end{aligned}$$

Therefore,

$$\int |\varphi_{n,j,\nu} * \varphi_{m,k,\mu}(x)|^s\, \mathrm{d}x \le C\,\frac{\delta^{2s+2}\, 2^{n(2+s)}\, 2^{m(2s+1)}}{(|\nu-\mu|+1)^{s-1}}.$$

Then with (i) and (ii) we can write that (5.7) is

$$\begin{aligned}
&\le C \sum_{\substack{n-m\le|\log_2\delta|\\ m\le n}} \Bigg\{ \sum_n \sum_{j,k} \frac{\delta^{-(s-1)/2}}{2^{2ns/q}\, 2^{2ms/q}}\\
&\quad\times \Bigg[\sum_{|\nu-\mu|+1\le 2^{n-m}} \left(\frac{2^{n-m}}{|\nu-\mu|+1}\right)^{s-1} \delta^{2s+2}\, 2^{3n}\, 2^{3ms}\, |a_{n,j,\nu}|^s\, |a_{m,k,\mu}|^s\\
&\qquad + \sum_{|\nu-\mu|+1\ge 2^{n-m}} \delta^{2s+2}\, 2^{n(2+s)}\, 2^{m(2s+1)}\, \frac{|a_{n,j,\nu}|\, s\, |a_{m,k,\mu}|^s}{(|\nu-\mu|+1)^{s-1}} \Bigg]\Bigg\}^{1/s}\\
&\le C \sum_{\substack{n\le m\le|\log_2\delta|\\ m\le n}} \Bigg\{ \sum_n \sum_{j,k} \frac{\delta^{-(s-1)/2}\, \delta^{2s+2}\, 2^{m(2s+1)}\, 2^{n(2+s)}}{2^{2ns/q}\, 2^{2ms/q}}\\
&\qquad \times \sum_{\nu,\mu} \frac{|a_{n,j,\nu}|^s\, |a_{m,k,\mu}|^s}{(|\nu-\mu|+1)^{s-1}} \Bigg\}^{1/s},
\end{aligned}$$

and by a fractional integration argument,

$$\begin{aligned}
&\sum_{\nu,\mu} \frac{|a_{n,j,\nu}|^s\, |a_{m,k,\mu}|^s}{(|\nu-\mu|+1)^{s-1}}\\
&\qquad \le C\Big(\sum_\nu |a_{n,j,\nu}|^{2s/(3-s)}\Big)^{(3-s)/2} \Big(\sum_\mu |a_{m,k,\mu}|^{2s/(3-s)}\Big)^{(3-s)/2}.
\end{aligned}$$

Hence, (5.7) is

$$
\le C \sum_{\substack{n-m\le|\log_2\delta| \\ m\le n}} \delta^{(3s-3)/s}\, 2^{-|n-m|(s-1)/s}
$$

$$
\cdot\Big\{\sum_n \Big(\sum_{j,\nu} |a_{n,j,\nu}|^{2s/(3-s)}\, \delta^2\, 2^{3n}\Big)^{(3-s)/2} \Big(\sum_{k\mu} |a_{m,k,\mu}|^{2s/(3-s)}\, \delta^2\, 2^{3m}\Big)^{(3-s)/2}\Big\}^{1/s}
$$

$$
\le C\delta^{(3s-3)/s} \Big(\sum_{\substack{n-m\le|\log_2\delta| \\ m\le n}} 2^{-|n-m|(s-1)/s}\Big) \Big(\sum_{n,j,\nu} |a_{n,j,\nu}|^{2s/(3-s)}\, \delta^2\, 2^{3n}\Big)^{(3-s)/s},
$$

but

$$
\sum_{\substack{n-m\le|\log_2\delta| \\ m\le n}} 2^{-|n-m|(s-1)/s} \le C,
$$

and then we have proved that

$$
\sum_{r\ge 0} A_r \le C\delta^{(3s-3)/s} \Big(\sum_{n,j,\nu} |a_{n,j,\nu}|^{2s/(3-s)}\, \delta^2\, 2^{3n}\Big)^{(3-s)/s}.
$$

Now, observing that $(3s-3)/2s = 1/q$ and $(3-s)/2s = 1/q'$, we can conclude finally

$$
\Big\| \sum_{n,j,\nu} \frac{a_{n,j,\nu}}{2^{2n/q}}\, \widehat{\varphi}_{n,j,\nu} \Big\|_{p'} \le C\Big(\sum_r A_\tau\Big)^{1/2} \le C\delta^{1/q} \Big\| \sum_{n,j,\nu} a_{n,j,\nu}\, \varphi_{n,j,\nu} \Big\|_{q'},
$$

which is inequality (5.5). □

The restriction theorem has been extended to the case of curves in $\mathbb{R}^n$ (see, for instance, [114, 125]), and there are also several partial results for hypersurfaces when $n \ge 3$. But their proofs, see [99, 127, 128], do not fit in the "suprematistic" point of view of this chapter.

5.7 Hilbert transform along plane curves

The real variable theory of singular integrals started by Calderón and Zygmund has been extended in several different ways.

One such extension is the singular integral operator

$$
Hf(x) = \text{p.v.} \int_{-\infty}^{\infty} f(x - \Gamma(t))\, \frac{\mathrm{d}t}{t},
$$

where $\Gamma\colon \mathbb{R} \to \mathbb{R}^n$, $n \ge 2$, is a curve passing through the origin: $\Gamma(0) = 0$. This object, called the Hilbert transform along the curve Γ, appears in the study of parabolic differential operators as well as in the harmonic analysis of homogeneous spaces.

These objects are also multipliers of the Fourier transform and, therefore, can be defined by the formula

$$Hf(\xi) = \widehat{f}(\xi)\, m(\xi),$$

where

$$m(\xi) = \text{p.v.} \int_{-\infty}^{\infty} \exp(i\xi \cdot \Gamma(t)) \, \frac{\mathrm{d}t}{t}.$$

Similarly to the case of the ordinary Hilbert transform, where we have the Hardy–Littlewood maximal function as the positive operator controlling its behaviour, one can define maximal operators associated to these singular integrals:

$$Mf(x) = \sup_{h>0} \frac{1}{2h} \int_{-h}^{h} |f(x - \Gamma(t))| \, \mathrm{d}t, \tag{5.8}$$

$$H^* f(x) = \sup_{\varepsilon>0} \left| \int_{|t|\geq\varepsilon} f(x - \Gamma(t)) \, \frac{\mathrm{d}t}{t} \right|. \tag{5.9}$$

The first natural question one may ask is which are the conditions to be imposed on the curve Γ so that H, H^* and M are bounded on some, or all, spaces $L^p(\mathbb{R}^n)$. In a series of papers, A. Nagel, E. Stein, S. Wainger and others [120, 121] have obtained results involving some kind of "positiveness" of the curvature of Γ, taking advantage of the decay of the multiplier $m(\xi)$ that it implies. The proofs are completed by an ingenious use of the method of complex interpolation, together with arguments of a Tauberian type involving certain square functions. In [121], it has been observed that, for convex curves $\Gamma(t) = (t, \gamma(t))$ in $\mathbb{R}^2$, with $\gamma(0) = \gamma'(0) = 0$, there exist necessary and sufficient conditions for H to be bounded in $L^2(\mathbb{R}^2)$: when γ is even (respectively, odd), the condition is that $\gamma'(t)$ (respectively, $h(t) = t\gamma'(t) - \gamma(t)$) must have the doubling time property, i.e., $\gamma'(Ct) \geq 2\gamma'(t)$ (respectively, $h(Ct) \geq 2h(t)$) for some fixed constant C and for every $t > 0$. The range of L^p-estimates has been extended from L^2 to L^p, $4/3 < p < 4$, in [111].

Let us consider a C^1 curve in $\mathbb{R}^2$, $\Gamma = \{(t, \gamma(t)) : -\infty < t < +\infty\}$, $\gamma(0) = \gamma'(0) = 0$, with the following properties:

(a) γ is *biconvex*, i.e., $|\gamma'(t)|$ is decreasing in $(-\infty, 0)$ and increasing in $(0, +\infty)$.
(b) γ has *doubling time*, i.e., there exists a constant $C > 1$ such that $|\gamma'(Ct)| \geq 2|\gamma'(t)|$.
(c) γ is *balanced*, by which we mean the following: there exists $k > 1$ such that $|\gamma(k^{-1}t)| \leq |\gamma(-t)| \leq |\gamma(kt)|$ for every $t > 0$.

Theorem 5.21. *Under the assumptions* (a), (b) *and* (c) *on the curve, the singular integral H and the maximal functions M and H^* are bounded operators in $L^p(\mathbb{R}^2)$ for every p, $1 < p < \infty$.*

Immediate consequences of the boundedness of M and H^* are:

Corollary 5.22. *For every $f \in L^p(\mathbb{R}^2)$, $1 < p < \infty$,*

$$\lim_{\varepsilon\to 0} \frac{1}{2\varepsilon} \int_{-\varepsilon}^{\varepsilon} f(x - \Gamma(t))\, \mathrm{d}t = f(x) \quad \text{a.e.},$$

$$\lim_{\varepsilon\to 0} \int_{|t|>\varepsilon} f(x - \Gamma(t))\, \frac{\mathrm{d}t}{t} = Hf(x) \quad \text{a.e.}$$

Theorem 5.21 gives a fairly complete analysis of the properties of Hilbert transforms along convex curves in the plane. For every fixed p, $1 < p < \infty$, the doubling time property is both necessary and sufficient for L^p-boundedness if the curve is even. Likewise, for a biconvex curve, the condition of being balanced is necessary for the L^2-boundedness of H, as it is proved in Section 5.7.5 below. The fact that for such curves one cannot obtain boundedness of the Hilbert transform in some $L^p(\mathbb{R}^2)$ without obtaining the full range $1 < p < \infty$ explains, in retrospect, the success of the theory, in contrast with the situation in the analysis of the spherical summation multipliers.

5.7.1 Decomposition of the operators and sketch of proof

All the L^p estimates will be proved for $f \in \mathcal{S}(\mathbb{R}^2)$, and then extended by continuity. Let us define the singular measures M_k and h_k $(k \in \mathbb{Z})$ such that

$$\widehat{M}_k(\xi,\eta) = \int_{1\le|t|\le 2} \exp\left\{i(2^k t\xi + \gamma(2^k t)\eta)\right\} \mathrm{d}t,$$

$$\widehat{h}_k(\xi,\eta) = \int_{1\le|t|\le 2} \exp\left\{i(2^k t\xi + \gamma(2^k t)\eta)\right\} \frac{\mathrm{d}t}{t},$$

and consider the following basic decompositions of our operators:

$$Mf(x) \simeq \sup_k |M_k * f(x)|, \tag{5.10}$$

$$Hf(x) = \sum_{k=-\infty}^{\infty} h_k * f(x), \tag{5.11}$$

$$H^* f(x) \simeq \sup_k \Big|\sum_{n>k} h_n * f(x)\Big|. \tag{5.12}$$

We shall also use the following square function:

$$Gf(x) = \Big(\sum_{k=-\infty}^{\infty} |\widetilde{M}_k f(x)|^2\Big)^{1/2}, \tag{5.13}$$

where

$$\widetilde{M}_k f(\xi,\eta) = \left[\widehat{M}_k(\xi,\eta) - \widehat{\Phi}(2^k\xi)\, \widehat{M}_k(0,\eta)\right] \widehat{f}(\xi,\eta)$$

for a suitable function Φ in the Schwartz class satisfying $\widehat{\Phi}(0) = 1$.

The proof of the boundedness of the maximal function Mf is based on the following strategy: we have the pointwise majorization $Mf(x) \le Gf(x) + f^*(x)$, where f^* denotes the strong maximal function of f. We start by proving the L^2-boundedness of Gf by means of Plancherel's theorem. This implies the same result for Mf which, in turn, implies the L^p estimate, $4/3 < p < 4$, for the square function Gf by a nowadays familiar argument based on a vector valued inequality and an ad-hoc Littlewood–Paley decomposition of frequency space. In general, an L^p estimate for M implies an L^q estimate for G, with $|1/q - 1/2| < 1/(2p)$, and therefore, a bootstrap argument will yield the whole range of p's, $1 < p < \infty$.

The doubling time condition of the curve will be used to define angular Littlewood–Paley decompositions. For the sake of simplicity, we shall assume that the doubling time constant is $C = 2$. Then, we consider the angular regions

$$\Delta_k^{\pm} = \left\{(\xi,\eta) : \frac{1}{2}|\gamma'(\pm 2^k)| < \frac{|\xi|}{|\eta|} < 2|\gamma'(\pm 2^{k+1})|\right\}, \qquad \text{for } k \in \mathbb{Z}.$$

It turns out that $\{\Delta_k^+\}_{-\infty}^{\infty}$ is a lacunary decomposition of $\mathbb{R}^2$ with finite overlapping: $\Delta_h^+ \cap \Delta_k^+ = \varnothing$ if $|k-h| \ge 3$, and the same is true for $\{\Delta_k^-\}$.

5.7.2 The maximal operator and the square function

Let $\Delta_k = \Delta_k^+ \cup \Delta_k^-$, and define the projection operators T_k by

$$\widehat{(T_k f)} = \widehat{f}\chi_{\Delta_k}, \qquad \text{for } k \in \mathbb{Z}.$$

Lemma 5.23. *For every p, $1 < p < \infty$, we have*

$$\left\|\Big(\sum_k |T_k f|^2\Big)^{1/2}\right\|_p \le c_p \|f\|_p, \tag{5.14}$$

$$\left\|\sum_k T_k f_k\right\|_p \le c_p \left\|\Big(\sum_k |f_k|^2\Big)^{1/2}\right\|_p, \tag{5.15}$$

$$\left\|\sup_k \Big|\sum_{n\ge k} T_n f\Big|\right\|_p \le c_p \|f\|_p. \tag{5.16}$$

Inequality (5.14) is nowadays well known (see [98] for references), and (5.15) is its dual form; (5.16) follows from (5.14) in the same way that one proves the lacunary convergence of Fourier series.

Lemma 5.24. *The operator*

$$\Gamma f(x) = \sup_{h>0} \frac{1}{2h}\int_{-h}^{h} |f(x-\gamma(t))|\,dt$$

is a bounded operator in $L^p(\mathbb{R})$, $1 < p < \infty$.

The proof is obvious due to the monotonicity of γ':

$$\frac{1}{h}\int_0^h f(x-\gamma(t))\,\mathrm{d}t = \frac{1}{h}\int_0^{\gamma(h)} f(x-u)\,\frac{du}{\gamma'(\gamma^{-1}(u))}$$
$$= \int_0^c f(x-u)\,\varphi(u)\,\mathrm{d}u \le f^*(x),$$

(Hardy–Littlewood maximal function), because $\varphi(u)$ is non-negative, decreasing and satisfies $\int_0^c \varphi(u)\mathrm{d}u = 1$. The same argument holds for the average over the interval $[-h,0]$.

Lemma 5.25. *Let $\Phi \in \mathcal{S}(\mathbb{R})$ be such that $\chi_{[-1,1]} \le \widehat{\Phi} \le \chi_{[-2,2]}$. Then*

$$\left|\widehat{M}_k(\xi,\eta) - \widehat{\Phi}(2^k\xi)\,\widehat{M}_k(0,\eta)\right| \le C\min\left(2^k|\xi|, (2^k|\xi|)^{-1}\right),$$

for all $(\xi,\eta) \notin \Delta_k$. The same estimates hold for $\widehat{h}_k(\xi,\eta)$ in place of $\widehat{M}_k(\xi,\eta)$.

Proof. The estimate for the case $2^k|\xi| \le 1$ is trivial, because M_k is a measure supported in $\{(x_1,x_2) : |x_1| \le 2^{k+1}\}$. For $2^k|\xi| > 1$, we use the Van der Corput lemma:

$$\widehat{M}_k(\xi,\eta) = \int_1^2 e^{ig_k(t)}\,\mathrm{d}t + \int_1^2 e^{il_k(t)}\,\mathrm{d}t,$$

with

$$|g_k'(t)| = |2^k(\xi + \gamma'(2^k t)\eta| \quad \text{and} \quad |l_k'(t)| = |2^k(\xi + \gamma'(-2^k t)\eta)|.$$

If $|\xi| \ge 2|\eta|\cdot|\gamma'(2^{k+1})|$, then we have $|g_k'(t)| \ge 2^{k-1}|\xi|$. If $|\xi| \le 1/2|\eta|\cdot|\gamma'(2^k)|$, then also $|g_k'(t)| \ge 2^{k-1}|\gamma'(2^k t)|\cdot|\eta| \ge 2^k|\xi|$, and the situation for l_k' is similar. Finally, the complement of these four cases corresponds to points $(\xi,\eta) \in \Delta_k$. The proof for $\widehat{h}_k(\xi,\eta)$ is essentially the same. □

Suppose that $U_k f = U_k * f$ is a sequence of positive operators uniformly bounded in $L^\infty(\mathbb{R}^n)$, and consider $U^* f(x) = \sup_k |U_k * f(x)|$.

Lemma 5.26. *If $U^* f(x) = \sup_k |U_k * f(x)|$ is bounded in $L^q(\mathbb{R}^n)$, then*

$$\left\|\Big(\sum_k |U_k f_k|^2\Big)^{1/2}\right\|_p \le c_p \left\|\Big(\sum_k |f_k|^2\Big)^{1/2}\right\|_p$$

for $|1/p - 1/2| = 1/(2q)$.

This is a well-known result, and it is stated here only to clarify the presentation of the proof of Theorem 5.21. The proof of the estimates for the maximal function depends on the following proposition, of a conditional nature.

Lemma 5.27. *Suppose that* $\|Mf\|_q \leq C\|f\|_q$ *for some* q. *Then, for all* p *such that* $|1/p - 1/2| < 1/(2q)$, *we have the inequality*

$$\|Gf\|_p \leq C_p \, \|f\|_p,$$

where G *is defined in* (5.13).

Proof. Recall that $T_k f = \chi_{\Delta_k} \cdot \widehat{f}$, and let us write

$$f^k = f - T_k f$$

and

$$S_k f = \widehat{f} \cdot \chi_{\{(\xi,\eta):2-k \leq |\xi| \leq 2-k+1\}}.$$

Then

$$Gf \leq \Big(\sum_k |\widetilde{M}_k * T_k f|^2\Big)^{1/2} + \Big(\sum_k |\widetilde{M}_k * f^k|^2\Big)^{1/2} = \mathrm{I} + \mathrm{II}.$$

Now, observe that $\sup_k (|\widetilde{M}_k| * f)$ is a bounded operator in L^q due to the hypothesis and Lemma 5.24. Therefore

$$\|\mathrm{I}\|_p \leq C_p \Big\| \Big(\sum_k |T_k f|^2\Big)^{1/2} \Big\|_p \leq C_p \, \|f\|_p$$

by Lemmas 5.26 and 5.23.

We have

$$\mathrm{II} = \Big(\sum_k \Big|\widetilde{M}_k * \Big(\sum_{j=-\infty}^{\infty} S_{j+k} f^k\Big)\Big|^2\Big)^{1/2} \leq \sum_j \Big(\sum_k |\widetilde{M}_k * S_{j+k} f^k|^2\Big)^{1/2} = \sum_j \mathrm{II}_j.$$

We shall prove:

$$\|\mathrm{II}_j\|_2 \leq C\, 2^{-|j|} \, \|f\|_2 \quad \text{and} \quad \|\mathrm{II}_j\|_{p_0} \leq C \, \|f\|_{p_0} \quad \text{for} \quad \Big|\frac{1}{2} - \frac{1}{p_0}\Big| = \frac{1}{2q}.$$

Interpolating, we shall obtain

$$\|\mathrm{II}\|_p \leq \sum_j \|\mathrm{II}_j\|_p \leq \sum_j C\, 2^{-\theta|j|} \, \|f\|_p \leq C_p \, \|f\|_p \quad \text{with} \quad \frac{1}{p} = \frac{1-\theta}{p_0} + \frac{\theta}{2}.$$

We use Plancherel's theorem for the L^2-inequality:

$$\|\mathrm{II}_j\|_2^2 = \sum_k \int_{\substack{|\xi| \sim 2^{-j-k} \\ (\xi,\eta) \notin \Delta_k}} |\widehat{M}_k(\xi,\eta) - \widehat{\Phi}(2^k \xi)\, \widehat{M}_k(0,\eta)|^2 \, |\widehat{f}(\xi,\eta)|^2 \, d\xi \, d\eta$$

$$\overset{\text{(by Lemma 5.25)}}{\leq} C \int |\widehat{f}(\xi,\eta)|^2 \, (2^{-|j|})^2 \, d\xi \, d\eta = C\, 2^{-2|j|} \, \|f\|_2^2.$$

Finally,

$$\begin{aligned}
\|\mathrm{II}_j\|_{p_0} &\le C_{p_0} \Big\| \Big(\sum_k |S_{j+k} f^k|^2 \Big)^{1/2} \Big\|_{p_0} \\
&\le C_{p_0} \Big\{ \Big\| \Big(\sum_k |S_{j+k} f|^2 \Big)^{1/2} \Big\|_{p_0} + \Big\| \Big(\sum_k |S_{j+k} (T_k f)|^2 \Big)^{1/2} \Big\|_{p_0} \Big\} \\
&\le C_{p_0} \Big\{ \|f\|_{p_0} + \Big\| \Big(\sum_k |T_k f|^2 \Big)^{1/2} \Big\|_{p_0} \Big\} \le C_{p_0} \|f\|_{p_0}.
\end{aligned}$$

To finish the proof of the theorem for the maximal operator, we use the majorization

$$Mf(x) \le 2 \sup_k |M_k * f(x)| \le 2 \sup_k |\widetilde{M}_k * f(x)| + 2Nf(x) \le 2Gf(x) + 2Nf(x),$$

where N is the Hardy–Littlewood operator in the x_1-variable and the operator Γ of Lemma 5.24 in the x_2-variable. Now, the bootstrap argument sketched in Section 5.7.1 can be applied and yields $\|Mf\|_p \le C_p \|f\|_p$ for $1 < p \le \infty$. □

5.7.3 The singular integral

The estimate for Hf will be based on an argument similar to that of Lemma 5.27, together with the fact that the operators $h_k * f$ are controlled by the maximal operator Mf.

We first consider the case of even $\gamma(t)$, so that $\widehat{h}_k(0, \eta) = 0$. The notation used is as in Lemma 5.27:

$$Hf = \sum_k h_k * T_k f + \sum_k h_k * f^k = H_{\mathrm{I}} f + H_{\mathrm{II}} f.$$

Using Lemmas 5.23 and 5.26, and the fact that $\sup_k |h_k| * f \le Mf$, we obtain

$$\begin{aligned}
\|H_{\mathrm{I}} f\|_p &= \Big\| \sum_k T_k (h_k * T_k f) \Big\|_p \le C_p \Big\| \Big(\sum_k |h_k * T_k f|^2 \Big)^{1/2} \Big\|_p \\
&\le C_p \Big\| \Big(\sum_k |T_k f|^2 \Big)^{1/2} \Big\|_p \le C_p \|f\|_p.
\end{aligned}$$

On the other hand,

$$|H_{\mathrm{II}} f| = \Big| \sum_k h_k * \sum_{j=-\infty}^{\infty} S_{j+k} f^k \Big| \le \sum_j \Big| \sum_k h_k * S_{j+k} f^k \Big| = \sum_j |\mathrm{II}_j|.$$

Since $h_k(0,\eta) = 0$, Lemma 5.25 and Plancherel's theorem yield

$$\begin{aligned}\|\mathrm{II}_j\|_2^2 &= \sum_k \int_{\substack{|\xi|\sim 2^{-j-k} \\ (\xi,\eta)\notin\Delta_k}} |\widehat{f}(\xi,\eta)|^2\, |\widehat{h}_k(\xi,\eta)|^2 \,\mathrm{d}\xi\,\mathrm{d}\eta \le C \int |\widehat{f}(\xi,\eta)|^2\, 2^{-2|j|}\,\mathrm{d}\xi\,\mathrm{d}\eta \\ &\le C\, 2^{-2|j|}\, \|f\|_2^2.\end{aligned}$$

In L^p, $1 < p < \infty$, we have a uniform estimate

$$\begin{aligned}\|\mathrm{II}_j\|_p &= \Big\| \sum_k S_{j+k}(h_k * S_{j+k} f^k) \Big\|_p \le C_p \Big\| \Big(\sum_k |h_k * (S_{j+k} f^k)|^2 \Big)^{1/2} \Big\|_p \\ &\le C_p \Big\| \Big(\sum_k |S_{j+k} f^k|^2 \Big)^{1/2} \Big\|_p \le C_p \|f\|_p\end{aligned}$$

and the proof ends by interpolation and summing a geometric series.

For the case of a general curve, we decompose

$$h_k = \widetilde{h}_k + m_k,$$

where, with Φ as above,

$$\widehat{m}_k(\xi,\eta) = \widehat{\Phi}(2^k\xi)\cdot \widehat{h}_k(0,\eta).$$

Then, the Fourier transform of $\widetilde{h}_k$ satisfies the same estimates as in the case on an even curve, namely

$$|\widehat{(\widetilde{h}_k)}(\xi,\eta)| \le C \min(2^k|\xi|, (2^k|\xi|)^{-1}) \quad \text{if } (\xi,\eta)\notin\Delta_k,$$

and $\sup |\widetilde{h}_k| * f \le Mf + Nf$ is bounded in $L^p(\mathbb{R}^2)$ for all $p > 1$. Therefore, the previous argument gives

$$\Big\| \sum_k \widetilde{h}_k * f \Big\|_p \le C_p\, \|f\|_p, \quad 1 < p < \infty.$$

Next, we need to estimate the decay of $h_k(0,\eta)$. It is precisely here where the hypothesis of having a balanced curve plays a role. Let us assume for simplicity that the balancing constant is $k \le 2$. Define $\lambda_k = |\gamma(2^k)|$, $k \in \mathbb{Z}$, so that $\{\lambda_k\}_{k\in\mathbb{Z}}$ is (by convexity) a lacunary sequence, and the balancing condition implies $\lambda_{k-1} \le |\gamma(-2^k)| \le \lambda_{k+1}$, $k \in \mathbb{Z}$. Now, for small η, we have

$$\begin{aligned}|\widehat{h}_k(0,\eta)| &= \Big| \int_{1\le|t|\le 2} [e^{i\eta\gamma(2^k t)} - 1]\, \frac{\mathrm{d}t}{t} \Big| \le (\log 2)\, |\eta|\, (|\gamma(2^{k+1})| + |\gamma(-2^{k+1})|) \\ &\le C\cdot \lambda_{k+2}\cdot |\eta|.\end{aligned}$$

For large η, we once again use Van der Corput's lemma. Since

$$\left|\frac{\mathrm{d}}{\mathrm{d}t}(\eta\gamma(2^k t))\right| = |2^k\eta\gamma'(2^k t)| \geq \lambda_{k-1}|\eta|, \quad 1 \leq |t| \leq 2,$$

we obtain

$$|\widehat{h}_k(0,\eta)| \leq C\,(\lambda_{k-1}|\eta|)^{-1}.$$

We now apply the same argument of the even case replacing the operators S_j by $\tilde{S}_j$ defined by

$$\widehat{(\tilde{S}_j f)}(\xi,\eta) = \widehat{f}(\xi,\eta)\cdot\chi_{I_j}(\eta),$$

where $I_j = [\lambda_{j+1}^{-1}, \lambda_j^{-1}] \cup [-\lambda_j^{-1}, -\lambda_{j+1}^{-1}]$. Then

$$\sum_k m_k * f = \sum_{j,k} m_k * \tilde{S}_{j+k} f = \sum_{j=-\infty}^{\infty} A_j f.$$

The uniform estimate $\|A_j f\|_p \leq C_p\|f\|_p$, $1 < p < \infty$ follows as in the even case.

In L^2, we have

$$\|A_j f\|_2^2 = \sum_k \int_{\mathbb{R}}\int_{I_{j+k}} |\widehat{f}(\xi,\eta)|^2\,|m_k(\xi,\eta)|^2\,\mathrm{d}\xi\,\mathrm{d}\eta.$$

If $j > 1$ and $\eta \in I_{j+k}$,

$$|m_k(\xi,\eta)| \leq \frac{\lambda_{k+2}}{\lambda_{k+j}} \leq \rho^{2-j},$$

where $\rho > 1$ is the lacunarity constant of the sequence $\{\lambda_k\}_{k\in\mathbb{Z}}$. When $j < -1$ and $\eta \in I_{j+k}$,

$$|m_k(\xi,\eta)| \leq \frac{\lambda_{j+k+1}}{\lambda_{k-1}} \leq \rho^{2+j}.$$

Therefore,

$$\|A_j f\|_2 \leq C\rho^{-|j|}\,\|f\|_2 \quad \text{for } -\infty < j < \infty,$$

and this ends the proof.

5.7.4 The maximal Hilbert transform

We shall prove here the L^p-boundedness of the operator $H^* f(x)$ defined by (5.12) for all $1 < p < \infty$. Since

$$H^* f(x) \leq Mf(x) + \sup_k \left|\sum_{n\geq k} h_n * f(x)\right|,$$

it suffices to consider the maximal operator:

$$\sup_k \Big| \sum_{n \geq k} h_n * f \Big|.$$

For simplicity, we shall only do this in the case of an even curve; the necessary modifications for dealing with general balanced curves are essentially contained in the last computations of Section 5.7.3. Define $\Phi \in \mathcal{S}(\mathbb{R})$ as in Lemma 5.25, write $\widehat{\Phi}_k(\xi) = \widehat{\Phi}(2^k \xi)$, and denote by $\overset{(1)}{*}$ convolution in the first variable. Then, we decompose the truncated Hilbert transform as follows:

$$\begin{aligned}
&\sum_{n \geq k} h_n * f \\
&= \Phi_k \overset{(1)}{*} \Big\{ Hf - \sum_{n<k} h_n * f \Big\} + (\delta - \Phi_k) \overset{(1)}{*} \sum_{n \geq k} h_n * T_n f + (\delta - \Phi_k) \overset{(1)}{*} \sum_{n \geq k} h_n * f^n \\
&= A_k f + B_k f + C_k f,
\end{aligned}$$

and we need to estimate $\sup_k |A_k f|$, $\sup_k |B_k f|$ and $\sup_k |C_k f|$. In doing so, we shall use the following maximal operators:

(a) $g_1^*(x)$, the Hardy–Littlewood maximal function of $g(x_1, x_2)$ in the x_1-direction.
(b) $Ng(x)$, the Hardy–Littlewood maximal operator acting on the x_1 variable, and the operator Γ of Lemma 5.24 acting on x_2 (this operator appears at the end of Section 5.7.2).
(c) $T^* g(x) = \sup_k |T_k g(x)|$.
(d) $S^* g(x) = \sup_k |\Sigma_{n \geq k} T_n g(x)|$ (see Lemma 5.23).
(e) $Mf(x)$, the maximal operator along the curve Γ.

All of these are bounded in $L^p(\mathbb{R}^2)$ for every $1 < p < \infty$. Now, to estimate $A_k f$, we observe first that the action of the kernel $\Phi_k \overset{(1)}{*} \sum_{n<k} h_n$ is pointwise dominated by the action of the maximal operator N. It suffices to prove this at the point $x = 0$ and for $k = 0$; given a test function g, we have

$$\begin{aligned}
\Big| \Big\langle \Phi \overset{(1)}{*} \sum_{n<0} h_n, g \Big\rangle \Big| &= \Big| \int_{-\infty}^{\infty} \Big\{ \text{p.v.} \int_{-1}^{1} g(s+t, \gamma(t)) \, \frac{\mathrm{d}t}{t} \Big\} \Phi(s) \, \mathrm{d}s \Big| \\
&= \Big| \int_{-\infty}^{\infty} \int_{-1}^{1} [\Phi(s-t) - \Phi(s)] \, g(s, \gamma(t)) \, \frac{\mathrm{d}t}{t} \, \mathrm{d}s \Big| \\
&\leq C \int_{-1}^{1} \int_{-\infty}^{\infty} \frac{1}{1+s^2} \, |g(s, \gamma(t))| \, \mathrm{d}s \, \mathrm{d}t \leq Ng(0)
\end{aligned}$$

(we have used the fact that γ is even in the second equality). Therefore,

$$\sup_k |A_k f(x)| \leq C \, [(Hf)_1^*(x) + Nf(x)].$$

For the second term, since $\{T_n\}_{n\in\mathbb{Z}}$ are orthogonal projections, we can write $h_n * T_n f = T_n(\sum_j h_j * T_j f)$. Remember from Section 5.7.4 that Hf was decomposed as $Hf = H_{\mathrm{I}} f + H_{\mathrm{II}} f$, and that both pieces were bounded in L^p for all $1 < p < \infty$. Then,

$$\sup_k |B_k f(x)| = \sup_k \left|(\delta - \Phi_k) \overset{(1)}{*} \sum_{n\geq k} T_n(H_{\mathrm{I}} f)(x)\right| \leq C\,(S^* H_{\mathrm{I}} f)_1^*(x).$$

Finally, the last term can be decomposed as follows:

$$\sup_k |C_k f| = \sup_k \left|\sum_{j=0}^{\infty} (\delta - \Phi_k) \overset{(1)}{*} h_{j+k} * f^{j+k}\right| \leq \sum_{j=0}^{\infty} P_j f,$$

where

$$P_j f(x) = \sup_k |(\delta - \Phi_k) \overset{(1)}{*} h_{j+k} * f^{j+k}(x)|.$$

The operators P_j are uniformly bounded in $L^p(\mathbb{R}^2)$, $1 < p < \infty$, because

$$P_j f(x) \leq C\,(Mf + MT^* f)_1^*(x).$$

On the other hand, a more favorable estimate holds for $p = 2$, since

$$\begin{aligned} \|P_j f\|_2^2 &\leq \sum_k \iint_{(\xi,\eta)\notin\Delta_{j+k}} |1 - \widehat{\Phi}(2^k\xi)|^2\, |h_{j+k}(\xi,\eta)|^2\, |\widehat{f}(\xi,\eta)|^2\, \mathrm{d}\xi\, \mathrm{d}\eta \\ &\leq C \iint_{k:2-k\leq|\xi|} |2^{j+k}\xi|^{-2}\, |\widehat{f}(\xi,\eta)|^2\, \mathrm{d}\xi\, \mathrm{d}\eta \leq C\,2^{-2j}\, \|f\|_2^2. \end{aligned}$$

Again, interpolating and summing a geometric series shows that

$$\left\|\sum_{j=0}^{\infty} P_j f\right\|_p \leq C_p\, \|f\|_p, \quad 1 < p < \infty,$$

finishing the proof.

5.7.5 Balance is necessary

Suppose that $\Gamma = \{(t, \gamma(t))\}_{t\in\mathbb{R}}$ is a biconvex curve, i.e., $\gamma(0) = \gamma'(0) = 0$ and $|\gamma'(t)|$ increases in $(0, +\infty)$ and decreases in $(-\infty, 0)$, and let H be the Hilbert transform along Γ. We wish to prove that, if H is bounded in any L^p space, then Γ must be a balanced curve. In fact, since H is a translation invariant operator corresponding to the multiplier

$$m(\xi, \eta) = \text{p.v.} \int_{-\infty}^{\infty} \exp(i\xi t + i\eta\gamma(t))\, \frac{\mathrm{d}t}{t},$$

we have that

$$H \text{ bounded in } L^p(\mathbb{R}^2) \implies H \text{ bounded in } L^2(\mathbb{R}^2) \iff m \in L^\infty(\mathbb{R}^2)$$
$$\implies \Gamma \text{ is balanced.}$$

Only the last implication must be proved, but this follows from:

Lemma 5.28. *If* Γ *is biconvex and* $|\gamma(-t_0)| = |\gamma(t_1)|$ *with* $t_0, t_1 > 0$*, then*

$$|\log t_1/t_0| \le 8 + |m(0,\eta)|, \quad \textit{where } \eta = 1/|\gamma(t_1)|.$$

Proof. Suppose that $t_0 < t_1$. Then

$$\begin{aligned} m(0,\eta) &= \int_{t_1}^{\infty} e^{i\gamma(t)\eta}\,\frac{\mathrm{d}t}{t} + \int_{-\infty}^{t_0} e^{i\gamma(t)\eta}\,\frac{\mathrm{d}t}{t} + \int_{-t_0}^{t_1} [e^{i\gamma(t)\eta} - 1]\,\frac{\mathrm{d}t}{t} + \int_{t_0}^{t_1} \frac{\mathrm{d}t}{t} \\ &= I_1 + I_2 + I_3 + \log\frac{t_1}{t_0}. \end{aligned}$$

Let

$$F(t) = \int_{t_1}^{t} e^{i\gamma(s)\eta}\,\mathrm{d}s, \quad \text{for } t_1 < t < \infty.$$

For $s > t_1$, we have

$$|\gamma'(s)|\eta \ge |\gamma'(t_1)|\eta \ge \frac{|\gamma(t_1|}{t_1}\eta = \frac{1}{t_1},$$

and γ' is monotone. Therefore, $F(t) \le 3/t_1$, and integrating by parts yields

$$|I_1| = \left|\int_{t_1}^{\infty} F(t)t^{-2}\,\mathrm{d}t\right| \le 3.$$

Similarly, $|I_2| \le 3$. On the other hand,

$$|I_3| \le \int_{-t_0}^{t_1} |\gamma(t)|\,\eta\,\frac{\mathrm{d}t}{|t|} \le \eta \int_{-t_0}^{t_1} |\gamma'(t)|\,\mathrm{d}t = \eta(|\gamma(t_1)| + |\gamma(-t_0)|) = 2.$$

Putting everything together, $|I_1 + I_2 + I_3| \le 8$. □

Chapter 6
The ball multiplier on mixed norm spaces

These are my spaces. If you don't like them... well, I have others.
(Rephrasing Groucho)

6.1 Introduction

As in the previous chapter, let $\widehat{Tf}(\xi) = \chi_B(\xi)\,\widehat{f}(\xi)$ be the Fourier multiplier operator given by the indicator function of the unit ball B in $\mathbb{R}^n$.

The main purpose of this chapter is to present a proof of the boundedness of T in the critical range $2n/(n+1) < p < 2n/(n-1)$ if we substitute $L^p(\mathbb{R}^n)$ by the mixed norm space $L^p_{\text{rad}}L^2_{\text{ang}}(\mathbb{R}^n)$, where the L^p-norm is taken in the radial variable, and the L^2-norm applies to the angular part; that is,

$$T \text{ is bounded on } L^p_{\text{rad}}L^2_{\text{ang}}(\mathbb{R}^n) \text{ if and only if } \frac{2n}{n+1} < p < \frac{2n}{n-1},$$

where the norm in $L^p_{\text{rad}}L^2_{\text{ang}}(\mathbb{R}^n)$ is given by the integral

$$\left(\int_0^\infty \left(\int_{S^{n-1}} |f(r\theta)|^2 \,\mathrm{d}\sigma(\theta)\right)^{p/2} r^{n-1}\,\mathrm{d}r\right)^{1/p},$$

with $\mathrm{d}\sigma$ the uniform measure in the unit sphere, and where $r > 0$ and $\theta \in S^{n-1}$ are the polar coordinates in $\mathbb{R}^n$. This will be the content of Theorem 6.2.

The $L^p_{\text{rad}}L^2_{\text{ang}}(\mathbb{R}^n)$ are Banach spaces whose norms are invariant under rotations, but not under translations. The Schwartz space $\mathcal{S}(\mathbb{R}^n)$ is dense with continuous inclusion on them, and we have the duality relation

$$(L^p_{\text{rad}}L^2_{\text{ang}}(\mathbb{R}^n))^* = L^q_{\text{rad}}L^2_{\text{ang}}(\mathbb{R}^n) \quad \text{for } 1/p + 1/q = 1,\ 1 < p, q < \infty.$$

and the inclusions

$$L^p_{\text{rad}}L^2_{\text{ang}}(\mathbb{R}^n) \subset L^p(\mathbb{R}^n) \quad \text{if } p \le 2, \quad \text{and} \quad L^p_{\text{rad}}L^2_{\text{ang}}(\mathbb{R}^n) \supset L^p(\mathbb{R}^n) \quad \text{if } p \ge 2.$$

A. Córdoba, *Suprematism in Harmonic Analysis*, Progress in Mathematics 356,
https://doi.org/10.1007/978-3-031-73737-4_6

6.2 Estimates of Bessel functions

The well-known decomposition of $L^2(\mathbb{R}^n)$ into subspaces invariant under the Fourier transform associated to spherical harmonics will be crucial and, in this chapter, Bessel functions are very relevant. Therefore, it will be convenient, for the sake of completeness, to describe their main properties in this section. Fortunately, there are excellent references for the task, namely [16] and [147].

Let

$$\widehat{F}(\xi) = \int F(x)\exp(-2\pi i x\cdot\xi)\,\mathrm{d}x$$

be the Fourier transform in the Euclidean space $\mathbb{R}^n$, which, in the case of a radial function $F(x) = f(|x|)$, reduces to the well-known formula

$$\widehat{F}(\xi) = 2\pi|\xi|^{-(n-2)/2}\int_0^\infty f(s)\,J_{(n-2)/2}(2\pi|\xi|s)\,s^{n/2}\,\mathrm{d}s. \tag{6.1}$$

Here J_ν denotes the Bessel function of the first kind and order ν, given by

$$\begin{aligned} J_\nu(x) &= \frac{1}{\pi}\int_0^\pi \cos(\nu t - x\sin t)\,\mathrm{d}t - \frac{\sin\pi\nu}{\pi}\int_0^\infty e^{-\nu t - x\sinh t}\,\mathrm{d}t \\ &= \frac{1}{\pi}\,\mathrm{Re}\int_0^\pi e^{i(\nu t - x\sin t)}\,\mathrm{d}t + O\Big(\frac{1}{x+\nu}\Big), \qquad x\in(0,\infty) \end{aligned} \tag{6.2}$$

(see [147], page 176 , formula 4). It satisfies the second order differential equation

$$x^2 J_\nu''(x) + xJ_\nu'(x) + (x^2-\nu^2)J_\nu(x) = 0 \tag{6.3}$$

and admits the expansions

$$J_\nu(x) = \sum_{m=0}^\infty \frac{(-1)^m (x/2)^{\nu+2m}}{\Gamma(\nu+m+1)} \sim \frac{(x/2)^\nu}{\Gamma(\nu+1)}(1+\Theta), \qquad \text{as } x\to 0,$$

where $|\Theta| \le e^{x^2/4-1}/(\nu+1)$, and

$$J_\nu(x) \sim \Big(\frac{2}{\pi x}\Big)^{1/2}\cos\Big(x - \frac{\nu\pi}{2} - \frac{\pi}{4}\Big), \qquad \text{as } x\to\infty.$$

We also have the two following important identities:

(a) Hankel inversion formula:

$$\int_0^\infty J_\nu(sr)\,J_\nu(tr)\,r\,\mathrm{d}r = \delta(s-t), \qquad \text{in the distributional sense.}$$

(b) Mellin transform:

$$MJ_\nu(s) = \int_0^\infty t^{-is} J_\nu(t)\,\frac{\mathrm{d}t}{t} = \frac{2^{-is}\Gamma(\nu-is)/2}{(\nu+is)\Gamma((\nu+is)/2)}.$$

References [16] and [147] contain proofs of all statements above, and different results where estimates uniform on the order of Bessel functions are relevant. In general, to find those estimates outside the "critical range" $\nu/2 \le x \le 3\nu/2$ is the easy part but inside that critical range one needs a rather careful analysis.

Without loss of generality, we may assume that $\nu > 0$. Observe that if $0 < x < \nu/2$ then the phase of the oscillatory integral (6.2) has derivative equal to $\nu - x\cos(t) \ge \nu/2$, implying the estimate $|J_\nu(x)| \lesssim 1/\nu$.

In the case $x > 3\nu/2$, the critical point of the phase is $\xi = \cos^{-1}(\nu/x)$, and the second derivative of the phase at that point is $x\sin(\xi) = x\sqrt{1-\nu^2/x^2}$, which is bigger than or equal to $\sqrt{5}x/3$.

Therefore, for $\delta \le 1$, the second derivative of the phase is bigger than $x/3$ at points inside the interval $(\xi - \delta, \xi + \delta)$. On the other hand, outside that interval the first derivative is bigger than $x\delta$. We are then in the conditions to apply Van der Corput's estimates, taking $\delta = 1/\sqrt{x}$, to obtain the inequality $|J_\nu(x)| \lesssim 1/\sqrt{x}$, when $x > 3\nu/2$.

Theorem 6.1. *The following estimates hold*:

$$\int_{\nu/2}^{3\nu/2} |J_\nu(x)|^p \, dx \lesssim \begin{cases} \nu^{(1-p)/3} & \text{if } p > 4, \\ \nu^{-1}\log\nu & \text{if } p = 4, \\ \nu^{1-p/2} & \text{if } 1 \le p < 4, \end{cases} \qquad \textit{and}$$

$$\sup_{\substack{x \ge 0 \\ \nu \text{ integer}}} \left| \int_0^x J_\nu(t)\, dt \right| \le C < \infty.$$

Proof. In the region $\nu + k\nu^{1/3} \le x \le \nu + 2k\nu^{1/3}$, with $k \le \frac{1}{4}\nu^{2/3}$, the critical point is $\xi = \cos^{-1}(\nu/x)$, and the second derivative of the phase is equal to $x\sin\xi = x\sqrt{1-\nu^2/x^2} \gtrsim k^{1/2}(\nu)^{2/3}$. Observe also that the absolute value of the phase derivative is bigger than $k^{1/2}\nu^{2/3}\delta$ if x is not contained in the interval $(\xi-\delta, \xi+\delta)$, with δ small enough independently of k. Therefore, $|J_\nu(x)| \lesssim k^{-1/4}\nu^{-1/3}$.

Next we search for estimates in the region $\nu - 2k\nu^{1/3} \le x \le \nu - k\nu^{1/3}$, with $k \le \frac{1}{4}\nu^{2/3}$. Here the derivative of the phase is equal to $\nu - x\cos t$, and it is always positive, while its second derivative is $x\sin(t)$. Therefore, the first derivative gets its minimum value, $\nu - x$, at $t = 0$, implying $|J_\nu(x)| \lesssim k^{-1}\nu^{-1/3}$.

Both estimates above yields the first part of the theorem.

To show that the Bessel functions of integer order have a uniformly bounded set of primitives, let us observe that for $\nu = 1, 2, 3, \ldots$ we have

$$J_\nu(x) = \frac{1}{2\pi}\int_{-\pi/2}^{\pi/2} e^{i(\nu\theta - x\sin\theta)}\, d\theta + (-1)^\nu \frac{1}{2\pi}\int_{-\pi/2}^{\pi/2} e^{i(\nu\theta + x\sin\theta)}\, d\theta.$$

Both oscillatory integrals will be treated with the same tools, namely stationary phase and Van der Corput's estimates, but the first one is more complicated because in the second the derivative of the phase is always greater than ν, and this fact simplifies the computations. Therefore, in the following we will discuss only the details involved in the estimation of the first integral.

We have

$$\int_0^x \int_{-\pi/2}^{\pi/2} e^{i(\nu\theta - t\sin\theta)}\, d\theta\, dt = \text{p.v.} \int_{-\pi/2}^{\pi/2} e^{i\nu\theta} \frac{i}{\sin\theta} [e^{-ix\sin\theta} - 1]\, d\theta$$
$$= -i\ \text{p.v.} \int_{-\pi/4}^{\pi/4} \frac{e^{i\nu\theta}}{\sin\theta}\, d\theta + i\ \text{p.v.} \int_{-\pi/4}^{\pi/4} \frac{e^{i(\nu\theta - x\sin\theta)}}{\sin\theta}\, d\theta + O(1),$$

uniformly on ν and $x > 0$; and

$$\text{p.v.} \int_{-\pi/4}^{\pi/4} \frac{e^{i\nu\theta}}{\sin\theta}\, d\theta = \text{p.v.} \int_{-1/\nu}^{1/\nu} \frac{e^{i\nu\theta} - 1}{\sin\theta}\, d\theta + \text{p.v.} \int_{1/\nu \le |\theta| \le \pi/4} \frac{e^{i\nu\theta}}{\sin\theta}\, d\theta,$$

implying that

$$\text{p.v.} \left| \int_{-\pi/4}^{\pi/4} \frac{e^{i\nu\theta}}{\sin\theta}\, d\theta \right| \lesssim 1 + \nu \sup_{\alpha} \left| \int_{1/\nu}^{\alpha} e^{i\nu\theta}\, d\theta \right|,$$

which is bounded uniformly in x and ν.

To estimate the other term, we shall consider first the region $x \le \nu$. In this case, the phase of the oscillatory integral is monotone, having a first derivative equal to $\nu - x\cos(\theta) > \nu - x > 0$, while its third derivative, $\nu\cos\theta$, is greater than $\sqrt{2}x/2$.

Then Van der Corput's estimates give us a uniform estimate of the oscillatory integrals on the regions where either $x^{-1/3} < |\theta| < \pi/4$, or $1/(\nu - x) < |\theta| < \pi/4$.

That is, denoting $m = \min(x^{-1/3}, 1/(\nu - x))$, and since

$$|e^{i(\nu\theta - x\sin\theta)} - 1| \lesssim |\nu - x||\theta| + x|\theta|^3,$$

we get

$$\left| \text{p.v.} \int_{-m}^{m} \frac{e^{i(\nu\theta - x\sin\theta)}}{\sin\theta}\, d\theta \right| < \infty$$

uniformly in the region $0 < x < \nu$.

For the remainder region, we write $x = \nu + k\nu^{1/3}$, $k > 0$, and observe that the third derivative of the phase, $x\cos(\theta)$, is greater than $\sqrt{2}\nu/2$. Therefore Van der Corput's estimates yield

$$\left| \int_{x^{-1/3}}^{\pi/4} \frac{e^{i(\nu\theta - x\sin\theta)}}{\sin\theta}\, d\theta \right| \lesssim \frac{x^{1/3}}{\nu^{1/3}} \lesssim 1.$$

Next observe that $|\theta| < x^{-1/3}$ implies that

$$|e^{i(\nu\theta - x\sin\theta)} - 1| \lesssim (|\theta||\nu - x| + x|\theta|^3).$$

Let us now define the sets U, V and W as follows:

$$U = \{|\theta| \le x^{-1/3}\}, \quad V = \{|\theta| \le x^{-1/3}/k\} \quad \text{and} \quad W = U - V.$$

We have

$$\left|\text{p.v.}\int_U \frac{e^{i(\nu\theta - x\sin\theta)}}{\sin\theta}\,d\theta\right| \le \left|\text{p.v.}\int_V \frac{e^{i(\nu\theta - x\sin\theta))}}{\sin\theta}\,d\theta\right| + \left|\text{p.v.}\int_W \frac{e^{i(\nu\theta - x\sin\theta))}}{\sin\theta}\,d\theta\right|.$$

Next we write the integral on the set V in the form

$$\int_V \frac{e^{i(\nu\theta - x\sin\theta)-1}}{\sin\theta}\,d\theta,$$

and observe that the absolute value of the integrand is controlled by

$$\frac{\nu|\theta|^3 + k|\theta|\nu^{1/3}}{|\sin\theta|}.$$

Therefore, the integral is uniformly bounded on x, k and ν.

Finally, we estimate the integral on W. There we have

$$e^{i(\nu\theta - x\sin\theta)} = e^{ik\theta x^{1/3}}(1 + O(x|\theta|^3)),$$

and since $\int_W \frac{O(x|\theta^3|)}{|\sin\theta|}\,d\theta$ is clearly bounded uniformly on k and ν, we are left with the oscillatory term

$$\int_W \frac{e^{ik\theta x^{1/3}}}{\sin\theta}\,d\theta.$$

And we observe again that in W the function $1/\sin\theta$ is monotonic and bounded by

$$\frac{1}{\sin(1/(kx^{1/3}))},$$

while the integrals

$$\int_\beta^\alpha e^{ik\theta x^{1/3}}\,d\theta$$

are bounded by $1/(kx^{1/3})$ independently of α, allowing us to conclude the proof. □

6.3 The disc multiplier on $L^p_{\text{rad}}L^2_{\text{ang}}(\mathbb{R}^n)$

With the same notation used in the introduction of this chapter, let us consider

$$\widehat{Tf}(\xi) = \chi_B(\xi)\,\widehat{f}(\xi)$$

for rapidly decreasing smooth functions f, where B is the unit ball in $\mathbb{R}^n$, $n \ge 2$.

Theorem 6.2. *The operator T is bounded on $L^p_{\text{rad}}L^2_{\text{ang}}(\mathbb{R}^n)$ if and only if $2n/(n+1) < p < 2n/(n-1)$,*

The proof of this theorem was first given in [134] (see also [139] for an independent proof, and [133] for a weighted version).

The "only if" part follows easily taking f to be the inverse Fourier transform of a C^∞-function, $\widehat{f}$, such that $\widehat{f} \equiv 1$ when $|\xi| \le 1$ and $\widehat{f} \equiv 0$ if $|\xi| \ge 2$. Then

$$Tf(x) = C_n\,|x|^{-n}\,J_{n/2}(2\pi|x|),$$

which belongs to $L^p_{\mathrm{rad}}L^2_{\mathrm{ang}}(\mathbb{R}^n)$ if and only if $p > 2n/(n+1)$. The other bound $p < 2n/(n-1)$ follows by duality.

Given $f \in \mathcal{S}(\mathbb{R}^n)$, it has an expansion

$$f(x) = \sum_{k,\ell} f_k^\ell(|x|)\,Y_k^\ell\Big(\frac{x}{|x|}\Big),$$

where the functions f_k^ℓ are defined on $[0,\infty)$ and $\{Y_k^\ell\}$ is an orthonormal basis of the d_k-dimensional space of spherical harmonic polynomials of degree k in $\mathbb{R}^n$. Here, the index k takes non-negative integer values and, for each k, the index l takes values in the interval

$$1 \le \ell \le d_k = \binom{n+k-1}{k} - \binom{n+k-3}{k-2}.$$

It is a well-known fact (see [14]) that the Fourier transform preserves that expansion:

$$\widehat{f}(\xi) = \sum_{k,\ell} i^k\,F_k^\ell(|\xi|)\,Y_k^\ell\Big(\frac{\xi}{|\xi|}\Big),$$

where

$$F_k^\ell(r) = r^{-n/2+1}\int_0^\infty f_k^\ell(s)\,s^{n/2}\,J_{k-1+n/2}(rs)\,\mathrm{d}s.$$

Continuing with the proof of the theorem, we note that

$$Tf(x) = \sum_{k=0}^{\infty} T_k^n f_k^\ell(|x|)\,Y_k^\ell\Big(\frac{x}{|x|}\Big),$$

where T_k^n are integral operators given by the formula

$$T_k^n f(t) = (-1)^k t^{-(n-1)/2}\int_0^\infty f(r)\,r^{(n-1)/2}\,K_{k-1+n/2}(t,r)\,\mathrm{d}r,$$

and

$$K_\nu(t,r) = \sqrt{tr}\int_0^1 J_\nu(rs)\,J_\nu(ts)\,s\,\mathrm{d}s.$$

Let us begin decoding the kernels. Consider the ODE satisfied (6.3) by the Bessel functions:

$$x^2J_\nu''(x) + xJ_\nu'(x) + (x^2-\nu^2)\,J_\nu(x) = 0,$$

which written in terms of the functions

$$\mathcal{U}_r(s) = \sqrt{rs}\, J_\nu(rs) \quad \text{and} \quad q_r(s) = \frac{1}{4}s^{-2} + r^2 - \nu^2 s^2,$$

becomes

$$\mathcal{U}_r''(s) + q_r(s)\,\mathcal{U}_r(s) = 0.$$

Then

$$\begin{aligned}(t^2 - r^2)\sqrt{rt}\int_0^1 J_\nu(rs)\,J_\nu(st)s\,\mathrm{d}s &= \int_0^1 (q_r(s) - q_t(s))\,\mathcal{U}_r(s)\,\mathcal{U}_t(s)\,\mathrm{d}s \\ &= \int_0^1 \left(\mathcal{U}_r(s)\,\mathcal{U}_t''(s) - \mathcal{U}_r''(s)\,\mathcal{U}_t(s)\right)\mathrm{d}s = \mathcal{U}_r(1)\,\mathcal{U}_t'(1) - \mathcal{U}_r'(1)\,\mathcal{U}_t(1).\end{aligned}$$

Therefore,

$$\begin{aligned}K_\nu(t,r) &= \sqrt{rt}\,\frac{tJ_\nu'(t)J_\nu(r) - rJ_\nu'(r)J_\nu(t)}{t^2 - r^2} \\ &= \frac{\sqrt{t}J_\nu'(t)J_\nu(r)\sqrt{r}}{2(t-r)} + \frac{\sqrt{t}J_\nu'(t)J_\nu(r)\sqrt{r}}{2(t+r)} + \frac{\sqrt{t}J_\nu(t)J_\nu'(r)\sqrt{r}}{2(r-t)} + \frac{\sqrt{t}J_\nu(t)J_\nu'(r)\sqrt{r}}{2(r+t)} \\ &= \sum_{j=1}^4 K_\nu^j(t,r).\end{aligned}$$

Thus we have obtained four families of integral operators:

$$K_\nu^j f(t) = \int_0^\infty K_\nu^j(t,r)\,f(r)\,\mathrm{d}r, \qquad \text{for } j = 1,\ldots,4.$$

However taking into account the asymptotics of Bessel functions (see [147], p. 199),

$$\begin{aligned}J_\nu(z) \sim \left(\frac{2}{\pi z}\right)^{1/2}\Bigg[&\cos\left(z - \frac{2\nu+1}{4}\pi\right)\sum_{m=0}^\infty \frac{(-1)^m(\nu,2m)}{(2z)^{2m}} \\ &- \sin\left(z - \frac{2\nu+1}{4}\pi\right)\sum_{m=0}^\infty \frac{(-1)^m(\nu,2m+1)}{(2z)^{2m+1}}\Bigg],\end{aligned}$$

where

$$(\nu,k) = \frac{\Gamma(\nu+k+1/2)}{k!\,\Gamma(\nu-k+1/2)},$$

one may infer heuristically that the kernels $j = 1$ and $j = 3$ correspond to Hilbert transforms, while the cases $j = 2$ and $j = 4$ are less singular and produce Hardy integral operators. Since the order ν of the Bessel functions is increasing, the asymptotic estimate above does not produce by brute force uniform bounds, and the more precise analysis of the following lemma is needed in the critical range $\nu/2 \le r \le 2\nu$.

Lemma 6.3. *The following estimates hold uniformly on* $\nu \geq 1$:

(i) $|J_\nu(r)| \lesssim \dfrac{1}{r^{1/2}}$ *and* $|J'_\nu(r)| \lesssim \dfrac{1}{r^{1/2}}$ *when* $r \geq \frac{3}{2}\nu$.

(ii) $|J_\nu(r)| \lesssim \dfrac{1}{1+\nu}$ *and* $|J'_\nu(r)| \lesssim \dfrac{1}{(1+\nu)^2}$ *when* $r \leq \frac{3}{2}\nu$.

(iii) $|J_\nu(\nu + \rho\nu^{1/3})| \lesssim \dfrac{1}{\rho^{1/4}\,\nu^{1/3}}$ *and* $|J'_\nu(\nu + \rho\nu^{1/3})| \lesssim \dfrac{\rho^{1/4}}{\nu^{2/3}}$ *when* $1 \leq \rho < \frac{1}{2}\nu^{2/3}$.

(iv) $|J_\nu(\nu - \rho\nu^{1/3})| \lesssim \dfrac{1}{\rho\,\nu^{1/3}}$ *and* $|J'_\nu(\nu - \rho\nu^{1/3})| \lesssim \dfrac{1}{\rho^2\,\nu^{2/3}}$ *when* $1 \leq \rho < \nu^{2/3}$.

As an indication for the reader, we sketch the proof of (iii). The stationary phase method, Van der Corput estimates, and integration by parts allow us to control the size of $J_\nu(x)$ in the critical interval $\nu/2 \leq x \leq 2\nu$. We have

$$J_\nu(x) = \operatorname{Re}\frac{1}{\pi}\int_0^\pi e^{i(\nu t - x\sin t)}\,\mathrm{d}t + O\Big(\frac{1}{\nu}\Big) = \operatorname{Re}\frac{1}{\pi}\int_0^\pi e^{ix(\theta t - \sin t)}\,\mathrm{d}t + O\Big(\frac{1}{\nu}\Big),$$

where $\theta = \nu/x \approx \nu/(\nu + \rho\nu^{1/3}) \approx 1 - \rho\nu^{-2/3}$. The critical point corresponds to

$$\frac{d}{dt}(\theta t - \sin t) = 0$$

i.e., $t_\theta = \frac{1}{\pi}\cos^{-1}\theta$. Then

$$\frac{\mathrm{d}^2}{\mathrm{d}t^2}(\theta t - \sin t)\Big|_{t=t_\theta} = \sqrt{1-\theta^2} \approx \sqrt{2\rho\nu^{-2/3}} \approx \rho^{1/2}\,\nu^{-1/3}.$$

Therefore, the stationary phase method yields

$$|J_\nu(x)| \lesssim \frac{1}{\sqrt{x}} \cdot \frac{1}{\sqrt{\rho^{1/2}\,\nu^{-1/3}}} \lesssim \frac{1}{\rho^{1/4}\,\nu^{1/3}}$$

because $x \approx \nu + \rho\nu^{1/3}$ and $\rho < \nu^{2/3}$. The other cases follow by similar arguments (see [130, 131, 147]).

In the proof of Theorem 6.2, we will get kernels of the form $\sqrt{t}\,J'_\nu(t)\,J_\nu(r)\sqrt{r}$, allowing us to make use of the extra decay hidden there to achieve the estimates.

Proposition 6.4. *Given* $4/3 < p < 4$, *there exists a finite constant* C_p *such that, for* $j = 1, 2, 3, 4$, *any function* $\nu\colon \mathbb{N} \to \mathbb{R}_+$, *and every sequence of rapidly decreasing smooth functions* $\{g_\ell\}$,

$$\int_0^\infty \Big(\sum_\ell |K^j_{\nu(l)} g_\ell(r)|^2\Big)^{p/2}\,\mathrm{d}r \leq C_p \int_0^\infty \Big(\sum_\ell |g_\ell(r)|^2\Big)^{p/2}\,\mathrm{d}r.$$

Proof. We will consider the operator K_ν^1 with kernel

$$K_\nu^1(t,r) = \frac{\sqrt{t}\,J'_\nu(t)\,J_\nu(r)\,\sqrt{r}}{2(t-r)}.$$

The estimate for K_ν^3 will follow as a result of duality; the cases K_ν^2 and K_ν^4, being less singular, are easier to handle, and the details will be left to the reader.

Fixing $\nu > 0$, we consider the partition

$$[0,\infty) = [0,\nu/2) \cup [\nu/2, 2\nu) \cup [2\nu,\infty) = I_0^\nu \cup I_c^\nu \cup I_\infty^\nu,$$

and the corresponding splitting of the kernel. Let us denote by $T^\nu_{\alpha,\beta}$ the integral operator with the kernel

$$\chi_{I_\alpha^\nu}(t)\,K_\nu^1(t)\,\chi_{I_\beta^\nu}(r), \qquad \alpha,\beta = 0, c, \infty.$$

We consider four cases.

Case 1. $\alpha,\beta = 0,\infty$, or $\{\alpha,\beta\} = \{0,c\}$.

The cases $\alpha,\beta = 0,\infty$ are easier, because in $I_0^\nu \cup I_\infty^\nu$ both $\sqrt{t}\,J'_\nu(t)$ and $\sqrt{t}\,J_\nu(t)$ are uniformly bounded, and the estimate in the proposition is reduced to a well-known inequality for the Hilbert transform.

Similarly, when $\alpha = 0$ and $\beta = c$, or when $\alpha = c$ and $\beta = 0$, we have that

$$\sup_{t\in I_0^\nu}\left|\sqrt{t}\,J'_\nu(t)\right| \sup_{r\in I_c^\nu}\left|\sqrt{r}\,J_\nu(r)\right| + \sup_{t\in I_0^\nu}\left|\sqrt{t}\,J_\nu(t)\right| \sup_{r\in I_c^\nu}\left|\sqrt{r}\,J'_\nu(r)\right| \lesssim 1$$

uniformly in ν (Lemma 6.3).

Case 2. $\alpha = c$, $\beta = \infty$.

Let us consider the partition of the interval I_c^ν through the sets

$$G_\tau^+ = [\nu + \tau\nu^{1/3}, \nu + (\tau+1)\nu^{1/3}) \quad \text{and} \quad G_\tau^- = [\nu - (\tau+1)\nu^{1/3}, \nu - \tau\nu^{1/3})$$

for $\nu = 0, 1, 2, \ldots, [\frac{1}{2}\nu^{2/3}]$.

We have

$$\begin{aligned}
&\sum_\tau \chi_{G_\tau^+}(t)\,|\sqrt{t}\,J'_\nu(t)|\left|\int_{2\nu}^\infty \frac{\sqrt{r}\,J_\nu(r)}{t-r}\,f(r)\,\chi_{I_\infty^\nu}(r)\,\mathrm{d}r\right| \\
&\quad + \sum_\tau \chi_{G_\tau^-}(t)\,|\sqrt{t}\,J'_\nu(t)|\left|\int_{2\nu}^\infty \frac{\sqrt{r}\,J_\nu(r)}{t-r}\,f(r)\,\chi_{I_\infty^\nu}(r)\,\mathrm{d}r\right| \\
&\quad \lesssim \nu^{-1/6}\sum_\tau (\tau^{1/4}\,\chi_{G_\tau^+}(t) + \tau^{-2}\,\chi_{G_\tau^-}(t))\,\left|H(f(r)\sqrt{r}J_\nu(r)\chi_{I_\infty^\nu}(r))(t)\right| \\
&\quad \lesssim |H(f(r)\sqrt{r}J_\nu(r)\chi_{I_\infty^\nu}(r))(t)| = |H(f(r)\theta_\nu(r))(t)|,
\end{aligned}$$

where $\theta_\nu(r) = |\sqrt{r}\,J_\nu(r)\chi_{I_\infty^\nu}(r)| \lesssim 1$ uniformly in ν.

Case 3. $\alpha = \infty, \beta = c$.

Since $|\sqrt{r}\, J_\nu(r)| \lesssim 1$ uniformly in ν when $r \geq \frac{3}{2}\nu$, the part of the estimate corresponding to that region trivializes.

For the remainder we have, with the notation of the previous case,

$$\chi_{[2\nu,\infty)}(t)\Big(\sum_\tau \frac{1}{|\nu|}\int_{G_\tau^+} \frac{|f(r)|}{(1+\tau)^{1/4}\nu^{1/3}}\,\mathrm{d}r + \sum_\tau \frac{1}{|\nu|}\int_{G_\tau^-} \frac{|f(r)|}{(1+\tau)\nu^{1/3}}\,\mathrm{d}r\Big)$$
$$\lesssim \chi_{[2\nu,\infty)}(t)\, Mf(t),$$

where as before, M denotes the Hardy–Littlewood maximal function.

Putting all the estimates together for the Cases 1, 2 and 3, we get

$$\int_0^\infty \Big|\sum_\ell |T^\nu_{\alpha,\beta} f_\ell|^2\Big|^{p/2} \lesssim \int_0^\infty \Big|\sum_\ell |H(f_\ell \theta_\nu)|^2\Big|^{p/2} + \int_0^\infty \Big|\sum_\ell |M(f_\ell)|^2\Big|^{p/2}$$
$$\lesssim \int_0^\infty \Big|\sum_\ell |f_\ell|^2\Big|^{p/2}.$$

Case 4. $\alpha = c, \beta = c$.

It will be convenient to modify the splitting, and for a fixed ν define

$$I_k^+ = [\nu + 2^k \nu^{1/3}, \nu + 2^{k+1}\nu^{1/3}) \quad \text{and} \quad I_k^- = [\nu - 2^{k+1}\nu^{1/3}, \nu - 2^k \nu^{1/3}),$$

where $0 \leq k \leq [\frac{2}{3}\log_2 \nu]$ and $I_0 = [\nu - \nu^{1/3}, \nu + \nu^{1/3})$. Define also

$$I^+ = \bigcup_k I_k^+ \quad \text{and} \quad I^- = \Big(\bigcup_k I_k^-\Big) \cup I_0.$$

Then we have the kernels

$$\chi_{I^\pm}(t)\,\frac{\sqrt{t}\,J_\nu'(t)\,J_\nu(r)\,\sqrt{r}}{t-r}\,\chi_{I^\pm}(r).$$

Taking into account once more the estimate of Lemma 6.3, it is obvious that the most dangerous is the ++ term. Therefore, in the following, we shall present the details of the proof corresponding to that case, leaving the other three as exercises.

To estimate the action of the operator with kernel $\chi_{I^+}(t)K_\nu^1(t,r)\chi_{I^+}(r)$ on a function f_ℓ at a point $t \in I_k^+$, we divide the integration in three parts:

$$A_k(t) = \chi_{I_k^+}(t)\int \sum_{j\leq k-1} K_\nu^1(t,r)\,\chi_{I_j^+}(r)\,f_\ell(r)\,\mathrm{d}r,$$
$$B_k(t) = \chi_{I_k^+}(t)\int \sum_{j\geq k+1} K_\nu^1(t,r)\,\chi_{I_j^+}(r)\,f_\ell(r)\,\mathrm{d}r,$$
$$C_k(t) = \chi_{I_k^+}(t)\int K_\nu^1(t,r)\,\chi_{I_k^+}(r)\,f_\ell(r)\,\mathrm{d}r.$$

First we estimate A. We have

$$\begin{aligned}|A_k(t)| &\lesssim \sum_{j\le k-1} |\sqrt{t}\,J_\nu'(t)| \left|\int_{I_j^+} \frac{\sqrt{r}\,J_\nu(r)}{t-r} f_\ell(r)\,\mathrm{d}r\right| \\ &\lesssim \sum_{j\le k-1} 2^{-3k/4} \int_{I_j^+} |J_\nu(r)\, f_\ell(r)|\,\mathrm{d}r,\end{aligned}$$

that gives

$$\begin{aligned}|A_k(t)| &\lesssim 2^{-3k/4} \int_\nu^{\nu+2^k\nu^{1/3}} |f_\ell(r)| \sum_{j\le k-1} \frac{2^{-j/4}}{\nu^{1/3}} \chi_{I_j^+}(r)\,\mathrm{d}r \\ &\lesssim 2^{-3k/4}\,\nu^{-1/3}\Big(\int_\nu^{\nu+2^k\nu^{1/3}} |f_\ell(r)|^s\,\mathrm{d}r\Big)^{1/s}\Big(\sum_{j\le k-1} 2^{-js'/4}\,2^j\,\nu^{1/3}\Big)^{1/s'}\end{aligned}$$

where s and s' are conjugate Hölder exponents. Observe that if $s > 4/3$, then $s' < 4$, and we have

$$\Big(\sum_{j\le k-1} 2^{j(1-s'/4)}\Big)^{1/s'} \approx 2^{k(1-s'/4)/s'},$$

and therefore,

$$|A_k(t)| \lesssim \Big(2^{-k}\,\nu^{-1/3} \int_\nu^{\nu+2^k\nu^{1/3}} |f_\ell(r)|^s\,\mathrm{d}r\Big)^{1/s} \lesssim (M f_\ell^s(t))^{1/s} \quad \text{for } s > \frac{4}{3}.$$

Next we estimate B:

$$\begin{aligned}|B_k(t)| &\lesssim \sum_{j\ge k+1} |\sqrt{t}\,J_\nu'(t)| \left|\int_{I_j^+} \frac{\sqrt{r}\,J_\nu(r)}{t-r} f_\ell(r)\,\mathrm{d}r\right| \\ &\lesssim \sum_{j\ge k+1} \nu\,\frac{2^{k/4}}{\nu^{2/3}}\,\frac{1}{2^j\nu^{1/3}} \int_{I_j^+} |f_\ell(r)|\,\frac{2^{-j/4}}{\nu^{1/3}}\,\mathrm{d}r,\end{aligned}$$

that gives

$$|B_k(t)| \lesssim \sum_{j\ge k+1} 2^{-|j-k|/4}\,\frac{1}{2^j\nu^{1/3}} \int_{\nu+2^j\nu^{1/3}}^{\nu+2^{j+1}\nu^{1/3}} |f_\ell(r)|\,\mathrm{d}r \lesssim Mf(t),$$

where we have applied several times the estimates of Lemma 6.3.

Finally, one needs to control the diagonal terms C:

$$C_k(t) = \sqrt{t}\,J_\nu'(t)\,H(\chi_{I_k^+}(r)\,f_\ell(r)\sqrt{r}\,J_\nu(r))(t)\,\chi_{I_k^+}(t)$$

We have

$$\Big(\sum_k |C_k(t)|^2\Big)^{1/2} \le \Big(\sum_k |\chi_{I_k^+}(t)\,\nu^{-1/6}\,2^{k/4}\,|H(\chi_{I_k^+}(r)\,f_\ell(r)\sqrt{r}\,J_\nu(r))(t)|^2\Big)^{1/2}$$
$$\le \Big(\sum_k |H(\chi_{I_k^+}(r)\,f_\ell(r)\,\nu^{-1/6}\,2^{k/4}\sqrt{r}\,J_\nu(r))(t)|^2\Big)^{1/2}$$
$$= \Big(\sum_k |H(\chi_{I_k^+}(r)\,f_\ell(r)\,\theta_\nu^k(r))(t)|^2\Big)^{1/2},$$

where $|\theta_\nu^k(r)| \lesssim 1$ uniformly in ν and k.

That is, for each α and β, we have obtained the pointwise estimate

$$\Big(\sum_\ell |T_{\alpha\beta}\,f_\ell|^2\Big)^{1/2} \lesssim \Big(\sum_\ell |H(f_\ell\cdot\theta_\nu)|^2\Big)^{1/2} + \Big(\sum_\ell |M f_\ell^s|^{2/s}\Big)^{1/2}$$
$$+ \Big(\sum_\ell\sum_k |H(f_\ell\cdot\theta_\nu^k\cdot\chi_{I_k^+})|^2\Big)^{1/2}$$

for $s > 4/3$, where the functions θ^ν and θ_k^ν are uniformly bounded.

So we can finish the proof of the proposition with the help of some well-known estimates for the Hilbert transform and the Hardy–Littlewood maximal functions (see [4, 139, 141]), namely,

$$\int \Big(\sum_\ell |H f_\ell(x)|^s\Big)^{p/s} \omega(x)\,\mathrm{d}x + \int \Big(\sum_\ell |M f_\ell(x)|^s\Big)^{p/s} \omega(x)\,\mathrm{d}x$$
$$\lesssim C_{p,s}(\omega) \int \Big(\sum_\ell |f_\ell(x)|^s\Big)^{p/s} \omega(x)\,\mathrm{d}x$$

so long as the weight ω belongs to the class A_p, that is,

$$\sup_I \Big(\frac{1}{|I|}\int_I \omega\Big)\Big(\frac{1}{|I|}\int_I \omega^{-1/(p-1)}\Big)^{p-1} = |\omega|_p < \infty$$

Furthermore, the constant $C_{p,s}(\omega)$ above depends only upon p, s and $|\omega|_p$. □

Proof of Theorem 6.2. Let us note that, in order to prove the theorem, it is enough to show the following inequality:

(6.4)
$$\int_0^\infty \Big|\sum_\ell |K_\nu\,g_\ell(r)|^2\Big|^{p/2} r^{(n-1)(1-p/2)}\,\mathrm{d}r \lesssim \int_0^\infty \Big|\sum_\ell |g_\ell(r)|^2\Big|^{p/2} r^{(n-1)(1-p/2)}\,\mathrm{d}r$$

so long as $2n/(n+1) < p < 2n/(n-1)$.

But this will be a consequence of Proposition 6.4, together with some observations about weights in the class A_p.

Since it happens that $|x|^\beta \in A_p$ if and only if $-1 < \beta < p-1$, we can conclude that $|x|^{(n-1)(1-p/2)} \in A_p$ if and only if $2n/(n+1) < p < 2n/(n-1)$. Furthermore, if $2 \le p < 2n/(n-1)$, we have that $|x|^{(n-1)(1-p/2)} \in A_q$, for every $q \ge 1$. However, in the interval $2n/(n+1) < p \le 2$ we have that $|x|^{-p/s'+(n-1)(1-p/2)} \in A_{p/s}$ under the hypothesis $s < 4$, $1/s + 1/s' = 1$.

The part of the inequality (6.4) which corresponds to the critical intervals,

$$K_\nu^c(t,r) = \chi_{I_c^\nu}(t)\,\frac{\sqrt{t}\,J_\nu'(t)\,J_\nu(r)\sqrt{r}}{t-r}\,\chi_{I_c^\nu}(t),$$

is a direct consequence of Proposition 6.4 for the following reasons:

$$\begin{aligned}\int_0^\infty \Big(\sum_\ell |K_\nu^c g_\ell(t)|^2\Big)^{p/2} t^\alpha\,dt &\lesssim \sum_{n=0}^\infty 2^{n\alpha}\int_{2^{n-1}}^{2^{n+1}} \Big(\sum_{\ell:\nu\sim 2^n} \big|K_\nu^c g_\ell(t)\big|^2\Big)^{p/2} dt\\ &\lesssim \sum_{n=0}^\infty 2^{n\alpha}\int_0^\infty \Big(\sum_{\ell:\nu\sim 2^n} |g_\ell(t)|^2\,\chi_{I_c^\nu}(t)\Big)^{p/2} dt\\ &\lesssim \sum_{n=0}^\infty \int_{2^{n-1}}^{2^{n+1}} \Big(\sum_{\ell:\nu\sim 2^n} |g_\ell(t)|^2\Big)^{p/2} t^\alpha\,dt \lesssim \int_0^\infty \Big(\sum_{\ell:\nu\sim 2^n} |g_\ell(t)|^2\Big)^{p/2} t^\alpha\,dt,\end{aligned}$$

and the result follows taking $\alpha = (n-1)(1-p/2)$.

The remainder terms can be controlled similarly, except for one of them which needs extra arguments, namely

$$\chi_{I_\infty^\nu}(t)\,K_\nu(t,r)\,\chi_{I_c^\nu}(t).$$

The case $2 \le p < 2n/(n-1)$ is easy because then $|x|^{(n-1)(1-p/2)} \in A_{p/s}$ for $4/3 < s < p$, and we have the estimate

$$\begin{aligned}&\int_0^\infty \Big(\sum_\ell |K_\nu g_\ell(x)|^2\Big)^{p/2} x^{(n-1)(1-p/2)}\,dx\\ &\qquad\lesssim \int_0^\infty \Big(\sum_\ell \big|M(g_\ell^s)(x)\big|^{2/s}\Big)^{p/2} x^{(n-1)(1-p/2)}\,dx\\ &\qquad\lesssim \int_0^\infty \Big(\sum_\ell |g_\ell(x)|^2\Big)^{p/2} x^{(n-1)(1-p/2)}\,dx.\end{aligned}$$

To treat the case $2n/(n+1) < p \le 2$, let us observe that for $t \ge 3\nu$, we have

$$\begin{aligned}\Big|\int_0^\infty K_\nu^1(t,r)\,g_\ell(r)\,\chi_{I_c^\nu}(r)\,dr\Big| &\lesssim \frac{1}{t}\int_{I_c^\nu} \sqrt{r}\,|J_\nu(r)|\,|g_\ell(r)|\,dr\\ &\lesssim \frac{1}{t}\Big(\int_{I_c^\nu} |g_\ell(r)|^s\,dr\Big)^{1/s} \nu^{1/s'} \lesssim \Big(\frac{\nu}{t}\Big)^{1/s'} (Mg_\ell^s(t))^{1/s} \qquad \text{for } s > 4/3.\end{aligned}$$

Then we get the integral

$$\int_0^\infty \Big(\sum_\ell \Big(\frac{\nu}{t}\Big)^{2/s'} (Mg_\ell^s(t))^{2/s}\Big)^{p/2} t^{(n-1)(1-p/2)}\,\mathrm{d}t,$$

and since $-1 < -p/s' + (n-1)(1-p/2) < p/s - 1$, it happens that

$$|t|^{-p/s'+(n-1)(1-p/2)} \in A_{p/s},$$

allowing us to obtain the estimate needed to finish the proof:

$$\begin{aligned}\int_0^\infty \Big(\sum_\ell \Big(\frac{\nu}{t}\Big)^{2/s'} (Mg_\ell^s(t))^{2/s}\Big)^{p/2} t^{(n-1)(1-p/2)}\,\mathrm{d}t \\ \lesssim \int_0^\infty \Big(\sum_\ell \nu^{2/s'} |g_\ell(t)|^2 \chi_{I_c^\nu}(t)\Big)^{p/2} t^{-p/s'+(n-1)(1-p/2)}\,\mathrm{d}t \\ \lesssim \int_0^\infty \Big(\sum_\ell |g_\ell(t)|^2\Big)^{p/2} t^{(n-1)(1-p/2)}\,\mathrm{d}t.\end{aligned}$$

□

6.4 Restriction properties of the Fourier transform on $L^p_{\mathrm{rad}} L^2_{\mathrm{ang}}(\mathbb{R}^n)$

With similar methods to those developed above, one can obtain the following restriction theorem.

Theorem 6.5. *In $\mathbb{R}^n$, $n \geq 2$, given $1 \leq p < 2n/(n+1)$, there exists a finite constant $C_{p,n}$ such that*

$$\|\widehat{f}\|_{L^2(S^{n-1})} \leq C_{p,n} \|f\|_{L^p_{\mathrm{rad}} L^2_{\mathrm{ang}}(\mathbb{R}^n)}$$

for every rapidly decreasing smooth function f.

Note that the interval $1 \leq p \leq 2n/(n+1)$ is bigger than $1 \leq p < 2(n+1)/(n+3)$, which corresponds to the Stein–Tomas theorem [1, 13, 128]:

$$\|\widehat{f}\|_{L^2(S^{n-1})} \lesssim \|f\|_{L^p(\mathbb{R}^n)}.$$

The proof of the restriction estimate above will be obtained by duality from the corresponding extension estimate

$$\|\widehat{f\,\mathrm{d}\sigma}\|_{L^q_{\mathrm{rad}} L^2_{\mathrm{ant}}(\mathbb{R}^n)} \lesssim_{q,n} \|f\|_{L^2(S^{n-1})} \qquad \text{for } q > \frac{2n}{n-1}.$$

Taking $f(\theta) \equiv 1$, it is easy to check that the estimate above cannot hold when $q \leq 2n/(n-1)$, which amounts to show that the range $1 \leq p < 2n/(n+1)$ is sharp in the statement of the theorem.

Given $f \in L^2(S^{n-1})$, we have an expansion $f = \sum_k a_k Y_k$, where Y_k is a spherical harmonic of degree k, normalized as $\|Y_{k_i}\|_{L^2(S^{n-1})} = 1$. Then we can invoke the formula

$$\widehat{Y_k \mathrm{d}\sigma}(\xi) = 2\pi\, i^k\, |\xi|^{1-n/2}\, J_{k-1+n/2}(2\pi|\xi|)\, Y_k\Big(\frac{\xi}{|\xi|}\Big). \tag{6.5}$$

Proof of Theorem 6.5. As we have mentioned, any $f \in L^2(S^{n-1})$ can be written in the form $f = \sum_k a_k Y_k$, where Y_k is in normalized harmonic polynomial of degree k, and it suffices to prove the following inequality:

$$\Big(\int_0^\infty \Big(\sum_k |a_k|^2 |J_{k-1+n/2}(r)|^2\Big)^{q/2} r^{(1-n/2)q+n-1}\, \mathrm{d}r\Big)^{1/q} \lesssim_q \Big(\sum_k |a_k|^2\Big)^{1/2}$$

for every $q > 2n/(n-1)$.

Since $|J_\nu(r)| \lesssim r^\nu$ when $0 \le r \le 1$, the contribution of that interval to the integral above trivializes. Let us then consider, for a fixed $M = 2^m$ $(m = 0, 1, \ldots)$ the integral I_M as above but restricted to $M \le r \le 2M$.

We shall first consider the case $n = 2$. Then we have

$$\begin{aligned} I_M &= \int_M^{2M} \Big(\sum_k |a_k|^2\, |J_k(r)|^2\Big)^{q/2} r\, \mathrm{d}r \\ &= \int_M^{2M} \Big(\sum_{k<M/2}\Big)^{q/2} r\, \mathrm{d}r + \int_M^{2M} \Big(\sum_{M/2\le k\le 4M}\Big)^{q/2} r\, \mathrm{d}r + \int_M^{2M} \Big(\sum_{k>4M}\Big)^{q/2} r\, \mathrm{d}r \\ &= I_M^1 + I_M^2 + I_M^3. \end{aligned}$$

To estimate I_M^1, we observe that $|J_k(r)| \lesssim 1/r^{1/2}$ uniformly in k when $k \le M/2 \le r/2$. Therefore,

$$I_M^1 \lesssim \int_M^{2M} \Big(\sum_{k<M/2} |a_k|^2\Big)^{q/2} r^{-q/2+1}\, \mathrm{d}r \lesssim M^{(4-q)/2} \Big(\sum |a_k|^2\Big)^{q/2}.$$

Regarding I_M^3, we have $k > 4M \ge 2r$ and $|J_k(r)| \lesssim 1/k$, which yields

$$I_M^3 \lesssim \int_M^{2M} \Big(\sum_{k>4M} k^{-2}\, |a_k|^2\Big)^{q/2} r\, \mathrm{d}r \lesssim M^{2-q} \Big(\sum |a_k|^2\Big)^{q/2}.$$

Finally let us write

$$I_M^2 \lesssim \sum_\alpha \int_{G_\alpha} \Big(\sum_{M/2\le k\le 4M} |a_k|^2 |J_k(r)|^2\Big)^{q/2} r\, \mathrm{d}r,$$

where

$$G_\alpha = \Big[\frac{M}{2} + \alpha M^{1/3}, \frac{M}{2} + (\alpha+1) M^{1/3}\Big]$$

is an interval in the real line and $\alpha = 0, 1, 2, \ldots, [\frac{3}{2}M^{2/3}]$. Let us also define

$$A_\alpha = \sum_{k\in G_\alpha} |a_k|^2.$$

Then we have

$$\begin{aligned} I_M^2 &\leq \sum_\beta \int_{G_\beta} \Big(\sum_{\alpha\leq\beta} A_\alpha \Big(\frac{1}{(|\alpha-\beta|+1)^{1/4} M^{1/3}} \Big)^2 \Big)^{q/2} r\, \mathrm{d}r \\ &\quad + \sum_\beta \int_{G_\beta} \Big(\sum_{\alpha\geq\beta} A_\alpha \Big(\frac{1}{(|\alpha-\beta|+1) M^{1/3}} \Big)^2 \Big)^{q/2} r\, \mathrm{d}r \\ &= I_M^{2,1} + I_M^{2,2}, \end{aligned}$$

and

$$\begin{aligned} I_M^{2,1} &\lesssim \sum_{\beta\leq\frac{3}{2}M^{2/3}} \Big(\sum_{\alpha\leq\beta} \frac{A_\alpha}{(|\alpha-\beta|+1)^{1/2}} \Big)^{q/2} M^{(4-q)/3} \lesssim M^{(4-q)/3} \Big(\sum_\gamma A_\gamma^s \Big)^{q/2s} \\ &\lesssim M^{(4-q)/3} \Big(\sum_\gamma A_\gamma \Big)^{q/2} \lesssim M^{(4-q)/3} \Big(\sum_k |a_k|^2 \Big)^{q/2}, \end{aligned}$$

where we have taken $2/q = 1/s - 1/2$ and used the fact that $q > 4$.

The term $I_M^{2,2}$ is controlled by the same argument. Therefore, when $n = 2$ and $q > 4$, adding all the estimates above over the dyadic intervals $[M, 2M]$, $M = 2^m$, we get

$$\sum_m \big(2^{m(4-q)/2} + 2^{m(4-q)/3}\big) \Big(\sum_k |a_k|^2 \Big)^{q/2} \lesssim \Big(\sum_k |a_k|^2 \Big)^{q/2},$$

and the theorem is proved in this case.

In the general case $n \geq 2$, we have to estimate

$$\Big(\int_{2^m}^{2^{m+1}} \Big(\sum_k |a_k|^2 \, |J_{k-1+n/2}(r)|^2 \Big)^{q/2} r^{(1-n/2)q+n-1}\, \mathrm{d}r \Big)^{1/q},$$

which, using the previous argument, is controlled by

$$\lesssim 2^{m(n-2)(-q/2+1)} \big(2^{m(4-q)/2} + 2^{m(4-q)/3}\big) \Big(\sum_k |a_k|^2 \Big)^{q/2}.$$

Finally, it is enough to note that the resulting exponent in the powers of 2 is negative for $q > 2n/(n-1)$. □

6.5 Singular integrals in $L^p_{\mathrm{rad}}L^2_{\mathrm{ang}}(\mathbb{R}^n)$

Knowing the relevance of the spaces $L^p_{\mathrm{rad}}L^2_{\mathrm{ang}}(\mathbb{R}^n)$ for the disc multiplier, it is interesting to uncover the action on them of Calderón–Zygmund singular integrals and directional Hilbert transforms in $\mathbb{R}^n$.

Consider the singular integral operator

$$Sf(x) = \text{p.v.} \int_{\mathbb{R}^n} f(x-y)\, \frac{\Omega(y)}{|y|^n}\, \mathrm{d}y,$$

where Ω is smooth enough, homogeneous of degree 0, and has zero mean value:

$$\int_{S^{n-1}} \Omega(y)\, \mathrm{d}\sigma(y) = 0.$$

Theorem 6.6. *The operator S extends to a bounded operator from $L^p_{\mathrm{rad}}L^2_{\mathrm{ang}}(\mathbb{R}^n)$ to itself when $1 < p < \infty$.*

Proof. It suffices to prove the theorem when $2 \le p < \infty$, because then the case $1 < p \le 2$ will follow by duality.

Given $f \in L^p_{\mathrm{rad}}L^2_{\mathrm{ang}}(\mathbb{R}^n)$ and $g \in \mathcal{S}(\mathbb{R})$, let us consider

$$I(f,g) = \int_0^\infty \int_{S^{n-1}} |Sf(r,\theta)|^2\, g(r)\, r^{n-1}\, \mathrm{d}\sigma(\theta)\, \mathrm{d}r = \int_{\mathbb{R}^n} |Sf(x)|^2\, g(|x|)\, \mathrm{d}x.$$

By [84], we have

$$I(f,g) \lesssim s \int_{\mathbb{R}^n} |f(x)|^2\, (Mg^s(x))^{1/s}\, \mathrm{d}x$$

for $1 < s < \infty$, where Mg^s denotes the Hardy–Littlewood maximal operator applied to the function $g^s(x) = (g(|x|))^s$. But

$$Mg^s(x) = \sup_{x \in B} \frac{1}{|B|} \int_B g^s(y)\, \mathrm{d}y$$

is also radial. Therefore

$$\begin{aligned} I(f,g) &\lesssim s \int_0^\infty \int_{S^{n-1}} |f(r,\theta)|^2\, (Mg^s(r))^{1/s} r^{n-1}\, \mathrm{d}\sigma(\theta)\, \mathrm{d}r \\ &\lesssim s \|f\|_{L^p_{\mathrm{rad}}L^2_{\mathrm{ang}}(\mathbb{R}^n)} \Big(\int_0^\infty (Mg^s(r))^{q/s}\, r^{n-1}\, \mathrm{d}r \Big)^{1/q} \end{aligned}$$

so long as $1 < s < q$, where $2/p + 1/q = 1$.

Taking the supremum over all such g so that $\int_0^\infty g^q(r) r^{n-1}\, \mathrm{d}r \le 1$, we conclude the proof. □

The theorem above is somehow the result that one may expect, but the following is, perhaps, more surprising. Let us consider the Hilbert transform H_ω in the direction of $\omega \in S^{n-1}$, defined in (5.2).

Theorem 6.7. *The operator H_ω is bounded on $L^p_{\rm rad}L^2_{\rm ang}(\mathbb{R}^n)$ if and only if $2n/(n+1) < p < 2n/(n-1)$.*

Proof. Because of the rotational symmetry, it suffices to show the conclusion for $\omega = e_n = (0, 0, \ldots, 1)$.

To see the "only if" part, we take f to be the indicator function of the unit cube and observe that

$$|H_{e_n} f(x_1, x_2, \ldots, x_{n-1}, y)| \gtrsim \frac{1}{|y|}$$

for $|x_j| \leq 1/2$, $j = 1, 2, \ldots, n-1$, and $|y| \geq 2$.

Then an elementary computation yields the inequality

$$\begin{aligned}\int_0^\infty \int_{S^{n-1}} (|H_{e_n} f(r,\theta)|^2 \, \mathrm{d}\sigma(\theta))^{p/2} \, r^{n-1} \, \mathrm{d}r &\geq \int_2^\infty \left(\left(\frac{1}{r}\right)^2 \frac{1}{r^{n-1}}\right)^{p/2} r^{n-1} \, \mathrm{d}r \\ &= \int_2^\infty r^{n-p(n+1)/2} \, \frac{\mathrm{d}r}{r},\end{aligned}$$

and therefore the boundedness of H_{e_n} implies that $p > 2n/(n+1)$. Again, the other bound $p < 2n/(n-1)$ follows by duality.

To prove the "if" part, let us observe first that the $L^p_{\rm rad}L^2_{\rm ang}(\mathbb{R}^n)$ norm is well behaved under dilations: if

$$f_\delta(x) = f(\delta x),$$

then

$$\|f_\delta\|_{L^p_{\rm rad}L^2_{\rm ang}(\mathbb{R}^n)} = \delta^{-n/p} \, \|f\|_{L^p_{\rm rad}L^2_{\rm ang}(\mathbb{R}^n)},$$

and it is preserved under multiplication by characters $e^{ix\cdot\eta}$.

Therefore, if $\widehat{T_R f}(\xi) = \chi_{B_R}(\xi)\widehat{f}(\xi)$, where B_R is the ball of radius R, we have

$$T_R f(x) = T_1 f_{1/R}(Rx),$$

which produces the estimate

$$\begin{aligned}\|T_R f\|_{L^p_{\rm rad}L^2_{\rm ang}(\mathbb{R}^n)} &= R^{-n/p} \, \|T_1 f_{1/R}\|_{L^p_{\rm rad}L^2_{\rm ang}(\mathbb{R}^n)} \lesssim R^{-n/p} \, \|f_{1/R}\|_{L^p_{\rm rad}L^2_{\rm ang}(\mathbb{R}^n)} \\ &\lesssim \|f\|_{L^p_{\rm rad}L^2_{\rm ang}(\mathbb{R}^n)}\end{aligned}$$

uniformly on $R > 0$, so long as $2n/(n+1) < p < 2n/(n-1)$.

Next we consider the multiplier, $i\,\mathrm{sign}(\xi \cdot \omega)$, corresponding to the Hilbert transform H_ω, and we observe that its boundedness properties are equivalent to those of the Fourier multipliers given by hyperplanes $\{\xi : \xi \cdot \omega \leq t\}$.

Given a function f (chosen so that its Fourier transform $\widehat{f}$ has compact support), the expression

$$e^{-2\pi i x\cdot\xi_k}\,T_{R_k}(e^{2\pi i y\cdot\xi_k} f(y))(x)$$

corresponds to the Fourier multiplier given by the indicator function of the ball of radius R_k centered at ξ_k. Choosing a convenient sequence of points ξ_k and radii R_k, we can write

$$\chi_{\{\xi\cdot\omega\le 0\}}(\xi)\,\widehat{f}(\xi) = \lim_{k\to\infty} \chi_{\|\xi-\xi_k\|\le R_k}(\xi)\,\widehat{f}(\xi)$$

Then the dominated convergence theorem, together with the uniform bounds of T_{R_k} in $L^p_{\mathrm{rad}}L^2_{\mathrm{ang}}(\mathbb{R}^n)$ for the range $2n/(n+1) < p < 2n/(n-1)$, allow us to finish the proof. □

A different proof of Theorem 6.7. We start with the integral

$$I(f,g) = \int_0^\infty \int_{S^{n-1}} |H_\omega f(r,\theta)|^2\, g(r)\, r^{n-1}\, \mathrm{d}\sigma(\theta)\, \mathrm{d}r.$$

We have

$$I(f,g) = \int_{\mathbb{R}^n} |H_\omega f(x)|^2\, g(|x|)\, \mathrm{d}x \lesssim_s \int_{\mathbb{R}^n} |f(x)|^2\, (M_\omega g^s(x))^{1/s}\, \mathrm{d}x,$$

where M_ω denotes the one-dimensional Hardy–Littlewood maximal function in the direction of ω, and $s > 1$.

As $g^s(x) = g^s(|x|)$ is radial, one may consider its universal Kakeya maximal function, defined as

(6.6)
$$\mathcal{U}f(x) = \sup_{\substack{a,b>0\\ \omega\in S^{n-1}}} \frac{1}{a+b}\int_{-a}^{b} |f(x+t\omega)|\, \mathrm{d}t,$$

to get the upper bound

$$I(f,g) \lesssim_s \int_{\mathbb{R}^n} |f(x)|^2\, (\mathcal{U}g^s(x))^{1/s}\, \mathrm{d}x,$$

and so,

$$I(f,g) \lesssim \left(\int_0^\infty \int_{S^{n-1}} (|f(r,\theta)|^2\, \mathrm{d}\sigma(\theta))^{p/2}\, r^{n-1}\, \mathrm{d}r\right)^{2/p} \left(\int_{\mathbb{R}^n} (\mathcal{U}g^s(x))^{q/s}\, \mathrm{d}x\right)^{1/q},$$

where $2/p + 1/q = 1$. To finish, we make use of the fact that $\mathcal{U}$ is bounded on $L^p(\mathbb{R}^n)$, $p > n$, when restricted to radial functions [132]: observe that $2n/(n+1) < p < 2n/(n-1)$ yields

$$\frac{1}{q} = 1 - \frac{2}{p} < 1 - \frac{n-1}{n} = \frac{1}{n}\cdot$$

Therefore, given $q > n$, we choose $s > 1$ so that $q/s > n$ to obtain

$$\int_{\mathbb{R}^n} (\mathcal{U}g^s(x))^{q/s}\,\mathrm{d}x \lesssim_s \int_{\mathbb{R}^n} |g(x)|^q\,\mathrm{d}x = \int_0^\infty |g(r)|^q\, r^{n-1}\,\mathrm{d}r,$$

which coincides with $\|f\|_{L^p_{\mathrm{rad}}L^2_{\mathrm{ang}}(\mathbb{R}^n)}$. □

Corollary 6.8 (Meyer's lemma). *Given $\{\theta_j\}$, a countable family of directions in $\mathbb{R}^n$, and the corresponding directional Hilbert transforms, $H_jf = H_{\theta_j}f$, the following inequality holds*:

$$\Big\|\Big(\sum_j |H_jf_j|^2\Big)^{1/2}\Big\|_{L^p_{\mathrm{rad}}L^2_{\mathrm{ang}}(\mathbb{R}^n)} \lesssim \Big\|\Big(\sum_j |f_j|^2\Big)^{1/2}\Big\|_{L^p_{\mathrm{rad}}L^2_{\mathrm{ang}}(\mathbb{R}^n)},$$

where $2n/(n+1) < p < 2n/(n-1)$.

Proof. Once more, it is enough to prove it when $2 \le p < 2n/(n-1)$ and then use duality to cover the other cases.

Given $g \in \mathcal{S}(\mathbb{R})$, we consider the integrals

$$\sum_j \int_0^\infty \int_{S^{n-1}} |H_jf_j(r,\theta)|^2\, g(r)\, r^{n-1}\,\mathrm{d}\sigma(\theta)\,\mathrm{d}r = \sum_j \int_{\mathbb{R}^n} |H_jf_j(x)|^2\, g(|x|)\,\mathrm{d}x,$$

that is

$$\lesssim_s \sum_j \int_{\mathbb{R}^n} |f_j(x)|^2\, (\mathcal{U}g^s(x))^{1/s}\,\mathrm{d}x,$$

and then

$$\lesssim_s \left|\int_0^\infty \int_{S^{n-1}} \Big(\Big|\sum_j f_j(r,\theta)\Big|^2 \mathrm{d}\sigma(\theta)\Big)^{p/2} r^{n-1}\,\mathrm{d}r\right|^{2/p} \left|\int_{\mathbb{R}^n} (\mathcal{U}g^s(x))^{q/s}\,\mathrm{d}x\right|^{1/q}.$$

The proof follows for the same reasons given in Theorem 6.7. □

We finish this section by presenting a new proof, using old techniques, of the boundedness of the universal Kakeya maximal function acting on radial functions.

Related with these problems are maximal functions associated to vector fields in $\mathbb{R}^n$. Given a continuous field of directions $v(x)$, and a positive valued real function $p(x)$, let us consider

$$M_vf(x) = \sup_{0<r,t<p(x)} \frac{1}{r+t}\int_{-r}^{t} |f(x+sv(x))|\,\mathrm{d}s$$

As it is usual in differentiation theory, the behavior of those maximal operators produces quantitative versions of the Lebesgue differentiation theorem. In general, the problem is rather difficult, but E. Stein and S. Wainger [126] have created a theory which yields sufficient conditions for boundedness.

However, for our discussion the following three examples are of special relevance:

(1) $v(x) = x/\|x\|$, $p(x) = \infty$. In this case, M is bounded on $L^p(\mathbb{R}^n)$ if and only if $p > n$.
(2) $v(x) = x/\|x\|$, $p(x) = \frac{1}{2}\|x\|$. Then M is of weak type $(1,1)$.
(3) In $\mathbb{R}^2$, $v(x,y) = (-y,x)/\|(x,y)\|$, $p(x) = \infty$. Then M is bounded on $L^p(\mathbb{R}^2$, $1 < p < \infty$.

The proof of (1) uses polar coordinates together with the fact that $|x|^{n-1}$ is an A_p weight in $\mathbb{R}^n$ for $p > n$:

$$\int_{\mathbb{R}^n} |Mf(x)|^p \, dx = \int_{S^{n-1}} \int_0^\infty |Mf(r,\theta)|^p \, r^{n-1} \, dr \, d\sigma(\theta)$$
$$\lesssim_p \int_{S^{n-1}} \int_0^\infty |f(r,\theta)|^p \, r^{n-1} \, dr \, d\sigma(\theta)$$

so long as $p > n$. The proof of (2) is left as an exercise to the reader.

Regarding (3), let us consider in $\mathbb{R}^2$, identified with $\mathbb{C}$, the change of variables $w = \log z$ mapping the straight line $\rho e^{i\theta}(1+ir)$, $r \in \mathbb{R}$, into the curve Γ: $\log\rho + i\theta + \log(1+ir)$. But Γ is the result of translating the fixed curve $w = \log(1+ir)$ to the point $\log\rho + i\theta$. Therefore, in the w-plane, the maximal function is realized as

$$\widetilde{Mg}(w) = \sup_{t,r>0} \frac{1}{r+t} \int_{-r}^{t} |g(w+\gamma(s))| \, ds,$$

where $\gamma(s) = \log(1+is)$. In this setting, we can apply the results of E. Stein and S. Wainger [126] to finish the proof. □

The result for the vector field (3) can be extended to higher dimensions if we restrict our attention to radial functions, because, in that case, we can take any direction perpendicular to the position vector x without changing the value of the maximal function there.

That is, considering the product structure $\mathbb{R}^n = \mathbb{R} \times \mathbb{R}^{n-1}$, the evaluation of the maximal function at the point $(x_1, x_2, \ldots, x_n)$ can be obtained as the evaluation of its $(n-1)$-dimensional version acting on the slice $\{x_1\} \times \mathbb{R}^{n-1}$. Then an induction argument together with Fubini's theorem gives the boundedness for every $p > 1$.

Our proof for $\mathcal{U}$ will be then modeled upon these two extreme examples (1) and (3), and it will make use of the geometry of parallelepipeds in $\mathbb{R}^n$ to treat the intermediate case.

In $\mathbb{R}^n$, $n \geq 2$, the universal Kakeya maximal function (6.6) is unbounded on every $L^p(\mathbb{R}^n)$, $1 \leq p < \infty$, as an appropriate Kakeya set argument easily shows. However, acting on radial functions f, we have:

Proposition 6.9. $\|\mathcal{U}f\|_p \lesssim \|f\|_p$ *if* $p > n$ *and* f *is radial.*

Proof. Fixing a radial function f and a positive number $\lambda > 0$, let us consider the set

$$E_\lambda = \{x : \mathcal{U}f(x) > \lambda\}.$$

Given $x \in E_\lambda$, there is a set of directions $\Gamma_x \subset S^{n-1}$ so that

$$\mathcal{U}f(x) = \sup_{\substack{a,b>0 \\ \omega\in S^{n-1}}} \frac{1}{a+b}\int_{-a}^{b} |f(x+t\omega)|\, dt \geq \lambda$$

for every $\omega \in \Gamma_x$. Clearly, if $\|x\| = \|y\|$, then Γ_x and Γ_y coincide after a rotation. Furthermore, for a given x and $\omega \in \Gamma_x$, we will choose a and b satisfying the inequality above and such that its sum $a + b$ if the biggest possible. Then, among all such $\omega \in S^{n-1}$, a and b, we select those maximizing the projection of the segment $\{x + s\omega : -a \leq s \leq b\}$ into the radial direction.

Taking advantage of the radial symmetry and after some elementary geometrical considerations, the set E_λ is contained in an "almost disjoint" union of annuli,

$$E_\lambda = \bigcup_j \{R_j^1 \leq |x| \leq R_j^2\} = \bigcup_j C_j,$$

satisfying the following properties:

(1) For each annulus C_j, there exists a direction ω_j so that the average of $|f|$ on the segments of directions ω_j, starting at some point of the sphere $\|x\| = R_j^1$ and ending on the point of the sphere $\|y\| = R_j^2$, is the given value λ.
(2) We have that $C_j \cap C_k = \emptyset$ if $|k - j| \geq 2$.

Clearly, in order to estimate the size of $E_\lambda = \{\mathcal{U}f > \lambda\}$ it is enough to control, independently, the portion of that set contained in each annulus C_j.

We shall distinguish two classes of annuli:

(i) Thin shells: $\{R \leq |x| \leq R + \Delta : \Delta < \frac{1}{2}R\}$.
(ii) Thick shells: $\{R \leq |x| \leq R + \Delta : \Delta \geq \frac{1}{2}R\}$.

In both cases (thin or thick), we fix two poles N and S (North and South) in order to estimate the portion of E_λ near the Equator. Given the radial symmetry, we can select the direction ω in such a way that the ray from each point of the shell meets the NS axis, but we can also establish the convention that, among the two possible rays, the chosen one in pointing North.

For a thick annulus C, its volume is comparable to $(\Delta + R)^n$. Therefore one needs to show the inequality

$$(\Delta + R)^n \lesssim_p \frac{1}{\lambda^p}\int_C |f(x)|^p\, dx \qquad \text{for any } p > n.$$

Let us observe that for each point in the sphere $\|x\| = R + \Delta$, we are then given a straight line segment L_x starting at x; ending at the inner sphere $\|y\| = R$; tangent to a certain sphere $\|y\| = R_0 \leq R$, and such that

- the average of $|f|$ on L_x is λ,
- the straight line L_x intersects the NS axis.

Again given the fixed radial function f, one can enlarge the segment L_x (with some small positive number ε) to become a tube of radius ε. In such a way that the average of $|f|$ in that tube is bigger than, say, $\lambda/2$.

Next we cover E_λ (or a fixed portion of it close to the Equator) with a family of those tubes, but keeping them pairwise disjoint at the outer sphere $\|x\| = R+\Delta$. That is, we have obtained what is called a *brush configuration* of tubes, meeting the NS axis and being disjoint at the outer sphere of the shell.

Let us denote by $\{T^\varepsilon_\nu\}_\nu$ the collection of those tubes, and let us consider their overlapping function $\sum_\nu \chi_{T^\varepsilon_\nu}(x)$. Then some elementary geometric considerations ("brush argument") yield the following estimate:

$$\Big|\Big\{\sum_\nu \chi_{T^\varepsilon_\nu}(x) \geq d > 0\Big\}\Big| \lesssim \frac{(\Delta+R)^n}{d^{n/(n-1)}}. \tag{6.7}$$

Therefore

$$\begin{aligned}|C| &\lesssim \sum_\nu |T^\varepsilon_\nu| \lesssim \frac{1}{\lambda}\sum_\nu \int_{T^\varepsilon_\nu} |f(x)|\,\mathrm{d}x \leq \frac{1}{\lambda}\sum_\nu \int |f(x)| \sum_\nu \chi_{T^\varepsilon_\nu}(x)\,\mathrm{d}x \\ &\leq \frac{1}{\lambda}\|f\|_p \Big\|\sum_\nu \chi_{T^\varepsilon_\nu}\Big\|_q \lesssim \frac{1}{\lambda}\|f\|_p |C|^{1/q},\end{aligned}$$

with $1/p + 1/q = 1$, so long as $q < n/(n-1)$ (equivalently $p > n$), allowing us to finish the proof.

The elementary geometrical considerations are the following.

(1) For each θ in the Equator S^{n-2} of S^{n-1}, let us consider the two-dimensional plane H_θ determined by θ and the NS axis, and also the enlarged band $H^\varepsilon_\theta = \{x : \mathrm{dist}(x, H_\theta) \leq \varepsilon/2\}$. Choosing conveniently ε-spaced points $\{\theta_j\}$ in S^{n-2}, we may assume that the sets $H^\varepsilon_{\theta_j} \cap \{|x| = R+\Delta\}$ are pairwise disjoint and that their union covers $\{|x| = R+\Delta\}$. Then their overlapping where $|x| < R+\Delta$ is easily controlled: at distance $R_0 < r < R+\Delta$, we have

$$\sum_j \chi_{H^\varepsilon_{\theta_j}}(x) \lesssim \Big(\frac{\Delta+R}{r}\Big)^{n-2}.$$

Then, the collection of tubes $\{T^\varepsilon_\nu\}$ is divided into disjoint classes by the inclusion relation $T^\varepsilon_\nu \subset H^\varepsilon_{\theta_j}$.

(2) The tubes inside $H^\varepsilon_{\theta_j}$ meet at the tangential inner sphere $\|x\| = R_0 \leq R$, producing a collection of *bush configurations* (that is, sets of tubes meeting at a common point in R_0) such that each tube inside $H^\varepsilon_{\theta_j}$ belong to, at most, two of those bushes. Then their overlapping function is easily controlled by an elementary

calculation ("bush argument"):

$$\sum_{T_\nu^\varepsilon \subset H_{\theta_j}^\varepsilon} \chi_{T_\nu^\varepsilon}(x) \lesssim \frac{\Delta + R}{r},$$

when $R_0 < |x| = r < R + \Delta$.

These two estimates together imply that in the sphere $|x| = r$, $R_0 < r < R+\Delta$, the overlapping is bounded by $(R+\Delta)^{n-1}/r^{n-1}$ uniformly in $\varepsilon > 0$. That is, we get (6.7).

To treat thin shells, we proceed in the same manner, but here the estimate is reduced easily to each plane H_θ. That is, we only have to consider the two-dimensional case, and observe that the ε-rectangles $R_{\theta,j}^\varepsilon$ are arranged in disjoint families of bushes satisfying the overlapping estimate

$$\left|\left\{\sum \chi_{R_{\theta,j}^\varepsilon}(x) \geq d\right\}\right| \lesssim \frac{\Delta R}{d^2},$$

which implies

$$\int_{S^{n-2}} \left|\left\{\sum \chi_{R_{\theta,j}^\varepsilon}(x) \geq d\right\}\right| R^{n-2}\,\mathrm{d}\theta \lesssim \frac{\Delta R^{n-1}}{d^2}. \qquad \square$$

6.6 Miscellanea: one-dimensional integral inequalities

In the following, we shall consider several integral inequalities in one dimension whose proofs, nevertheless, rely upon certain properties of Bessel functions, see Section 6.2, and upon the formula (6.1) for the Fourier transform of radial functions in $\mathbb{R}^n$.

Theorem 6.10. *For every measurable function f on the real line, the following estimate holds*:

$$\left\{\int \left[\int |f(s)|\,|s|^{n/2} \min(|s/t|, |t/s|)^{(n-2)/2}\,\mathrm{d}s\right]^q t^{n+q-1-(qn/2)}\,\mathrm{d}t\right\}^{1/q} \leq C(p,n)\left(\int |f(r)|^p r^{n-1}\,\mathrm{d}r\right)^{1/p},$$

where $C(p,n)$ is a universal (independent of f) finite constant, $n \geq 3$ is an integer, and $1 < p < n/2$ and $1/q = 1/p - 2/n$.

Let us observe that it will be enough to prove the particular case where the function f is non-negative and the integrals are in $[0,\infty]$.

Proof of Theorem 6.10. For $\nu > 0$, we consider the function

$$G_\nu(t) = \int_0^\infty J_\nu(ts) J_\nu(s)\,\frac{\mathrm{d}s}{s},$$

and observe that $G_\nu(t) = G_\nu(1/t)$. It is well known that the integral can be explicitly evaluated and is equal to $\frac{1}{2\nu}\min(t, 1/t)^\nu$.

However, for the sake of completeness, we shall present two arguments to sustain the identity above.

(1) Observe that for $t > 0$, we have

$$\int_0^\infty J_\nu(tr) r^{-is} \frac{\mathrm{d}r}{r} = t^{is} \int_0^\infty J_\nu(r) r^{-is} \frac{\mathrm{d}r}{r} = t^{is} M J_\nu(s),$$

and apply Parseval's theorem for the Mellin transform:

$$\begin{aligned} G_\nu(t) &= \frac{1}{2\pi} \int_{-\infty}^{\infty} e^{is\log t}\, |MJ_\nu(s)|^2 \,\mathrm{d}s = \frac{1}{2\pi} \int_{-\infty}^{\infty} e^{is\log t} \frac{1}{\nu^2 + s^2} \,\mathrm{d}s \\ &= \frac{1}{2\nu} \left[\min\left(t, \frac{1}{t}\right) \right]^\nu. \end{aligned}$$

(2) The continuous function G_ν satisfies the differential equation

$$t^2 \frac{\mathrm{d}^2 y(t)}{\mathrm{d}t^2} + t \frac{\mathrm{d}y(t)}{\mathrm{d}t} - \nu^2 y(t) = 0$$

inside the domain $(0, 1) \cup (1, \infty)$, which is consequence of the differential equation satisfied by Bessel functions together with Hankel's inversion formula.

Since $G_\nu(0) = 0, G_\nu(1) = \int_0^\infty r^{-1}(J_\nu(r))^2 \,\mathrm{d}r = 1/(2\nu)$, by uniqueness it must coincides with the other solution $\frac{1}{2\nu}[\min(t, 1/t)]^\nu$.

Next, with a given f (as in the hypothesis of Theorem 6.10), let us consider $F(x) = f(|x|)$ defined on $\mathbb{R}^n$, and $\widehat{F}(\xi) = g(|\xi|)$, where

$$g(r) = 2\pi\, r^{-(n-2)/2} \int_0^\infty f(s) J_{(n-2)/2}(2\pi r s)\, s^{n/2} \,\mathrm{d}s$$

and apply the Riesz potential ($n \geq 3$):

$$\widehat{I^2 F}(\xi) = |\xi|^{-2} g(|\xi|)$$

Denoting

$$h(|x|) := I^2 F(x),$$

we have that

$$\begin{aligned} h(t) &= 4\pi^2 t^{-(n-2)/2} \int_0^\infty r^{-(n-2)/2-2} \Big(\int_0^\infty f(s) J_{(n-2)/2}(2\pi r s)\, s^{n/2} \,\mathrm{d}s \Big) \\ &\qquad \times J_{(n-2)/2}(2\pi r t)\, r^{n/2} \,\mathrm{d}r \\ &= 4\pi^2 t^{-(n-2)/2} \int_0^\infty f(s)\, s^{n/2} \Big(\int_0^\infty J_{(n-2)/2}(2\pi r s)\, J_{(n-2)/2}(2\pi r t) \frac{\mathrm{d}r}{r} \Big) \,\mathrm{d}s \\ &= \frac{2\pi^2}{\nu} t^{-(n-2)/2} \int_0^\infty f(s)\, s^{n/2} \min\left(\frac{s}{t}, \frac{t}{s}\right)^{(n-2)/2} \mathrm{d}s. \end{aligned}$$

The theorem follows from well-known estimates for the Riesz potentials in $\mathbb{R}^n$, see [12], namely,

$$\|I^2 F\|_q \lesssim \|F\|_p, \quad \text{for } \frac{1}{q} = \frac{1}{p} - \frac{2}{n}, \quad 1 < p < \frac{n}{2}. \qquad \square$$

As was mentioned above, the proof of Theorem 6.10 presented here relays upon estimates of higher dimensional Fourier analysis. However, being a one-dimensional result, it is natural to ask for a strictly one-dimensional proof which, more likely, would shed light over the extensions to non-integer values of the parameter n.

Some examples of Theorem 6.10 follow.

(a) For $n = 3$, $p = 4/3$ and $q = 12$,

$$\Big[\int_0^\infty \Big(\int_0^\infty f(s)\, s^{3/2} \min((s/t, t/s))^{1/2}\, \mathrm{d}s\Big)^{12} t^{-4}\, \mathrm{d}t\Big]^{1/12} \lesssim \Big[\int_0^\infty f(r)^{4/3} r\, \mathrm{d}r\Big]^{3/4}.$$

(b) For $n = 4$, $p = 6/5$ and $q = 3$,

$$\Big[\int_0^\infty \Big(\int_0^\infty f(s)\, s^2 \min(s/t, t/s)\, \mathrm{d}s\Big)^3 \mathrm{d}t\Big]^{1/3} \lesssim \Big[\int_0^\infty f(r)^{6/5} r^3\, \mathrm{d}r\Big]^{5/6}.$$

(c) For $n = 6$, $p = 2$ and $q = 6$,

$$\Big[\int_0^\infty \Big(\int_0^\infty f(s)\, s^3 \min(s/t, t/s)^2\, \mathrm{d}s\Big)^6 t^{-7}\, \mathrm{d}t\Big]^{1/6} \lesssim \Big[\int_0^\infty f(r)^2 r^5\, \mathrm{d}r\Big]^{1/2}.$$

6.7 Fugue: band-limited functions

The method introduced in this chapter to estimate the ball multiplier in mixed norm spaces was motivated by the related work [131, 130] about band-limited functions.

There the convergence of series $\sum_{n=0}^{\infty} \lambda_n b_n(x)$ was investigated, where $\{b_n(x)\}$ are the prolate spheroidal wave functions which had been considered by several authors [143, 137, 138, 142] because of their relevance in communication theory and their role in the formulation of the uncertainty principle.

That is, when one considers in communication theory the impossibility of the simultaneous confinement of a given signal and of its amplitude spectrum.

Given a time interval $[-T, +T]$ and a frequency interval $[-\Omega, +\Omega]$, we may ask for signals $f(t)$ of total energy $\|f\|_2 = 1$ and such that

$$\tau^2 = \int_{-T}^{T} |f(t)|^2\, \mathrm{d}t \quad \text{and} \quad \omega^2 = \int_{-\Omega}^{\Omega} |\widehat{f}(\xi)|^2\, \mathrm{d}\xi$$

are as close to 1 as possible, where $\widehat{f}(\xi) = \int e^{-2\pi i x \xi} f(x)\, \mathrm{d}x$ denotes the Fourier transform.

- $\tau = 1$ means that the signal is confined to the period $|t| \leq T$, or that f is *time-limited*.
- $\omega = 1$ means that its power spectrum is confined to the frequency band $|\xi| \leq \Omega$, or that f is *band-limited*.

A well-known fact about the Fourier transform is that τ and ω cannot both be equal to 1 at the same time. In [137, 138], Landau and Pollak proved that the couple (τ, ω) corresponding to a signal f, $\|f\|_2 = 1$, describes the planar region defined by the inequality

$$\cos^{-1}(\tau) + \cos^{-1}(\omega) \geq \cos^{-1}\sqrt{\lambda_0}, \qquad \text{for } 0 \leq \tau \leq 1, 0 \leq \omega \leq 1,$$

where

$$\lambda_0 = \sup\left\{\int_{-T}^{T} |f(t)|^2\, \mathrm{d}t : f \in L^2(\mathbb{R}), \|f\|_2 = 1, \operatorname{supp}(\widehat{f}) \subset [-\Omega, +\Omega]\right\}$$

(with the proviso that if τ or $\omega = 1$ $(= 0)$, then the other is > 0 (< 1)).

Another relevant result where the functions $b_n(x)$ play an important role is the so-called dimension theorem. Roughly speaking, it says that the set of signals band-limited to the interval $[-\Omega, +\Omega]$ and "concentrated" in $[-T, +T]$ has dimension $4\Omega T$. Landau, Pollak and Slepian give precise statements of this theorem, of clear applications in the theory of signal.

In the proofs of those results, the operator

$$Pf(x) = \int_{-T}^{T} \frac{\sin 2\pi\Omega(x-t)}{\pi(x-t)} f(t)\, \mathrm{d}t$$

appears quite naturally. It is a positive operator, and by the standard theory, it has a countable family of eigenvalues: $1 > \lambda_0 > \lambda_1 > \cdots$, decreasing to 0 . The associated family of normalized eigenfunctions are the b_n's. The functions b_n can be chosen to be real valued, and they constitute a complete orthonormal system in the space of band-limited functions:

$$B_2(\Omega) = \left\{f \in L^2(\mathbb{R}) : \operatorname{supp}(\widehat{f}) \subset [-\Omega, +\Omega]\right\}.$$

Furthermore, the family $\{\lambda^{-1/2} b_n(x)\chi_{[-T,+T]}(x)\}$ is also a complete orthonormal system in the space of time-limited functions, i.e., $L^2[-T, +T]$. Further information of these statements are included in the articles of Landau, Pollak and Slepian mentioned above.

If we normalize P by considering the integration in the interval $[-1, +1]$ and taking $c = 2\pi\Omega T$, the eigenfunctions of the corresponding operator P acting on $L^2[-1, +1]$ are exactly those of the following Sturm–Liouville problem:

(6.8) $$(1 - x^2)\, y'' - 2xy' + (x - c^2x^2)\, y = 0,$$

imposing as boundary conditions that the solutions must be finite or continuous at the extreme points. This equation appears in the study of the three-dimensional wave equation by the method of separation of variables, using a prolate spheroidal coordinate system. This point of view is developed in [136] and [144].

Problem (6.8) only has solutions for a discrete set of real values $\chi_n(c)$, $n = 0, 1, 2, \ldots$ If we denote by $S_n(x)$ the corresponding eigenfunctions, they satisfy the following relations:

$$2c\pi^{-1}\,[R_n(1)]^2\,S_n(x) = \int_{-1}^{1} \frac{\sin c(x-t)}{\pi(x-t)}\,S_n(t)\,\mathrm{d}t$$

and

$$2i^n R_n(1) S_n(x) = \int_{-1}^{1} e^{icxt}\,S_n(t)\,\mathrm{d}t,$$

where $R_n(t)$ is a solution of (6.8), with $\chi = \chi_n(c)$ normalized in such a way that its behavior at infinity is given by $(1/ct)\cos[ct - (n+1)\pi/2]$. Obviously, we can extend $S_n(t)$ as an entire function in the complex plane.

Taking

$$\lambda_n(c) = \lambda_n = 2c\pi^{-1}\,[R_n(1)]^2$$

and

$$b_n(x, c) = b_n(x) = \sqrt{\lambda_n T^{-1}}\,u_n^{-1}\,S_n(xT^{-1}),$$

where

$$u_n^2 = \int_{-1}^{1} [S_n(t)]^2\,\mathrm{d}t,$$

we find another characterization of those functions $b_n(x)$, and their orthogonality relations

$$\int_{-T}^{T} b_n(t)\,b_m(t)\,\mathrm{d}t = \lambda_n \delta_{nm} \quad \text{and} \quad \int_{-\infty}^{\infty} b_n(t)\,b_m(t)\,\mathrm{d}t = \delta_{nm},$$

implying that they constitute also a complete orthogonal system in $L^2([-T, +T])$, which follows easily from the analogous statement for the family of functions $S_n(x)$ on $L^2([-1, +1])$.

We have

$$\widehat{b}_n(x) = i^{-n}\sqrt{T\Omega^{-1}\lambda_n^{-1}}\;b_n\left(T\Omega^{-1}x\right)\chi_{[-\Omega,+\Omega]}(x),$$

and the Plancherel theorem shows that $\{b_n\}$ is a complete orthonormal system in $B_2(\Omega)$, as was stated above.

The prolate spheroidal wave functions are closely related with Bessel functions of order "integer + one half". More concretely, with the spherical Bessel functions

$$j_n(x) = \sqrt{(\pi/2)x^{-1}}\,J_{n+1/2}(x),$$

which belong to the space $L^p(\mathbb{R})$, $p > 1$, and which behavior at infinity is given by $x^{-1}\sin[x - (n\pi)/2]$. They satisfy the orthogonality relation

$$\int_{-\infty}^{\infty} j_n(t)\, j_m(t)\, \mathrm{d}t = \pi(2n+1)^{-1}\delta_{nm},$$

and their Fourier transforms are given by

$$\widehat{j_n}(x) = (-i)^n \pi\, p_n(2\pi x)\, \chi_{[+1,-1]}(2\pi x),$$

where p_n is the Legendre polynomial of order n. This formula, together with Plancherel's theorem, shows that the family of functions $\sqrt{(2n+1)\pi^{-1}}\, j_n(x)$ also constitutes a complete orthonormal system in $B_2(\Omega)$. Problem (6.8) can be considered as a perturbation of the Sturm–Liouville problem for Legendre's equation, and their solutions $S_n(x)$ can be expanded as a series in the normalized Legendre polynomials $q_n(x) = \sqrt{(2n+1)/2}\, p_n(x)$. Therefore, we can express the functions $b_n(x)$ in terms of the spherical Bessel functions $j_n(x)$. More concretely, if

$$S_n(x) = \sum_{[-(n-1)/2] \le k \le \infty} d(k,n)\, q_{n+2k}(x),$$

then

$$b_n(x) = \sqrt{2\Omega}\, u_n^{-1} \sum_{[-(n-1)/2] \le k \le \infty} (-1)^k\, d(k,n)\, \sqrt{2n+4k+1}\, j_{n+2k}(2\pi\Omega x).$$

In this series expansion, the "leading" coefficient is the one corresponding to $k = 0$. In fact, when n is big enough, we have the asymptotic

$$b_n(x) \sim \sqrt{(2n+1)\pi^{-1}}\, j_n(x) \qquad (\Omega = 1/2\pi).$$

In [13], this theory is extended to the case $L^p(\mathbb{R})$, $p \neq 2$, defining the spaces

$$B_p(\Omega) = \left\{ f \in L^p(\mathbb{R}) : \operatorname{supp}(\widehat{f}) \subset [-\Omega, +\Omega] \right\}, \quad 1 \le p \le \infty.$$

They are Banach spaces with the induced topology of $L^p(\mathbb{R})$. If $p < q$, then $B_p(\Omega) \subsetneq B_q(\Omega)$, and the inclusion is continuous and dense. We have that $b_n(x) \in B_p(\Omega)$ for every $p > 1$, and if $\Omega = 1/2\pi$, then $j_n(x) \in B_p(\Omega)$, $p > 1$, for every n.

The dual space $(B_p(\Omega))^*$ is equal to $B_q(\Omega)$, with $1/p + 1/q = 1$, $1 < p < \infty$, in the standard sense, that is, given $T \in (B_p(\Omega))^*$, there exists a unique $g \in B_q(\Omega)$ such that $T(f) = \langle f, g\rangle$ for every $f \in B_p(\Omega)$. Also, we have the norm equivalence: there exists a positive constant $A > 0$ so that $A\|g\|_q \le \|T\|_{B_p(\Omega)^*} \le \|g\|_q$ (the value of the constant A depends on the norm of the Hilbert transform on $L^q(\mathbb{R})$).

Given $f \in B_p(\Omega)$, one can consider its Fourier coefficients

$$\langle f, b_n\rangle = \int f(x)\, b_n(x)\, \mathrm{d}x$$

and its partial sums

$$\sigma_N f(x) = \sum_{0 \le n \le M} \langle f, b_n \rangle \, b_n(x).$$

Similarly, given $f \in B_p(1/2\pi)$, we may consider

$$S_N f(x) = \sum_{0 \le n \le N} (2n+1)\pi^{-1} \langle f, j_n \rangle \, j_n(x).$$

The main result contained in [131] is the following.

Theorem 6.11. (a) $\lim_{N\to\infty} \|\sigma_N f - f\|_{L^p(\mathbb{R})} = 0$ *for every* $f \in B_p(\Omega)$ *if and only if* $4/3 < p < 4$.

(b) $S_N f$ *converges to* f *in* $B_p(1/2\pi)$ *if and only if* $4/3 < p < 4$.

As it happens with ordinary Fourier series, the convergence of partial sums in L^p, $p \neq 2$, is more delicate than the L^2 case. The proof follows, nevertheless, the standard strategy, namely: first show that the linear combinations of $\{b_n\}$ (respectively, $\{j_n\}$) are dense on $B_p(\Omega)$ (respectively, $B_p(1/(2\pi))$), and second, prove that the partial sums operators σ_N (respectively, S_N) are uniformly bounded. The first part is easy, and the heart of the proof lies in the second.

The kernels involved are

$$K_N(x,y) = \sum_{n=0}^{N} b_n(x)\, b_n(y) \quad \text{and} \quad H_N(x,y) = \sum_{n=0}^{N} (2n+1)\, j_n(x)\, j_n(y).$$

Then, after a certain amount of manipulation and decoding, the searched uniform boundedness of those partial sums operators is reduced to that of the operators given by the kernels

$$\frac{\sqrt{x}\, J'_\nu(x)\, J_\nu(y)\, \sqrt{y}}{x-y}$$

uniformly on $\nu \ge 0$.

But these are precisely the kernels involved in the analysis of the ball multiplier on mixed-norm spaces $L^p_{\text{rad}}(L^2_{\text{ang}})$ that were treated in previous sections.

Chapter 7
The Poisson summation formula

The Poisson summation formula changes counting lattice points into counting waves.

Fernando Chamizo

7.1 Introduction

Given a function f in the Schwartz class $\mathcal{S}(\mathbb{R}^n)$ with Fourier transform

$$\widehat{f}(\xi) = \int_{\mathbb{R}^n} e^{-2\pi i x\cdot\xi} f(x)\,\mathrm{d}x,$$

there are two standard ways of producing a periodic function out of f, namely,

$$F(x) = \sum_{\nu\in\mathbb{Z}^n} f(x+\nu) \quad \text{and} \quad G(x) = \sum_{\nu\in\mathbb{Z}^n} \widehat{f}(\nu)\, e^{2\pi i \nu\cdot x}.$$

Poisson's summation formula is just the statement that both are the same: $F \equiv G$. In particular, we have the identity

$$\sum_{\nu\in\mathbb{Z}^n} f(\nu) = \sum_{\nu\in\mathbb{Z}^n} \widehat{f}(\nu), \tag{7.1}$$

which has applications in many different areas ranging from number theory to crystallography.

The Poisson summation formula can be extended to functions beyond the Schwartz class. For example, to the case where they satisfy the estimate $|f(x)| + |\widehat{f}(x)| \lesssim (1+|x|)^{-n-\varepsilon}$, for some $\varepsilon > 0$. This is because in this case both members of the identity

$$\sum_{\nu\in\mathbb{Z}^n} f(\nu) = \sum_{\nu\in\mathbb{Z}^n} \widehat{f}(\nu)$$

are absolutely convergent. In general, if one of the members, say $\sum_\nu f(\nu)$, has a clear meaning (even if f is not in the Schwartz class but can be well approximated by functions $f_\varepsilon \in \mathcal{S}(\mathbb{R}^n)$)), then we obtain also the Poisson identity, where the sum $\sum_\nu \widehat{f}(\nu)$ is defined by $\lim_{\varepsilon\to 0}\sum_\nu \widehat{f_\varepsilon}(u)$.

A. Córdoba, *Suprematism in Harmonic Analysis*, Progress in Mathematics 356,
https://doi.org/10.1007/978-3-031-73737-4_7

If δ_x denotes the Dirac delta function at the point x, then the formula can be rephrased saying that the tempered distribution $\sum_{\nu\in\mathbb{Z}^n}\delta_\nu$ coincides with its own Fourier transform. Such distributions are called *Dirac combs* in crystallography, and they have an interesting interpretation in terms of X-ray diffraction: a periodic structure produces *Bragg peaks* – a periodic structure on the frequency side.

7.2 An inverse result

If $A\colon \mathbb{R}^n \to \mathbb{R}^n$ is a linear map such that $\det(A) = 1$, then an elementary calculation yields

$$\widehat{f\circ A}(\xi) = \widehat{f}((A^{-1})^{\mathsf{T}}\xi).$$

In particular, if we denote $x_\nu = A\nu$ and $y_\nu = (A^{-1})^{\mathsf{T}}\nu$, we get the "generalized" Poisson formula

$$\widehat{\sum_{\nu\in\mathbb{Z}^n}\delta_{x_\nu}} = \sum_{\mu\in\mathbb{Z}^n}\delta_{y_\mu}. \tag{7.2}$$

It is then natural to ask ourselves about the possibility of other such formulas. The answer is the following.

Theorem 7.1. *Suppose that $X = \{x_\nu\}_{\nu\in\mathbb{Z}^n}$ and $Y = \{y_\mu\}_{\mu\in\mathbb{Z}^n}$ are two sets of points in Euclidean space $\mathbb{R}^n$ satisfying that*

$$\min_{\nu\neq\nu'}\|x_\nu - x_{\nu'}\| \geq d \quad \textit{and} \quad \min_{\mu\neq\mu'}\|y_\mu - y_{\mu'}\| \geq d$$

for some positive constant d, and such that

$$\widehat{\sum_{\nu\in\mathbb{Z}^n}\delta_{x_\nu}} = \sum_{\mu\in\mathbb{Z}^n}\delta_{y_\mu}.$$

Then there exists a linear map $A\colon \mathbb{R}^n \to \mathbb{R}^n$ with $\det(A) = 1$ so that

$$x_\nu = A\nu \quad \textit{and} \quad y_\mu = (A^{-1})^{\mathsf{T}}\mu \quad \textit{for all } \nu,\mu\in\mathbb{Z}^n.$$

Proof. Step 1. A set of testing functions.

We shall make use of a continuous non-negative function ϕ that is supported in the cube $Q = \{x = (x^1, x^2, \ldots, x^n) : |x^j| \leq 1\} = [-1,1]^n$, and is such that $\phi(0) = 1$, $\widehat{\phi}(0) > 0$ and $0 \leq \widehat{\phi}(\xi) \leq c\prod_{j=1}^n(1+|\xi^j|)^{-1-\delta}$ for some constants $\delta > 0$ and $C < \infty$. An example of such a function is given by

$$\phi(x^1,\ldots,x^n) = \prod_{j=1}^n(1-|x^j|)_+, \qquad \text{for which} \quad \widehat{\phi}(\xi) = \prod_j\left(\frac{\sin\pi\xi^j}{\pi\xi^j}\right)^2.$$

Next, we shall consider the family

$$\phi_{a,b,\varepsilon}(x) = e^{2\pi i a\cdot x}\, \phi\Big(\frac{x-b}{\varepsilon}\Big),$$

where $a, b \in \mathbb{R}^n$ and ε is a positive real number. We have

$$\widehat{\phi_{a,b,c}}(\xi) = e^{-2\pi i b\cdot \xi}\, \varepsilon^n\, \widehat{\phi}(\varepsilon(\xi - a)).$$

Step 2. $0 \in X \cap Y$.

Given the symmetrical character of Poisson's formula, it is enough to show that $0 \in X$. We will proceed by contradiction.

Let us assume that $0 \notin X$. Then, for every $\varepsilon < d$, we have

$$0 = \sum_{\nu} \phi(\varepsilon^{-1} x_\nu) = \varepsilon^n \sum_{\nu\in\mathbb{Z}^n} \prod_{l=1}^{n} \Big(\frac{\sin \pi\varepsilon y_\nu^l}{\pi\varepsilon y_\nu^l}\Big)^2.$$

But this is not possible since $(\frac{\sin \pi t}{\pi t})^2$ vanishes only in the discrete set $t \in \mathbb{Z} - \{0\}$. Therefore, without loss of generality, we may assume that $x_0 = y_0 = 0 \in \mathbb{Z}^n$.

Step 3. For all $\nu, \mu \in \mathbb{Z}^n$,

$$x_\nu \cdot y_\mu = \sum_{j=1}^{n} x_\nu^j\, y_\mu^j \in \mathbb{Z}.$$

Let us take $b = x_\nu$, $a \in \mathbb{R}^n$ arbitrary, and $0 < \varepsilon < \delta$. Since x_ν is the only element of X in the support of $\phi(\varepsilon^{-1}(x - x_\nu))$, we have

$$e^{2\pi i a\cdot x_\nu} = \sum_{\mu\in\mathbb{Z}^n} e^{-2\pi i (y_\mu - a)\cdot x_\nu}\, \varepsilon^n\, \widehat{\phi}(\varepsilon^{-1}(y_\mu - a)).$$

That is to say

$$1 = \varepsilon^n \sum_{\mu\in\mathbb{Z}^n} \widehat{\phi}(\varepsilon^{-1}(y_\mu - a))\, e^{-2\pi i\, x_\nu\cdot y_\mu} \tag{7.3}$$

for every $x_\nu \in X$ and $a \in \mathbb{R}^n$. In particular, taking $x_0 = 0$, we get also

$$1 = \varepsilon^n \sum_{\mu\in\mathbb{Z}^n} \widehat{\phi}(\varepsilon^{-1}(y_\mu - a)).$$

But since $\widehat{\phi} \geq 0$, $\widehat{\phi}(0) = 1$, and $\widehat{\phi}$ vanishes only on a countable set, the identity (7.3), valid for every $\varepsilon > 0$ and $a \in \mathbb{R}^n$, implies that $x_\nu \cdot y_\mu \in \mathbb{Z}$.

Step 4. Neither of the sets X, Y can be contained on a hyperplane of $\mathbb{R}^n$.

Again, it is enough to prove this for the set X, and we shall argue by contradiction. Assume not: then, after an appropriate rotation A, we may suppose that X is contained in the hyperplane $\{x^n = 0\}$. Let f_1 and f_2 be arbitrary functions in the Schwartz class, $f_1 \in \mathcal{S}(\mathbb{R})$ and $f_2 \in \mathcal{S}(\mathbb{R}^{n-1})$, so that

$$\begin{aligned}\sum_{\nu\in\mathbb{Z}^n} f_1(x_\nu^n)\, f_2(x_\nu^1,\dots,x_\nu^{n-1}) &= \sum_{\nu\in\mathbb{Z}^n} f_1(0)\, f_2(x_\nu^1,\dots,x_\nu^{n-1})\\ &= \sum_{\mu\in\mathbb{Z}^n} \widehat{f_1}(y_\mu^n)\, \widehat{f_2}(y_\mu^1,\dots,y_\mu^{n-1}),\end{aligned}$$

which is impossible by the following argument: let $\chi \in C^\infty(\mathbb{R})$ be such that $\chi(t) = 0$ if $|t| \le 1$ and $\chi(t) = 1$ if $|t| \ge 2$, and take

$$f_1^\varepsilon(t) = f_1(t)\,\chi(\varepsilon^{-1}t), \quad f_1^\varepsilon(0) = 0, \qquad \text{for every } \varepsilon > 0.$$

Then

$$0 = \lim_{\varepsilon\to 0} \sum_{\nu\in\mathbb{Z}^n} f_1^\varepsilon(0)\, f_2(x_\nu^1,\dots,x_\nu^{n-1}) = \lim_{\varepsilon\to 0} \sum_{\mu\in\mathbb{Z}^n} \widehat{f_1^\varepsilon}(y_\mu^n)\, \widehat{f_2}(y_\mu^1,\dots,y_\mu^{n-1}).$$

That is to say

$$0 = \sum_{\mu\in\mathbb{Z}^n} \widehat{f_1}(y_\mu^n)\, \widehat{f_2}(y_\mu^1,\dots,y_\mu^{n-1})$$

for every $f_1 \in \mathcal{S}(\mathbb{R})$, $f_2 \in \mathcal{S}(\mathbb{R}^{n-1})$, which is absurd.

Step 5. End of the proof.

Let us take n linearly independent points $x_{\nu_1},\dots,x_{\nu_n}$ in X. Then the relations $y_\mu \cdot x_{\nu_j} \in \mathbb{Z}$, $j = 1,2,\dots,n$, imply that the set Y is contained in the dual lattice of the one generated by the vectors $x_{\nu_1},\dots,x_{\nu_n}$. Similarly, the set X is contained in a certain lattice in $\mathbb{R}^n$.

Next, let us consider all systems of n-linearly independent vectors of X and then choose one of them so that the absolute value of its determinant takes its minimum value. Let $\{x_{\nu_1},\dots,x_{\nu_n}\}$ be such a system for which, without loss of generality, we may assume that

$$\Delta = \det\begin{pmatrix} x_{\nu_1}^1 & \cdots & x_{\nu_1}^n \\ \vdots & \cdots & \vdots \\ x_{\nu_n}^1 & \cdots & x_{\nu_n}^n \end{pmatrix} > 0.$$

Because of the minimality of Δ, 0 is the only element of X contained in the parallelepiped $\{x = \sum_{j=1}^n s_j x_{\nu_j} : 0 \le |s_j| < 1, j = 1,\dots,n\}$, and there exists a linear mapping A satisfying

$$\det A = 1, \qquad \begin{cases} Ax_{\nu_1} = te_1 = t(1,0,\dots,0), \\ \qquad\vdots \\ Ax_{\nu_n} = te_n = t(0,0,\dots,1), \end{cases} \qquad \text{with } t = \Delta^{1/n},$$

and 0 is the only element of $A(X) = \{Ax_\nu\}$ inside the cube $Q_t = \{x = (x^1,\dots,x^n) : |x^j| \le t\}$.

Let us introduce the notation $\bar{x}_\nu = Ax_\nu$ and $\bar{y}_\mu = (A^{-1})^\mathsf{T} y_\mu$. Then, using again the function $\phi(u) = \prod_{j=1}^n (1 - |u^j|)_+$, we can write

$$1 = \sum_{\nu \in \mathbb{Z}^n} \phi(t^{-1}\bar{x}_\nu) = t^n \sum_{\mu \in \mathbb{Z}^n} \widehat{\phi}(t\bar{y}_\mu) = t^n \Big(1 + \sum_{\mu \neq 0} \widehat{\phi}(t\bar{y}_\mu)\Big),$$

implying that $0 < t \leq 1$, since we know the positivity $\widehat{\phi} \geq 0$. Furthermore, we have $\bar{x}_\nu \cdot \bar{y}_\mu \in \mathbb{Z}$ for every ν and μ. Therefore, $te_j \cdot \bar{y}_\mu \in \mathbb{Z}$ for every $j = 1, \ldots, n$, allowing us to conclude the identification $\bar{y}_\mu \in t^{-1}\mathbb{Z}^n$.

Next, let us use the identity

$$1 = t^n \sum_{\mu \in \mathbb{Z}^n} \widehat{\phi}(t(\bar{y}_\mu - a)),$$

and integrate it over a in the cube $Q = [-1/2, 1/2]^n$. We obtain

$$\begin{aligned}
1 &= \sum_{\mu \in \mathbb{Z}^n} t^n \int_{-1/2}^{1/2} \cdots \int_{-1/2}^{1/2} \widehat{\phi}(t(\bar{y}_\mu - a))\, da^1 \ldots da^n \\
&= \sum_{\mu \in \mathbb{Z}^n} \int_{t((\bar{y}_\mu)^1 - 1/2)}^{t((\bar{y}_\mu)^1 + 1/2)} \cdots \int_{t((\bar{y}_\mu)^n - 1/2)}^{t((\bar{y}_\mu)^n + 1/2)} \widehat{\phi}(u^1, \ldots, u^n)\, du^1 \ldots du^n \\
&= \sum_{\mu \in \mathbb{Z}^n} \int_{\mu^1 - t/2}^{\mu^1 + t/2} \cdots \int_{\mu^n - t/2}^{\mu^n + t/2} \widehat{\phi}(u^1, \ldots, u^n)\, du^1 \ldots du^n = \sum_{\mu \in \mathbb{Z}^n} \int_{Q_\mu(t)} \widehat{\phi}(u)\, du,
\end{aligned}$$

where $Q_\mu(t) = [\mu^1 - t/2, \mu^1 + t/2) \times \cdots \times [\mu^n - t/2, \mu^n + t/2)$.

To finish, observe that the cubes $Q_\mu(t)$ are pairwise disjoint and they cover completely $\mathbb{R}^n$ only in the case $t = 1$. Since $1 = \phi(0) = \int_{\mathbb{R}^n} \widehat{\phi}(u)\, du$, $\widehat{\phi} \geq 0$, and the set $\{\widehat{\phi}(x) = 0\}$ has measure zero, the only possibility left is precisely $t = 1$ and $X = \mathbb{Z}^n$. □

The origin of this proof goes back to 1986, when I attended a talk concerning quasicrystals at the ETH in Zurich. It struck me that crystallographers were taking a version of Theorem 7.1 for granted: that Bragg peaks, on the frequency side, implied periodicity on the spatial side. Consequently, I asked myself if a direct proof could be given of such a fundamental statement.

7.3 Extension to more general coefficients

It also seems natural to ask what happens if our distribution gives different masses to different deltas. Can these distributions be characterized as finite superpositions of periodic structures?

The following theorem gives a positive solution to this question, but its proof needs some powerful tools of the harmonic analysis of locally compact abelian groups, namely the characterization obtained by P. Cohen for idempotent measures on such a group G: $\mu * \mu = \mu$ if and only if $\operatorname{supp}(\widehat{\mu})$ belongs to the coset ring of the dual group $\hat{G}$ (i.e., the smallest ring of subsets of $\hat{G}$ which contains all open cosets).

Let the set $\Lambda = \bigcup_{j=1}^{N} \Lambda_j$ be given as a finite disjoint union of subsets of $\mathbb{R}^n$ and such that there exists a positive lower bound for the distances between any two distinct points in Λ. We shall describe this situation by saying that the set Λ is strongly discrete.

Given different complex numbers $\{a_j\}_{j=1,\dots,N}$, we may consider the tempered distribution

$$\mu = \sum_{j=1}^{N} a_j \sum_{x \in \Lambda_j} \delta_x .$$

Let us assume that its Fourier transform $\widehat{\mu}$ can also be expressed in the form

$$\widehat{\mu} = \sum_{\alpha} b_\alpha \, \delta_{y_\alpha},$$

and that it satisfies that, for a certain constant $C > 0$,

$$\sum_{y_\alpha \in Q} |b_\alpha| \le C < \infty, \qquad \text{for every unit cube } Q \subset \mathbb{R}^n. \tag{7.4}$$

Then we have the following theorem.

Theorem 7.2. *Under the hypothesis above, each set Λ_j is a finite disjoint union of n-dimensional lattices.*

Prior to the presentation of the proof, we shall consider various auxiliary lemmas.

Lemma 7.3. *Let $\{x_j\} \subset \mathbb{R}^n$ be a discrete subset of $\mathbb{R}^n$ whose complement contains arbitrarily large balls, and suppose that $\mu = \sum_j a_j \delta_{x_j}$ is a tempered distribution whose Fourier transform $\widehat{\mu}$ can be expressed in the form $\widehat{\mu} = \sum_k b_k \delta_{y_k}$ and satisfies condition* (7.4). *Then necessarily $\mu \equiv 0$.*

Proof. The hypothesis implies that for every positive number t, one can find a ball $B(z_t; t)$ disjoint with the support set $\{x_j\}$. We choose a smooth function ψ satisfying $\operatorname{supp}(\psi) \subset B(0; 1)$ and $\widehat{\psi}(0) = 1$, and we consider, for a fixed y_k, the function

$$\phi(x) = \exp(2\pi i \, y_k \cdot x) \, \psi\Big(\frac{x - z_t}{t}\Big).$$

We have

$$0 = \mu(\Phi) = \widehat{\mu}(\widehat{\phi}) = t^n \sum_{\alpha} b_\alpha \widehat{\psi}(t(y_\alpha - y_k)) \exp(2\pi i \, z_t \cdot y_\alpha),$$

which yields

$$b_k = -\sum_{\alpha \neq k} b_\alpha \widehat{\psi}(t(y_\alpha - y_k)) \exp(2\pi i\, z_t \cdot (y_\alpha - y_k)).$$

In particular,

$$|b_k| \leq \sum_{\alpha \neq k} |b_\alpha| \left|\widehat{\psi}(t(y_\alpha - y_k))\right|.$$

Taking the limit $t \to \infty$, we get $b_k = 0$ as a consequence of property (7.4). □

Before presenting our second lemma, let us recall the fact that the discrete subgroups of $\mathbb{R}^n$ are lattices, and that a subgroup H of $\mathbb{R}^n$ is not discrete if and only if each neighborhood of 0 in $\mathbb{R}^n$ contains infinitely many elements of H.

Lemma 7.4. *Let $A = (x+H) - (x_1+H_1) - \cdots - (x_r+H_r) \neq \emptyset$ be a strongly discrete set, where $H, H_1, \ldots, H_r$ are subgroups of $\mathbb{R}^n$. Then H is a discrete group.*

Proof. By induction on r.

Case $r = 1$: $A = (x+H) - (x_1+H_1) \neq \emptyset$.

If $(x+H) \cap (x_1+H_1) = \emptyset$, then $A = x + H$, and the result is immediate.

If $(x+H) \cap (x_1+H_1) \neq \emptyset$, then there exist elements $h \in H$ and $h_1 \in H_1$ such that $x + h = x_1 + h_1$, and we have

$$(x+H) \cap (x_1+H_1) = x + h + H \cap H_1,$$

that is,

$$A = x + [H - (h + H \cap H_1)].$$

Since H is a disjoint union of cosets of $H \cap H_1$, it is clear that A discrete implies $H \cap H_1$ discrete.

Suppose that H is not discrete. Then each open ball $B(0;r)$ centered at the origin in $\mathbb{R}^n$ must contain an infinite number of points in H. Since $h + H \cap H_1$ is discrete, $B(0;r) \cap (h + H \cap H_1)$ is a finite set and, therefore, $A - x$ contains an infinity of points in such a ball $B(0;r)$. This is contrary to the hypothesis that A is a strongly discrete set. (The remainder of the induction argument is left to the reader.) □

Lemma 7.5. *Suppose that $\{e_1, \ldots, e_m\}$ are linearly independent vectors in $\mathbb{R}^n$. Then for every $r > 0$, the set*

$$D_r = \{v = k_1 e_1 + \cdots + k_m e_m : d(v, \mathbb{Z}^n) < r, k_j \in \mathbb{Z}, j = 1, \ldots, m\}$$

cannot be contained in a finite union of $(m-1)$-dimensional planes.

Proof. By induction in m and using the theorem of H. Weyl about the uniform distribution modulo 1 of $(k\theta_1, \ldots, k\theta_s)$, if the numbers $\{1, \theta_1, \ldots, \theta_s\}$ are linearly independent over the rationals. Observe that the conclusion of the lemma follows easily if all the vectors e_j have rational coordinates. The details are left to the reader. □

Proof of Theorem 7.2. Let ψ by chosen as in the proof of Lemma 7.3 and consider, for each positive integer M, the measure

$$\nu_M = M^{-n}\psi\left(\frac{\cdot}{M}\right)\cdot\widehat{\mu} = M^{-n}\sum_{\alpha} b_\alpha\,\psi\left(\frac{y_\alpha}{M}\right)\delta_{y_\alpha}.$$

Then condition (7.4) implies that for each M, ν_M is a measure of finite total variation in $\mathbb{R}^n$, uniformly on M: $\|\nu_M\| \leq c < \infty$.

The measures ν_M have a natural extension as finite measures in $\overline{\mathbb{R}}^n$, the Bohr compactification of $\mathbb{R}^n$, with uniformly bounded total variation. Therefore, by a standard weak-star compactness argument, there is a subsequence, which we shall also denote by $\{\nu_M\}$, converging in the weak star topology to a finite measure $\bar{\nu}$.

We have

$$\begin{aligned}\widehat{\bar{\nu}}(\xi) &= \lim_{M\to\infty}\widehat{\nu}_M(\xi) = \lim_{M}\widehat{\left(M^{-n}\psi\left(\frac{\cdot}{M}\right)\cdot\widehat{\mu}\right)}(\xi) = \lim_{M\to\infty}\widehat{\psi}(M\cdot)*\mu(\xi)\\ &= \lim_{M\to\infty}\sum_{j=1}^{N} a_j\sum_{x\in\Lambda_j}\widehat{\psi}(M(\xi-x)).\end{aligned}$$

Therefore,

$$\widehat{\bar{\nu}}(\xi) = \begin{cases} a_j, & \text{if } \xi\in\Lambda_j,\\ 0, & \text{if } \xi\notin\Lambda.\end{cases}$$

Our next step is to use this fact to show that for each set Λ_j, we can find a finite measure μ_j on the Bohr group $\overline{\mathbb{R}}^n$ satisfying that

$$\widehat{\mu}_j(\xi) = 1 \quad \text{if } \xi\in\Lambda_j, \qquad \text{and} \qquad \widehat{\mu}_j(\xi) = 0 \quad \text{if } \xi\notin\Lambda_j.$$

We prove this by an induction argument on the number of sets Λ_j. If $N = 1$, then the result is obvious and we can take $\mu_1 = (1/a_1)\bar{\nu}$. Suppose then that the result holds for $N-1$. Consider the measure $\bar{\nu}_k = \bar{\nu}*\bar{\nu} - a_k\,\bar{\nu}$. Then we have

$$\widehat{\bar{\nu}}_k(\xi) = \widehat{\bar{\nu}}(\xi)^2 - a_k\,\widehat{\bar{\nu}}(\xi) = \begin{cases} a_j^2 - a_k a_j, & \text{if } \xi\in\Lambda_j,\ j\neq k,\\ 0, & \text{if } \xi\notin\Lambda-\Lambda_k.\end{cases}$$

It is easy to see that there exists k such that there exists j with the property that

$$a_j^2 - a_k a_j \neq a_m^2 - a_k a_m \qquad \text{for } m\neq j.$$

By the induction hypothesis, we can find μ_j with the required properties with respect to Λ_j. Then the measure $\lambda = \bar{\nu} - a_j\mu_j$ satisfies that its Fourier transform is given by

$$\widehat{\lambda}(\xi) = \begin{cases} a_k - a_j, & \text{if } \xi\in\Lambda_k,\ k\neq j,\\ 0, & \text{if } \xi\notin\Lambda-\Lambda_j,\end{cases}$$

and the values $a_k - a_j$, $k = 1, \ldots, j-1, j+1, \ldots, N$ are all different. We can use the induction hypothesis again to claim the existence of the required measures μ_k, $k = 1, \ldots, N$.

Therefore, we have conditions to apply the theorem of Paul Cohen on idempotent measures, to conclude that each Λ_j belongs to the coset ring of the group $\mathbb{R}^n$ provided with the discrete topology, i.e., as a dual group to $\overline{\mathbb{R}}^n$.

That is, each Λ_j is a finite union of sets of the form

$$A = (x + H) - (x_1 + H_1) - \cdots - (x_r + H_r) \neq \emptyset,$$

and by Lemma 7.4, we know that H is a discrete subgroup of $\mathbb{R}^n$ (with respect to the ordinary topology), that is, a lattice. Furthermore, there exist elements $h_j \in H_j$ such that

$$A = x + \{H - (h_1 + H \cap H_1) - \cdots - (h_r + H \cap H_r)\}.$$

We can write such a set in the form

$$A = x + \{(y_1 + K_1) \cup \cdots \cup (y_s + K_s)\} - A_1,$$

where $y_j - K_j$ are cosets of discrete subgroups K_j of $\mathbb{R}^n$ and A_1 is contained in a finite union of hyperplanes.

To see this, we merely observe that, in $\mathbb{R}^n$, if G_1 is a subgroup of the discrete group G_2 and both have equal dimensions (as lattices in $\mathbb{R}^n$), then the index $[G_2 : G_1]$ must be finite, i.e., G_2 is a finite union of cosets of G_1, and therefore given $g \in G_2$, $G_2 - (g + G_1)$ is a finite union of cosets of G_1.

The identity

$$(X_1 - Y_1) \cup (X_2 - Y_2) = X_1 \cup X_2 - \{(X_2^c \cap X_1 \cap Y_1) \cup (X_1^c \cap X_2 \cap Y_2) \cup (Y_1 \cap Y_2)\},$$

valid for general sets X_1, X_2, Y_1, Y_2, yields the following: each set Λ_j can be written in the form

$$\Lambda_j = (x_1 + H_1) \cup \cdots \cup (x_s + H_s) \cup B_1 - B_2,$$

where the H_j are n-dimensional discrete subgroups of $\mathbb{R}^n$ and the discrete sets B_1 and B_2 are contained in a finite union of hyperplanes.

Next, we change the last expression in order to have disjoint lattices.

Suppose that $(x_j + H_j) \cap (x_k + H_k) \neq \emptyset$. Then there exist elements $h_j \in H_j$ and $h_k \in H_k$ such that $x_j + h_j = x_k + h_k$, that is,

$$(x_j + H_j) \cap (x_k + H_k) = x_j + h_j + (H_j \cap H_k),$$
$$(x_j + H_j) \cup (x_k + H_k) = x_j + h_j + (H_j \cup H_k).$$

The hypothesis that Λ_j is strongly discrete implies that $H_j \cap H_k$ has dimension n. If not, after an application of a convenient linear mapping A, we can assume that $H_j = \mathbb{Z}^n$ and

$$H_k = \{m_1 e_1 + \cdots + m_n e_n : m_j \in \mathbb{Z}\},$$

where $e_j = (\theta_1^j, \dots, \theta_n^j)$ and, at least, one of the θ_i^j is not rational. But then an application of Lemma 7.5 yields that the set

$$\{x \in H_k : \exists y \in H_j, \quad 0 < d(x, y) < r\}$$

cannot be contained in a finite union of hyperplanes, and therefore, $H_j \cup H_k - B$ cannot be strongly discrete for any set B contained on such a union. This is a contradiction with our hypothesis about Λ_j.

But if $H_j \cap H_k \neq \emptyset$ has dimension n, then $H_j \cup H_k$ is a finite disjoint union of cosets of $H_j \cap H_k$. A finite number of applications of this procedure allows us to write each set Λ_j in the form

$$\Lambda_j = (x_1^j + H_1^j) \cup \cdots \cup (x_{s(j)}^j + H_{s(j)}^j) \cup B_1^j - B_2^j,$$

where the n-dimensional lattices $x_k^j + H_k^j$ are disjoint, and the discrete sets B_1^j and B_2^j are contained in a finite union of hyperplanes.

With the family of sets

$$\Lambda_j^* = (x_1^j + H_1^j) \cup \cdots \cup (x_{s(j)}^j + H_{s(j)}^j),$$

we define the following measure:

$$\mu_1 = \sum_{j=1}^{n} a_j \sum_{x \in \Lambda_j^*} \delta_x$$

Then the Poisson summation formula implies that $\widehat{\mu}_1$ is a measure satisfying condition (7.4). In that case, $\mu - \mu_1$ is also a measure whose Fourier transform satisfies (7.4), but whose support is not relatively compact, being contained in a finite union of hyperplanes, and by Lemma 7.3 we must have $\mu = \mu_1$, which implies that $\Lambda_j^* = \Lambda_j$, for each $j = 1, \dots, N$. □

7.4 An application: the phase problem for Gaussian sums

A relevant and well known application of the formula is the evaluation of Gaussian sums, which are of capital importance in several chapters of number theory, such as the quadratic reciprocity law.

Theorem 7.6. *For* $N \geq 1$,

$$G(N) = \sum_{n=0}^{N-1} e^{2\pi i n^2/N} = \frac{1}{2}\sqrt{N}\,(1+i)\,(1+e^{-\pi i N/2}),$$

that is,

$$G(N) = \begin{cases} \sqrt{N}, & \textit{if } N \equiv 1 \bmod (4), \\ 0, & \textit{if } N \equiv 2 \bmod (4), \\ i\sqrt{N}, & \textit{if } N \equiv 3 \bmod (4), \\ (1+i)\sqrt{N}, & \textit{if } N \equiv 0 \bmod (4). \end{cases}$$

Proof. We apply the Poisson summation formula to the function

$$f(x) = e^{2\pi i\, x^2/N}\, \chi_{[0,N]}(x).$$

Then

$$\widehat{f}(k) = \int_0^N e^{-2\pi i\, kx}\, e^{2\pi i\, x^2/N}\, dx = e^{-\pi i\, k^2 N/2} \sqrt{N} \int_{-k\sqrt{N}/2}^{-k\sqrt{N}/2+\sqrt{N}} e^{2\pi i\, u^2}\, du,$$

and the sum over even frequencies $k = 2m$ is equal to

$$\sum_m \widehat{f}(2m) = \lim_{M\to\infty} \sqrt{N} \int_{-M}^{M+\sqrt{N}} e^{2\pi i\, u^2}\, du = \sqrt{N}\, I,$$

while the sum over odd frequencies $k = 2m+1$ yields

$$\sum_m \widehat{f}(2m+1) = e^{-\pi i\, N/2} \sqrt{N}\, I.$$

To finish, let us observe that taking $N = 1$, the identity above gives

$$1 = (1 + e^{-\pi i/2})\, I,$$

that is, $I = \frac{1}{2}(1+i)$. □

In the case that $N = p$ is a prime number, we have the Legendre symbol $(\frac{n}{p})$, defined to be equal to $+1$ if n is a quadratic residue mod p, -1 if n is a quadratic non-residue, and 0 if $p \mid n$, giving the identity

$$\sum_{n=0}^{N-1} \left(\frac{n}{p}\right) e^{2\pi i\, n/p} = \sum_{n=0}^{N-1} e^{2\pi i\, n^2/p},$$

which will be useful in the following discussion of what we may call *the phase problem for Gaussian molecules.*

The spatial configurations of crystallized molecules are usually obtained via X-ray diffraction data. As was first suggested by M. von Laue, when the intensities of the diffracted rays are registered on a flat screen, high peaks appear in a discrete set, revealing the symmetries of the crystal. The standard interpretation assigns diffraction peak intensities to absolute values of the Fourier transform $\widehat{\rho}$ of the electron density ρ. The phase problem asks for the reconstruction of ρ from the knowledge of $|\widehat{\rho}|$. In certain interesting cases, this leads naturally to problems of factorization in suitable rings of polynomials (see [170]). For example, if we have a density $\rho = \sum \delta_{n_j}$, where δ_{n_j} denotes a Dirac delta function placed at the integer n_j, then $|\widehat{\rho}|$ determines ρ (modulo translations or reflections) if the polynomial $\sum x^{n_j}$ is irreducible in $\mathbb{Z}[x]$. This leads to the study of irreducible polynomials with coefficients 0 or 1. In [168], the conjecture that most of these polynomials are irreducible is stated, and some other related results are quoted. On the other hand, in general, if the polynomial $\sum x^{n_j}$ is not irreducible, there is a lack of uniqueness, showing that, in that generality, the phase problem is not well posed (the first practical example of non-uniqueness was considered in 1930 by Pauling and Shappell [169], who were studying crystals of bixbyite). A rather interesting question is which kind of "chemical", "geometric", or "arithmetic" information about ρ is relevant to ensure the reconstruction (see [160, 170]).

A plausible model for the electronic density of one-dimensional (periodic) crystals is given by infinite sums of Dirac delta functions (cf. [158]):

$$\rho = \sum_{j=1}^{N} b_j \sum_{n=-\infty}^{\infty} \delta_{x_j+n},$$

where $b_j \in \mathbb{Z}^+$are positive integers and $0 \leq x_j < 1$.

In this context, the phase problem seeks to locate the positions $\{x_j\}$ (modulo translations or reflections $x'_j = 1 - x_j$) knowing the absolute values

$$F(\nu) = \Big|\sum_{j=1}^{N} b_j\, e^{2\pi i x_j \nu}\Big|, \qquad \nu \in \mathbb{Z}.$$

The result presented below consists of a new observation about Gaussian sums, i.e., roughly speaking, they are determined by their absolute value among a class of exponential sums.

Notation. We shall write $e(x)$ as an abbreviation of $e^{2\pi i x}$.

Theorem 7.7. *Let $0 = x_1 < x_2 < \cdots < x_N < 1$ be real numbers and assume that there exists a prime number p such that the sum*

$$S(m) = \sum_{j=1}^{N} e(m x_j)$$

is of constant modulus $|S(m)| = \Gamma$ if p is not a divisor of m, and $|S(m)| = N$ otherwise. Then $p x_j \in \mathbb{Z}$ for $1 \leq j \leq N$, and either

$$S(m) = A T(m) + B\, e\Big(\frac{mk}{p}\Big) G(m) \quad \text{or} \quad S(m) = A T(m) + B\, e\Big(\frac{mk}{p}\Big),$$

where $A, B, k \in \mathbb{Z}$ and

$$T(m) = \sum_{n=0}^{p-1} e\left(\frac{mn}{p}\right) \quad \textit{and} \quad G(m) = \sum_{n=1}^{p-1} \left(\frac{n}{p}\right) e\left(\frac{mn}{p}\right).$$

The proof will be based on the following lemma.

Lemma 7.8. *If all the algebraic conjugates of $x \in \mathbb{Q}(\zeta)$, $\zeta = e(1/p)$, are complex numbers of equal modulus, and their arguments differ in algebraic multiples of π, then either*

$$x = B\zeta^k \sum_{n=1}^{p-1} \left(\frac{n}{p}\right) \zeta^n \quad \textit{or} \quad x = B\zeta^k$$

for some $B \in \mathbb{Q}$ and $k \in \mathbb{Z}$.

Proof. Let σ be a generator of the Galois group of the extension $\mathbb{Q}(\zeta)/\mathbb{Q}$. Using the hypothesis of the lemma, we can write $\sigma(x)/x = e(\alpha)$, for some $\alpha \in \mathbb{Q}$ (if $\alpha \notin \mathbb{Q}$, then it is well known that $e(\alpha)$ is not an algebraic number), i.e., $\sigma(x)/x = \zeta_b^a$, where $\zeta_b = e(1/b)$ for $a, b \in \mathbb{Z}^+$, $(a, b) = 1$.

Taking a^* such that $a^*a \equiv 1 \bmod (b)$, we get that $\zeta_b = (\zeta_b^a)^{a^*} \in \mathbb{Q}(\zeta)$. We have two cases

(i) If $p \mid b$, then $[\mathbb{Q}(\zeta) : \mathbb{Q}(\zeta_b)] = \phi(p)/\phi(b)$ yields $b = p$ or $b = 2p$.
(ii) If $p \nmid b$, then $\mathbb{Q}(\zeta) = \mathbb{Q}(\zeta, \zeta_b) = \mathbb{Q}(\zeta_{pb})$ yields $pb = p$ or $pb = 2p$.

Therefore we have that $b = 1, 2, p, 2p$ and $\zeta_b^a = \pm\zeta^l$ for some integer l, $0 \le l \le p-1$.

Let us assume that $\sigma(\zeta) = \zeta^g$, and take k such that $(g-1)k \equiv l \bmod p$, then since $\sigma(x)/x = \pm\zeta^l$, we get

$$\frac{\sigma(\zeta^{-k}x)}{\zeta^{-k}x} = \pm 1 \quad \text{and} \quad \frac{\sigma^2(\zeta^{-k}x)}{\sigma(\zeta^{-k}x)} = \pm 1.$$

Therefore $\sigma^2(\zeta^{-k}x) = \zeta^{-k}x$.

The subfield invariant under σ^2 is

$$M = \left\{a(\sigma^2(\zeta)+\sigma^4(\zeta)+\cdots+\sigma^{p-1}(\zeta))+b(\sigma(\zeta)+\sigma^3(\zeta)+\cdots+\sigma^{p-2}(\zeta)); a, b \in \mathbb{Q}\right\},$$

hence

$$\zeta^{-k}x = a\sum_{n\in\mathcal{R}} \zeta^n + b\sum_{n\in\mathcal{N}} \zeta^n, \quad a, b \in \mathbb{Q},$$

where $\mathcal{R}$ and $\mathcal{N}$ denote, respectively, the set of quadratic and non-quadratic residues mod p.

If $\sigma(\zeta^{-k}x) = \zeta^{-k}x$, then $\zeta^{-k}x \in \mathbb{Q}$. If $\sigma(\zeta^{-k}x) = -\zeta^{-k}x$, then we have $b = -a$ and that $\zeta^{-k}x$ is a rational multiple of a Gauss sum. □

Proof of the Theorem 7.7. The identity $|S(p)| = N$ implies $e(px_1) = e(px_2) = \cdots = e(px_N)$. Since we have fixed $x_1 = 0$, then we must have $x_r = n_r/p$ for some integers n_r, $0 \le n_r < p$. Therefore $x = S(1)$ is in the hypothesis of Lemma 7.8, and we get

$$\text{either} \quad S(1) = B\, e\Big(\frac{k}{p}\Big) G(1) \quad \text{or} \quad S(1) = B\, e\Big(\frac{k}{p}\Big).$$

For m prime with p, we obtain by conjugation in $\mathbb{Q}(\zeta)$ either

$$S(m) = B\, e\Big(\frac{mk}{p}\Big) G(m) \quad \text{or} \quad S(m) = B\, e\Big(\frac{mk}{p}\Big).$$

Finally, let us observe that $T(m)$ vanishes if and only if $p \nmid m$. Therefore there exists $A \in \mathbb{Q}$ such that either

$$S(m) = A\,T(m) + B\, e\Big(\frac{mk}{p}\Big) G(m) \quad \text{or} \quad S(m) = A\,T(m) + B\, e\Big(\frac{mk}{p}\Big),$$

for every $m \in \mathbb{Z}$. Identifying coefficients, we deduce that A and B are integers. □

7.5 Counting lattice points (a quick overview)

To count how many elements of the lattice $\mathbb{Z}^n$ are inside a given region is a problem appearing in many interesting contexts. A good example is provided by the arithmetical function

$$r_n(k) = \#\{v \in \mathbb{Z}^n : |v|^2 = k\},$$

i.e., $r_n(k)$ counts the number of lattice points in the surface of the sphere of radius $\sqrt{k}$ or, equivalently, the number of representations of the positive integer k as a sum of n squares.

When $n = 2$ or 3, the function $r_n(k)$ shows a rather irregular behavior and its value depends upon the prime factorization of the integer k. But for $n = 4$, we have the "more regular" formula obtained by Jacobi:

$$r_4(k) = 8 \sum_{m|k} m \quad \text{if } k \text{ is odd}, \quad \text{and} \quad r_4(k) = 24 \sum_{m|k,\, m \text{ odd}} m \quad \text{if } k \text{ is even}.$$

For $n \geqslant 5$, Hardy, making use of the circle method, proved [162] the following asymptotic expansion:

$$r_n(k) = \frac{\pi^{n/2}}{\Gamma(n/2)} k^{n/2-1} \sum_{m=1}^{\infty} \sum_{\substack{h \bmod (m) \\ (h,m)=1}} \left(\frac{G(h,m)}{m}\right)^n e^{-2\pi i\, hk/m} + O(k^{n/4}),$$

where

$$G(h,m) = \sum_{j=1}^{m} e^{2\pi i\, hj^2/k}.$$

A standard way of regularizing functions in analysis is taking its averages. Define

$$R_n(x) = \#\{\nu \in \mathbb{Z}^n : |\nu| \le x\} = r_n(0) + r_n(1) + \cdots + r_n([x^2]),$$

which, in this particular case, represents the number of lattice points inside the ball centered at the origin an radius $x > 0$ (we have used the standard notation $[y]$ for the integer part of the real number y).

It is credited to Gauss the observation that $R_n(x)$ is well estimated by the n-dimensional volume of the ball; more precisely,

$$R_n(x) = \omega_n x^n + E_n(x),$$

where $\omega_n = \frac{\pi^{n/2}}{\Gamma(n/2+1)}$ is the volume of the unit ball and $E_n(x) = O(x^{n-1})$.

To find the true order of magnitude (usually in the form $O(x^\alpha)$) of the error term, or discrepancy term, $E_n(x)$ is an outstanding problem in number theory. However, due to the extensive history of this subject, we can only phrase some of its highlights.

The quoted estimate of Gauss, $O(x^{n-1})$, follows easily from the fact that the region obtained as union of the lattice cubes Q_ν, $|\nu| \le x$, has volume equal to $R_n(x)$ and its boundary is contained between the balls centered at the origin and radii $x \pm \sqrt{n}$. Nevertheless, that crude estimate can be improved using the Poisson summation formula:

$$\begin{aligned} R_n(x) &= \sum_\nu \chi_{B_x(0)}(\nu) = \sum_\nu \widehat{\chi_{B_x(0)}}(\nu) \\ &= \frac{\pi^{n/2}}{\Gamma(n/2+1)} x^n + x^{n/2} \sum_{j\geqslant 1} \frac{r_n(j)}{j^{n/4}} J_{n/2}(2\pi x\sqrt{j}), \end{aligned}$$

which is the expression for the error term obtained by Hardy, and where the series of the second term has to be considered with a certain care.

The standard approach implies the substitution of the indicator function $\chi_{B_x(0)}$ by a smooth mollification, which allows, by integration by parts, the truncation of the series giving the discrepancy term. And this can be done in such a way that the approximate formulas of Hardy for $r_n(k)$, $n \geqslant 5$ (or Jacobi in the case $n = 4$) introduces, in both sides of Poisson formula, an error consistent with the expected $O(x^{n-2+\varepsilon})$, $\varepsilon > 0$. Finally, for the truncated series, we can use the asymptotic of Bessel functions together with the well-known Weyl estimates of trigonometric sums to conclude that for $n \geqslant 4$, the discrepancy term satisfies the estimate

$$E_n(x) = O(x^{n-2+\varepsilon}) \qquad \text{for every } \varepsilon > 0.$$

In fact, the case $n > 4$ is a consequence of the expression

$$\sum_{k<x^2} r_n(k) = \sum_{k<x^2} \sum_{-\sqrt{k}<\ell<\sqrt{k}} r_{n-1}(k-\ell^2) = \sum_{|\ell|\le x} \sum_{k\le x^2-\ell^2} r_{n-1}(k).$$

The error term in the inner summation is $O(x^{n-3+\varepsilon})$, where the principal term can be obtained via the Euler–Maclaurin summation formula.

Furthermore, using the expression for the discrepancy term given by Poisson's formula, it is easy to prove the estimate

$$\int_0^x |E_n(t)|^2 \, dt = c_n \, x^{2n-3} + \text{ lower order terms,}$$

implying that the estimate above is the best possible.

For balls in three dimensions, an early result of Landau showed that

$$R_3(x) - w_3 \, x^3 = O(x^{3/2}).$$

But this has been improved by several authors, among which we can mention Chamizo–Iwaniec [153] and Heath-Brown [163], where the present record for the exponent is $21/16 + \varepsilon$.

In the other direction, the method of estimating the L^2-norm of the discrepancy term implies here that the exponent 1 cannot be obtained, and this provides a limit and a conjecture: is $E_3(x) = O(x^{1+\varepsilon})$, for every $\varepsilon > 0$?

For the disc in dimension two, the exponent α provided in the first instance by the above mentioned application of Poisson's formula is $\alpha = 2/3$; this was obtained by Sierpiński and Voronoi, and the reader may consult the reference [151] for a detailed exposition. One can improve slightly this exponent by using Weyl's estimates and, up to now, the process culminated in a work of Kolesnik [159], which gave the exponent $\alpha = 278/429 + \varepsilon$, for any $\varepsilon > 0$. In the 1980's, there were results of Bombieri–Iwaniec [148] and Iwaniec–Mozzochi [165], which developed a new approach based in the estimates for Kloosterman sums. This method was exploited by Huxley ($\alpha = 131/208 + \varepsilon$) and Bourgain–Watt ($\alpha = 517/824$), but those exponents are "far" from the expected $\alpha = 1/2 + \varepsilon$.

One can also move the center of the disc. Given y in the unit square Q, consider the disc

$$D_x(y) = \{w \in \mathbb{R}^2 : \|w - y\| \leq x\}$$

and define

$$E_2(x; y) = \sum_{\nu} \chi_{D_x(y)}(\nu) - \pi x^2 = \sum_{\nu \neq 0} \widehat{\chi_{D_x(0)}}(\nu) \, e^{2\pi i \, \nu \cdot y}.$$

Then

$$\int_Q |E_2(x; y)|^2 \, dy = \sum_{\nu \neq 0} |\widehat{\chi}_{D_x}(\nu)|^2 = \frac{x}{\pi^2} \sum_{n=1}^{\infty} \frac{(r_2(n))^2}{n^{3/2}} \cos^2\left(2\pi x \sqrt{n} - \frac{3\pi}{4}\right) + O(1).$$

That is, if we choose at random a disc of radius x, then the expected number of lattice points contained in its interior is πx^2, and the standard deviation is $O(x^{1/2})$.

7.6 Random lattice points problems

Standard (two-dimensional) lattice point problems reduce to counting lattice points under positive graphs $y = g(x)$, with $g \in C_N^2$, where

$$C_N^2 = \{f \in C^2([0, N+1]) : 0 < C_1/N < -f'' < C_2/N,\ f(0) = f(N+1) = 0\},$$

for constants C_1 and C_2 and for $N \in \mathbb{Z}^+$. Well-known arguments allow to obtain asymptotic formulas in which the lattice error term is given by the difference between the area under g and the number of lattice points (lattice points on the real axis count one half). Variations on an old argument due to Jarník [166] prove that the bound $O(N^{2/3})$ is best possible (see Section 1 in [164]). After a century of advances in the method of exponential sums, this bound has been improved by M. N. Huxley [164] to $O(N^{46/73+\varepsilon})$ under stronger regularity conditions over g, namely three continuous derivatives. But this is still far from Hardy's conjecture $O(N^{1/2+\varepsilon})$ for the case of a circle [161]. There is some evidence of an extended conjecture for closed convex curves supported by average results over centers and radii. However, this latter kind of results are not completely satisfactory because they can be understood as an average over the lattice, rather than a real arbitrary variation of the curve. With this idea in mind, and taking Feynman path integral formulation of quantum mechanics as main motivation, we shall prove here that the integral of the squared lattice error term over all arcs $g \in C_N^2$ agrees with Hardy's conjecture, and hence that for "most of the arcs" belonging to C_N^2, the optimal $O(N^{1/2+\varepsilon})$ error bound holds without further regularity conditions. It is not the first time that the number theoretical problem of counting lattice points in convex region has been related to quantum mechanics. At the contrary, that connection appeared already in the pioneering work of H. Weyl: considering a potential V in Euclidean space $\mathbb{R}^n$, then the number of bounded states of $-\Delta + V$ (sometimes associated to lattice points) can be estimated, in a semiclassical approximation, by the volume $|\{(\xi, x) : |\xi|^2 + V(x) < \infty\}|$ in phase space, a question that will be mentioned in Chapter 9 regarding the Thomas–Fermi model of quantum mechanics of many interacting particles. On the other hand, there is the precedent of Ya. G. Sinai, who has also considered random lattice point problems in connection with physical models of phase transitions.

Summarizing the ideas exposed above, we want to give a sense to $\int |\Delta(f)|^2 \, d\mu(f)$, where $d\mu$ is a normalized measure on C_N^2 and $\Delta(f)$ is the lattice error function, which is given by

$$\Delta(f) = \int_0^{N+1} f(t)\, dt - \sum_{i=0}^{N+1} \left([f(i)] + \frac{1}{2}\right).$$

The Euler–Maclaurin summation formula or the truncation error formula for trapezoidal rule yields

$$\int_0^{N+1} f(t)\, dt = \sum_{i=0}^{N+1} f(i) + O(1).$$

Hence we infer that any reasonable definition of $\int |\Delta(f)|^2 \,\mathrm{d}\mu(f)$ should coincide, up to a negligible $O(1)$ term, with

$$\int |\Delta(x_1, x_2, \ldots, x_N)|^2 \,\mathrm{d}\mu_N(f)$$

where $x_i = f(i)$,

$$\Delta(x_1, x_2, \ldots, x_N) = \sum_{i=1}^{N} \psi(x_i), \qquad \text{with } \psi(x) = x - [x] - 1/2,$$

and $\mathrm{d}\mu_N$ factors into N measures supported on the vertical slices $x = x_1, \ldots, x = x_N$.

In this discrete setting, we can say that C_N^2 transforms into

$$\left\{(x_0, x_1, \ldots, x_{N+1}) \in \mathbb{R}^{N+2} : 0 < C_1/N < -f_i^{(2)} < C_2/N, x_0 = x_{N+1} = 0\right\},$$

where $f_i^{(2)}$ is the discrete second derivative,

$$f_i^{(2)} = \frac{(x_{i+1} - x_i) - (x_i - x_{i-1})}{2}.$$

Then, the most natural way to define the measure $\mathrm{d}\mu_N(f)$ is penalizing large or small values of the (discrete) second derivative. Hence we consider

$$(7.5) \quad \mathrm{d}\mu_N(f) = \mathrm{d}\mu(x_1, \ldots, x_N) = K_N \prod_{i=1}^{N} \phi\Big(\frac{(x_i - x_{i-1}) - (x_{i+1} - x_i)}{2} N\Big) \,\mathrm{d}x_i,$$

where $\phi \in C_0^\infty((C_1, C_2))$, $\phi > 0$ and K_N is a normalizing constant.

Definition 7.9. We shall call all arcs variance to the value of the integral

$$\mathcal{V} = \int_{\mathbb{R}^N} |\Delta(x_1, x_2, \ldots, x_N)|^2 \,\mathrm{d}\mu_N(x_1, x_2, \ldots, x_N),$$

where $\mathrm{d}\mu_N$ is as in (7.5), with $\int \phi = 1$ and $K_N = (N+1)(N/2)^N$.

Remark 7.10. We shall see later that with this value of K_N, the measure is actually normalized, i.e., $\int \mathrm{d}\mu_N = 1$.

Now we can state the main result.

Theorem 7.11. *Let $\mathcal{V}$ be the all arcs variance. Then for every $\varepsilon > 0$, there holds*

$$\mathcal{V} = O(N^{1+\varepsilon}).$$

In particular, Hardy's conjecture $|\Delta| = O(N^{1/2+\varepsilon})$ holds except in a set of μ_N-vanishing measure.

The scheme of the proof is as follows. We shall firstly smooth out Δ to get an analytic approximation in terms of some oscillatory sums. Secondly, we shall perform the integration over $\mathbb{R}^N$ to express all arcs variance as an exponential sum with certain coefficients. To finish, we have to estimate these coefficients and the value of the corresponding exponential sum.

Apart from the special notation already introduced, henceforth with ε we shall mean a small enough positive constant (not necessarily always the same), and the O-constants usually will depend on ε and collapse when $\varepsilon \to 0^+$.

7.6.1 Smoothing process

Firstly, we shall smooth the function $\psi(x)$. This can be done in several ways, but it will be convenient here to choose one leading to a finite Fourier series.

Lemma 7.12. *There exists $\sigma \in C_0^\infty((-1,1))$ such that the functions*

$$S_\pm(y) = -\sum_{h=1}^{\infty} \frac{\sin(2\pi h(y \pm \delta))}{\pi h}\, \sigma\Big(\frac{h}{N^{1/2}}\Big), \qquad \textit{with } \delta = N^{-1/2+\varepsilon},$$

satisfy

$$S_+(y) + O(\delta) \le \psi(y) \le S_-(y) + O(\delta).$$

Proof. By the periodicity, we can assume $y > 2\delta$. Let $s \in C_0^\infty((-1/2,1/2))$ be any real valued non-negative even function such that $\|s\|_2 = 1$ then we take $\sigma = s * s$. It is easy to prove that σ is even, compactly supported inside $(-1,1)$, and satisfies

$$\sigma\ \widehat{\sigma} \ge 0, \quad \widehat{\sigma}(x) = O((1+|x|)^{-1/(2\varepsilon)}), \quad \sigma(0) = \int \widehat{\sigma}(x)\,\mathrm{d}x = 1.$$

Consider the convolutions, say $p_y^\pm$, of $N^{1/2}\widehat{\sigma}(N^{1/2}x)$, and the characteristic function of the interval $[-y \mp \delta, y \pm \delta]$. Then we have

$$p_y^\pm(x) = \int_{I_\pm} \widehat{\sigma}(t)\,\mathrm{d}t, \qquad \text{with } I_\pm = [(x-y)N^{1/2} \mp N^\varepsilon, (x+y)N^{1/2} \pm N^\varepsilon].$$

The properties of $\widehat{\sigma}$ imply that $0 \le p_y^\pm(x) \le 1$ and that

$$p_y^-(x) = O\Big(\frac{\delta}{(1+|x|-y)^2}\Big) \quad \text{for } |x| \ge y,$$
$$p_y^+(x) = 1 + O\Big(\frac{\delta}{(1+y-|x|)^2}\Big) \quad \text{for } |x| \le y.$$

Hence

$$\sum_{h=-\infty}^{\infty} p_y^-(h) + O(\delta) \le 2[y] + 1 \le \sum_{h=-\infty}^{\infty} p_y^+(h) + O(\delta).$$

We apply Poisson's summation formula to obtain

$$y - \frac{1}{2}\sum_{h=-\infty}^{\infty} \widehat{p_y^+}(h) + O(\delta) \le \psi(y) \le y - \frac{1}{2}\sum_{h=-\infty}^{\infty} \widehat{p_y^-}(h) + O(\delta). \tag{7.6}$$

Then the definition of $p_y^\pm$and the properties of the convolution and Fourier transform allow us to deduce

$$\widehat{p_y^\pm}(\xi) = \frac{\sin(2\pi(y \pm \delta)\xi)}{\pi\xi}\,\sigma\Big(\frac{\xi}{N^{1/2}}\Big) \quad \text{for } \xi \ne 0,$$

and $\widehat{p_y^\pm}(0) = 2y \pm 2\delta$. Substituting in (7.6), the results follows. □

A consequence of the previous result is the following.

Lemma 7.13. *Let $\Delta = \Delta(x_1, \dots, x_N)$ be given as in the introduction of Section 7.6. Then*

$$|\Delta|^2 \le |\Delta_+|^2 + |\Delta_-|^2 + O(N^{1+\varepsilon}),$$

where

$$\Delta_\pm = \sum_{h=1}^{\infty}\sum_{m=1}^{N} \frac{e(hx_m \pm h\delta)}{h}\,\sigma\Big(\frac{h}{N^{1/2}}\Big), \qquad \text{with } \delta = N^{-1/2+\varepsilon}.$$

Proof. From Lemma 7.12, we obtain the estimate

$$\sum_{m=1}^{N} S_+(x_m) + O(N\delta) \le \Delta \le \sum_{m=1}^{N} S_-(x_m) + O(N\delta).$$

Therefore, taking absolute values and since $\sin(2\pi t) = \operatorname{Im} e(t)$, we get the inequality

$$|\Delta| \le \frac{1}{\pi}\,|\Delta_+| + \frac{1}{\pi}\,|\Delta_-| + O(N^{1/2+\varepsilon}),$$

and the proof follows by convexity. □

7.6.2 The exponential sum

Our next step in the proof of Theorem 7.11 is to estimate all arcs variance by an exponential sum. In this process, we shall need the following technical result.

Lemma 7.14. *Let J be the $N \times N$ Jacobi matrix*

$$J_{i,j} = \begin{cases} 1 & \text{if } i = j, \\ -1/2 & \text{if } |i-j| = 1, \\ 0 & \text{otherwise,} \end{cases}$$

and let $(b_{i,j})$ *be its inverse matrix* J^{-1}*. Then*

(a) $$\det J = \frac{N+1}{2^N}.$$

(b) $$b_{l,k} - b_{m,k} = \begin{cases} 2k(m-l)/(N+1) & \textit{if } k \le l \le m, \\ 2l - 2k(1-(m-l)/(N+1)) & \textit{if } l \le k \le m, \\ 2(m-l)(1-k/(N+1)) & \textit{if } l \le m \le k. \end{cases}$$

Proof. An elementary argument proves that the value of the determinant as a function of N, say d_N, satisfies the recursive formula

$$d_{N+2} = d_{N+1} - \frac{1}{4} d_N, \qquad \text{with } d_1 = 1, d_2 = \frac{3}{4},$$

and (a) follows by induction.

On the other hand, (b) is a direct consequence of

$$b_{i,j} = \frac{2\min(i,j)(N+1-\max(i,j))}{N+1}.$$

The proof of this formula reduces to check that

$$\sum_j J_{i,j} b_{j,k} = \delta_{i,k}.$$

If we define as zero the coefficients with non-positive subindices, we can write this as

$$-\frac{1}{2} b_{i-1,k} + b_{i,k} - \frac{1}{2} b_{i+1,k} = \delta_{i,k}. \tag{7.7}$$

For $k = i$, this is

$$-\frac{(i-1)(N+1-i)}{N+1} + \frac{2i(N+1-i)}{N+1} - \frac{i(N+1-i-1)}{N+1} = 1,$$

and for $k < i$,

$$-\frac{k(N+1-i+1)}{N+1} + \frac{2k(N+1-i)}{N+1} - \frac{k(N+1-i-1)}{N+1} = 0.$$

The case $k > i$ follows by the symmetry. Hence (7.7) holds. □

Remark 7.15. Note that (a) implies that $\int_{\mathbb{R}^N} \mathrm{d}\mu_N = 1$ (use the change of variables $\mathbf{y} = J\mathbf{x}$ and the fact that $\int \phi = 1$), i.e., the measure $\mathrm{d}\mu_N$ is normalized.

Remark 7.16. From the point of view of numerical analysis, the matrix J appears in approximating the second derivative, and its Jordan canonical form is a known object (see [157]). The formula for its eigenvalues (which are related to the stability of some algorithms) can be used to prove (a), and the whole canonical form to prove (b), but in both cases some auxiliary trigonometric identities are needed.

The function ϕ appearing in $d\mu_N$ is “off-centered”, which causes an oscillation on its Fourier transform, and it will be convenient to separate the non-oscillatory part. In order to do that, let us define

$$\rho(\xi) = e(\alpha\xi)\,\widehat{\phi}(\xi), \qquad \text{with } \alpha = \int x\phi(x)\,\mathrm{d}x.$$

Then $\phi \geq 0$ and $\int \phi = 1$ can be used to show that $\rho(0) = 1$, $\rho'(0) = 0$, $\rho''(0) < 0$, and $|\rho(\xi)| \leq 1$. Hence there exists a positive constant $0 < C_0 < 1$ such that

$$|\rho(x)| \leq \max(1 - C_0 x^2, 1 - C_0).$$

It turns out that all arcs variance is controlled by an exponential sum whose coefficients are related to ρ. They are precisely

$$a_{l,m}(h) = \prod_{k=1}^{N} \rho\Big(\frac{h}{N}(b_{l,k} - b_{m,k})\Big).$$

The explicit form of the exponential sum and its relation with our problem is the content of the following result.

Proposition 7.17. *Let $S(M,h)$ be the exponential sum*

$$S(M,h) = \sum_{M \leq l < m < 2M} a_{l,m}(h)\, e\Big(\frac{\alpha h}{N}(l-m)(m+l-N-1)\Big).$$

Then

$$\mathcal{V} = O\Big(N^{1+\varepsilon} + N^{\varepsilon} \sup_{h < N^{1/2}} \sup_{2M < N+1} |S(M,h)|\Big).$$

Proof. By Lemma 7.8,

$$\mathcal{V} \leq \int_{\mathbb{R}^N} |\Delta_+|^2\,\mathrm{d}\mu_N + \int_{\mathbb{R}^N} |\Delta_-|^2\,\mathrm{d}\mu_N + O(N^{1+\varepsilon}). \tag{7.8}$$

Dyadic subdivision on h and m (note that there are $O(N^{\varepsilon})$ dyadic blocks) and Cauchy's inequality give

$$\int_{\mathbb{R}^N} |\Delta_+|^2\,\mathrm{d}\mu_N = O\Big(N^{\varepsilon} \sup_{h<N^{1/2}} \sup_{2M<N+1} \int_{\mathbb{R}^N} \Big|\sum_{M \leq m < 2M} e(hx_m)\Big|^2 \mathrm{d}\mu_N\Big).$$

The same bound applies to $\int |\Delta_-|^2\,\mathrm{d}\mu_N$, and substituting in (7.8), we obtain

$$\mathcal{V} = O\Big(N^{1+\varepsilon} + N^{\varepsilon} \sup_{h<N^{1/2}} \sup_{2M<N+1} \Big|\sum_{M \leq l < m < 2M} \int_{\mathbb{R}^N} e(h(x_l - x_m))\,\mathrm{d}\mu_N\Big|\Big). \tag{7.9}$$

Using the definition of $d\mu_N$, the change of variables $\mathbf{y} = J\mathbf{x}$, and Lemma 7.14 (a),

$$\int_{\mathbb{R}^N} e(h(x_l - x_m))\,d\mu_N = K_N \int_{\mathbb{R}^N} e(h(x_l - x_m)) \prod_{k=1}^{N} \phi((J\mathbf{x})_k N)\,dx_k \tag{7.10}$$
$$= \prod_{k=1}^{N} \widehat{\phi}\Big(\frac{h}{N}(b_{lk} - b_{mk})\Big).$$

By Lemma 7.14 (b),

$$\sum_{k=1}^{N} (b_{l,k} - b_{m,k}) = (m - l)(m + l - N - 1)$$

and the result follows from (7.9) and (7.10). □

7.6.3 End of the proof

According to the last proposition, the bound $S(M, h) = O(N^{1+\varepsilon})$ suffices to complete the proof of our main result.

Proposition 7.18. *For $h \le N^{1/2}$ and $2M \le N + 1$, it holds that*

$$S(M, h) = O(N^{1+\varepsilon}).$$

Proof. Since $|\rho| \le 1$ and $l < m$, we have

$$|a_{l,m}(h)| = \prod_{k=1}^{N} \Big|\rho\Big(\frac{h}{N}(b_{l,k} - b_{m,k})\Big)\Big|$$
$$\le \prod_{1\le k\le l} \Big|\rho\Big(\frac{h}{N}(b_{l,k} - b_{m,k})\Big)\Big| \cdot \prod_{m\le k\le N} \Big|\rho\Big(\frac{h}{N}(b_{l,k} - b_{m,k})\Big)\Big|.$$

Note that the right-hand side of the inequality

$$|\rho(x)| \le \max(1 - C_0 x^2, 1 - C_0)$$

is non-increasing for $x > 0$.

Hence if $m - l > CN^{1/2+\varepsilon} h^{-1}$, with a small enough constant C, the first product is bounded by

$$\prod_{1\le k\le} \max\Big(1 - C_0 \frac{4k^2 h^2 (m - l)^2}{N^2 (N + 1)^2}, 1 - C_0\Big) \le \prod_{1\le k\le l} \Big(1 - C_0' \frac{k^2}{(N + 1)^{3-\varepsilon}}\Big),$$

and the second product is bounded by

$$\prod_{m\le k\le N} \max\left(1-4C_0\,\frac{h^2(m-l)^2}{N^2}\left(1-\frac{k}{N+1}\right)^2, 1-C_0\right) \le \prod_{m\le k\le N}\left(1-C_0'\,\frac{(N+1-k)^2}{(N+1)^{3-\varepsilon}}\right).$$

Noting that l and m belong to the same dyadic interval, we deduce from both bounds that $|a_{l,m}(h)|$ has an exponential decay with N when $m-l > CN^{1/2+\varepsilon}h^{-1}$. On the other hand, the terms with $0 < m-l < N^{\varepsilon}$ contribute trivially $O(N^{1+\varepsilon})$. Therefore we have

$$S(M,h) = S^*(M,h) + O(N^{1+\varepsilon}), \tag{7.11}$$

where $S^*(M,h)$ is the sum $S(M,h)$ but restricted to $N^{\varepsilon} \le m-l \le CN^{1/2+\varepsilon}h^{-1}$.

Assume that $a_{l,m}(h) \ne 0$ and that $\frac{h}{N}(b_{l,k}-b_{m,k})$ is small enough, say $O(N^{-\varepsilon/4})$, for $l < k \le m$. The definition of $a_{l,m}(h)$ implies that

$$1 + \frac{a_{l+1,m+1}(h) - a_{lm}(h)}{a_{l,m}(h)} = \frac{a_{l+1,m+1}(h)}{a_{l,m}(h)} = \prod_{k=l+1}^{m} \frac{\rho(\frac{h}{N}(b_{l+1,k}-b_{m+1,k}))}{\rho(\frac{h}{N}(b_{l,k}-b_{m,k}))}.$$

As $\rho'(0)=0$ and ρ'' is bounded, for $x = b_{l,k}-b_{mk}$, $l < k \le m$,

$$\frac{\rho(\frac{h}{N}(x+2))}{\rho(\frac{h}{N}x)} - 1 = \frac{\rho(\frac{h}{N}(x+2)) - \rho(\frac{h}{N}x) - \frac{2h}{N}\rho'(0)}{\rho(\frac{h}{N}x)} = O\Big(\frac{h^2}{N^2}\,x\Big).$$

Hence

$$\log\left(1 + \frac{a_{l+1,m+1}(h) - a_{l,m}(h)}{a_{l,m}(h)}\right) = O\Big(\sum_{k=l+1}^{m} \frac{h^2}{N^2}(m-l)\Big) = O(N^{-1+\varepsilon}),$$

and $|a_{l,m}(h)| \le 1$ implies

$$|a_{l+1,m+1}(h) - a_{l,m}(h)| = O(N^{-1+\varepsilon}).$$

This formula is valid in general, because if $a_{l,m}(h) = 0$ then $a_{l+1,m+1}(h) = 0$ except when $\rho(\frac{h}{N}(b_{l,k}-b_{m,k}))$ vanishes for some $l < k \le m$. But in this case Taylor expansion proves that $\frac{h}{N}(b_{l,k}-b_{m,k})$ is not $O(N^{-\varepsilon/4})$, and then $a_{l,m}(h)$ and $a_{l+1,m+1}(h)$ are exponentially small (in their definitions are at least N^{ε} terms smaller than $CN^{-\varepsilon/2}$). Therefore partial summation in $S^*(M,h)$ with the new variables $r = m-l$, $s = m$ gives

$$S^*(M,h) = O\Bigg(N^{\varepsilon} \sum_{N^{\varepsilon}\le r\le CN^{1/2+\varepsilon}h^{-1}} \sup_{M\le M'<2M} \Big|\sum_{M'\le s<2M} e\Big(\frac{2\alpha h}{N}\,rs\Big)\Big|\Bigg)$$

The innermost sum is bounded by $2\min(M, \|2\alpha hrN^{-1}\|^{-1})$ where $\|\cdot\|$ is the distance to the nearest integer. A final substitution in (7.11) allows us to conclude $S(M,h) = O(N^{1+\varepsilon})$, which is the desired result. □

Chapter 8
Encounters at the interface with number theory

In Harmonic Analysis we deal with nice kernels acting over complicated functions, while in Number Theory we get difficult kernels operating on very smooth functions.

Henryk Iwaniec

8.1 Questions from Riemann's tesis

In his Habilitationsschrift [194], B. Riemann states the following: "[. . .] the usefulness of Fourier series is not limited to research in physics; they have been successfully applied also to a field in pure mathematics, namely number theory, and here it seems to be of importance precisely to consider those functions whose representability by trigonometric series has not been yet investigated".

Riemann introduces

$$f(x) = \sum_{n=1}^{\infty} \frac{\{nx\}}{n^2},$$

and shows that f is discontinuous precisely at the rational points of the form $a/(2b)$, with $\gcd(a, 2b) = 1$, which constitute a dense subset of the real line.

Furthermore, using the formula

$$\sum_{n=1}^{\infty} \frac{1}{n^2} = \frac{\pi^2}{6} \quad \left(\text{or } \sum_{n=0}^{\infty} \frac{1}{(2n+1)^2} = \frac{\pi^2}{8}\right),$$

he proves that at those points of discontinuity there is a jump:

$$f^-\left(\frac{a}{2b}\right) - f^+\left(\frac{a}{2b}\right) = \frac{\pi^2}{8b^2}.$$

Since

$$|f(x)| \leq \frac{\pi^2}{6}$$

and the set of discontinuity is countable, it follows that f is Riemann-integrable, and its indefinite integral

$$F(x) = \int_0^x f(t)\,dt$$

A. Córdoba, *Suprematism in Harmonic Analysis*, Progress in Mathematics 356,
https://doi.org/10.1007/978-3-031-73737-4_8

turns out to be a continuous but non-differentiable function at those rational points:

$$\lim_{h\to 0^+} \frac{F\left(\frac{a}{2b}+h\right)-F\left(\frac{a}{2b}\right)}{h} = f^+\left(\frac{a}{2b}\right),$$
$$\lim_{h\to 0^-} \frac{F\left(\frac{a}{2b}+h\right)-F\left(\frac{a}{2b}\right)}{h} = f^-\left(\frac{a}{2b}\right).$$

In part IV of Riemann's Habilitationsschrift, one finds the function

$$f(x)=\sum_{n=1}^{\infty}\frac{\{nx\}}{n}.$$

We know that f is Lebesgue-integrable, but not Riemann-integrable, because its oscillation is unbounded on any interval.

It can be represented by the trigonometric series

$$f(x)=\frac{1}{\pi}\sum_{n=1}^{\infty}\frac{d_{\mathrm{o}}(n)-d_{\mathrm{e}}(n)}{n}\sin(\pi n x)$$

where $d_{\mathrm{o}}(n)$ (respectively, $d_{\mathrm{e}}(n)$) is the number of odd (respectively, even) divisors of n.

Riemann writes: "One can obtain similar examples with series of the form

$$\sum_{n=0}^{\infty} C_n\, e^{2\pi i n^2 x}$$

when the positive quantities C_n are decreasing to 0 but for which $\sum_n C_n=\infty$".

We find here a very interesting problem whose solution remains open.

Question 8.1. *If f is a Lebesgue-integrable function whose Fourier spectrum is contained in the set of the square numbers, does it follow that $\|f\|_p \lesssim \|f\|_1$, for every $p<4$?*

An equivalent formulation is the following. Let S be the Fourier multiplier operator given by the characteristic function of the set of square numbers, i.e., if f has the Fourier series $\sum \widehat{f}(k)\, e^{2\pi i k x}$, then the Fourier series of Sf is given by

$$\sum_{n=0}^{\infty}\widehat{f}(n^2)\, e^{2\pi i n^2 x}.$$

Question 8.2. *Is S a bounded operator from $L^2[0,1]$ to $L^p[0,1]$, for $2\le p<4$?*

By duality, it is equivalent to the boundedness of S from $L^q[0,1]$ to $L^2[0,1]$, for every $q>4/3$. A positive answer would have very interesting arithmetical consequences. It would imply, for example, that any arithmetic progression $\{a+br : 0\le r\le N-1\}$ of length N may contain, at most, $O(N^{1/2+\varepsilon})$ square numbers for every $\varepsilon>0$ (see reference [180] for more details about this problem).

If $G(x) = \sum_{r=0}^{N-1} e^{2\pi i(a+br)x}$, then we have $\|G\|_p \sim N^{(p-1)/p}$ for $p > 1$. On the other hand, if $n_1^2 < n_2^2 < \cdots < n_k^2$ are the square numbers contained in the arithmetic progression $\{a+br\}_{r=0,\dots,N-1}$, and assuming a positive answer to our question about S, we obtain

$$k^{1/2} = \|S(G)\|_2 \lesssim \|S(G)\|_p \lesssim N^{(p-1)/p}$$

for every $p > 4/3$, implying that $k = O(N^{1/2+\varepsilon})$ where $p = 1/(3/4 - \varepsilon)$.

Given $a \in \mathbb{Z}$ and $r \in \mathbb{Z}^+$, let us define

$$\sigma(N; a, r) = \#\{m^2 = a + nr; 0 \le r < N\} \quad \text{and} \quad \sigma(N) = \sup_{a,r} \sigma(N; a, r).$$

There is a conjecture, attributed to P. Erdős:

$$\sigma(N) = O(N^{1/2+\varepsilon}), \qquad \text{for all } \varepsilon > 0.$$

It is not difficult to see that $G(N) = O(N^{1/2})$ is not true, and the lines above show that a positive answer to the question about the multiplier S implies the Erdős conjecture. However, the best known result is $\sigma(N) = O(N^{3/5+\varepsilon})$, for all $\varepsilon > 0$. This is due to Bombieri–Zannier [174].

8.2 $B_2[\infty]$-sequences of square numbers

Let us denote by $r(n)$ the number of representations of the integer n as a sum of two squares. It is an outstanding arithmetical function, whose study is a familiar topic in number theory. This function has a rather "irregular" behavior, because the value $r(n)$ depends upon the decomposition of n in prime factors.

On the other hand, it is well known that its average satisfies

$$\sum_{n \le x} r(n) \sim \pi x \qquad \text{as } x \to \infty.$$

A particularly interesting expression of this "irregularity" is given by the fact that the order of magnitude of $\sum_{n \le x} r^2(n)$ is strictly greater than $\sum_{n \le x} r(n)$. More precisely,

$$\frac{\sum_{n \le x} r^2(n)}{\sum_{n \le x} r(n)} \gtrsim \log x.$$

It is then natural to consider the following class of sequences.

Definition 8.3. We say that a sequence of positive integers $\{n_k\}$ has the property $B_2[\infty]$, or that it is a $B_2[\infty]$-sequence, if

$$\limsup_{x \to \infty} \frac{\sum_{n \le x} r^2(n)}{\sum_{n \le x} r(n)} < +\infty,$$

where $r(n) = \#\{n = n_k + n_j, n_k \le n_j\}$.

This definition is also a natural extension of the concept of $B_2[g]$-sequences, i.e., those for which $r(n) \leq g$ for every n. We refer to [189] for details and applications.

As we have pointed out, the whole sequence of squares is not a $B_2[\infty]$-sequence. It seems that J. E. Littlewood was the first to observe that this fact is "atypical" among sequences with growth similar to the square numbers. In his Ph.D. Thesis, A. O. L. Atkin (see [171]) confirmed Littlewood's prediction proving the existence of a $B_2[\infty]$-sequence very close to the squares ($n_k = k^2 + O(\log k)$). He also observed that no sequence $\{n_k\}$ such that $n_k = k^2 + o(\sqrt{\log k})$ has the property $B_2[\infty]$.

Later on, P. Erdős and A. Rényi [185], using probabilistic methods, were able to show that, in a certain sense, almost every sequence close to the squares ($n_k = k^2(1 + o(1))$) is $B_2[\infty]$. In both [171] and [185], the sequences considered keep the growth of the squares but their terms are not square numbers. In the following, we shall proceed in the opposite direction, building subsequences of squares with the required property $B_2[\infty]$ and with a moderate rate of growth.

Theorem 8.4. *There exists an infinite sequence $a_1^2 < a_2^2 < \cdots < a_k^2 < \cdots$ of square numbers satisfying property $B_2[\infty]$ and such that $a_k^2 \lesssim k^2 \log^2 k$.*

Proof. We shall construct explicitly a family of such sequences. Let

$$I_j^\nu = \{2n \mid 2^j(1 + \nu/j) \leq 2n < 2^j(1 + (\nu + 1)/j)\}, \quad j = 1, \ldots, 0 \leq \nu \leq j - 1.$$

With a choice of ν_j, for each j, consider the sets

$$I^{(\nu)} = \bigcup_j I_j^{\nu_j} \quad \text{and} \quad A^{(\nu)} = \{a_k^2 \mid a_k \in I^{(\nu)}\}.$$

We claim that $A^{(\nu)}$ satisfies the conditions required by Theorem 8.4. (To simplify notation, we shall consider the case $\nu_j = 0$; the proof for general values of ν_j, $j = 1, 2, \ldots$, is identical.)

Clearly we have

$$a_{[2^{j-1}/j]} \leq 2^j + 2^j/j \leq 2^{j+1},$$

therefore, given k, there exists j such that $2^{j-1}/j \leq k < 2^j/(j+1)$. Then

$$a_k \leq a_{[2^j/(j+1)]} < 2^{j+2} \lesssim k \log k.$$

Let us consider

$$r(n) = \#\{n_j = a_k^2 + a_l^2 \mid a_k \leq a_l \text{ and } a_k, a_l \in I = \textstyle\bigcup_j I_j^0\}$$

We have the following inequality:

$$\sum_{n \leq x} r(n) = \sum_{\substack{a_k^2 + a_l^2 \leq x \\ a_k \leq a_l}} 1 \geq \frac{1}{2}\Big(\sum_{a_k^2 \leq x/2} 1\Big)^2 \gtrsim \frac{x}{\log^2 x}. \tag{8.1}$$

Therefore to prove the theorem, it is enough to show the following estimate:

$$\sum_{n\le x} r^2(n) \lesssim \frac{x}{\log^2 x}.$$

To see this, let us define, for every j, the function

$$r_j(n) = \#\{n = a_k^2 + a_l^2 \mid a_k \le a_l \text{ and } a_k, a_l \in I_j^0\}.$$

Then

$$\sum_{n\le x} r^2(n) \lesssim \Big[\sum_{j\le[\log_4 x]} \Big(\sum_n r_j^2(n)\Big)^{1/2}\Big]^2.$$

This is a consequence of the Littlewood–Paley theory of Fourier series, together with the natural interpretation of the sums $\sum_{n\le x} r^2(n)$ as L^4-norms of the corresponding trigonometric polynomials:

$$\begin{aligned}\sum_{n\le x} r^2(n) &\simeq \Big\|\sum_{j\le(\log_2 x)/2}\sum_{n\in I_j^0} e^{2\pi i n^2 x}\Big\|_4^4 \simeq \Big\|\Big(\sum_{j\le(\log_2 x)/2}\Big|\sum_{n\in I_j^0} e^{2\pi i n^2 x}\Big|^2\Big)^{1/2}\Big\|_4^4\\ &\lesssim \Big(\sum_{j\le(\log_2 x)/2}\Big\|\sum_{n\in I_j^0} e^{2\pi i n^2 x}\Big\|_4^2\Big)^2 \simeq \Big(\sum_{j\le(\log_2 x)/2}\Big(\sum_n r_j^2(n)\Big)^{1/2}\Big)^2.\end{aligned}$$

We use the standard notation (see the section of Notation) $A(z) \simeq B(z)$ if there exist constants $0 < C_1 \le C_2 < \infty$ such that $C_1 A(z) \le B(z) \le C_2 A(z)$ for every z.

We claim that

$$\sum_n r_j^2(n) \lesssim \Big(\frac{2^j}{j}\Big)^2. \tag{8.2}$$

Then we obtain

$$\sum_{n\le x} r^2(n) \lesssim \Big(\sum_{j\le(\log_2 x)/2}\frac{2^j}{j}\Big)^2 \lesssim \frac{x}{\log^2 x}.$$

To finish the proof, we have to show the estimate (8.2). We shall do this in two steps: first we state a lemma.

Lemma 8.5. *If $r_j(n) \ge 2$, then $r_j^2(n) \lesssim F(n)$, where $F(n)$ is the number of integer solutions (a_1, b_1, a_2, b_2) of the following system of inequalities:*

$$0 < \Big|\frac{a_1}{b_1}\Big| < \frac{1}{j}, \quad 0 < \Big|\frac{a_2}{b_2} - 1\Big| < \frac{1}{j}, \quad n = (a_1^2 + b_1^2)\cdot(a_2^2 + b_2^2).$$

Assuming this lemma, we can complete the proof of (8.2). With j fixed, we have

$$\sum_n r_j^2(n) \le \sum_{r_j(n)=1} r_j(n) + \sum_{r_j(n)\ge 2} r_j^2(n).$$

It is clear that the first term satisfies the estimate $\sum r_j(n) \lesssim (2^j/j)^2$.

By Lemma 8.5, the second term is bounded by the number of solutions (a_1, b_1, a_2, b_2) of the following inequalities:

$$2(2^j)^2 \le (a_1^2+b_1^2)(a_2^2+b_2^2) \le 2(2^j+2^j/j)^2, \quad 0<\left|\frac{a_1}{b_1}\right|<1/j, \quad 0<\left|\frac{a_2}{b_2}-1\right|<\frac{1}{j}.$$

Keeping a_1 and b_1 fixed, we shall estimate the number of pairs (a_2, b_2) satisfying

$$\sqrt{2}\,\frac{2^j}{\sqrt{a_1^2+b_1^2}} \le \sqrt{a_2^2+b_2^2} \le \frac{\sqrt{2}\,(2^j+2^j/j)}{\sqrt{a_1^2+b_1^2}}, \quad 0<\left|\frac{a_2}{b_2}-1\right|<\frac{1}{j}.$$

One can identify each solution (a_2, b_2) with a lattice point in the region described in the figure.

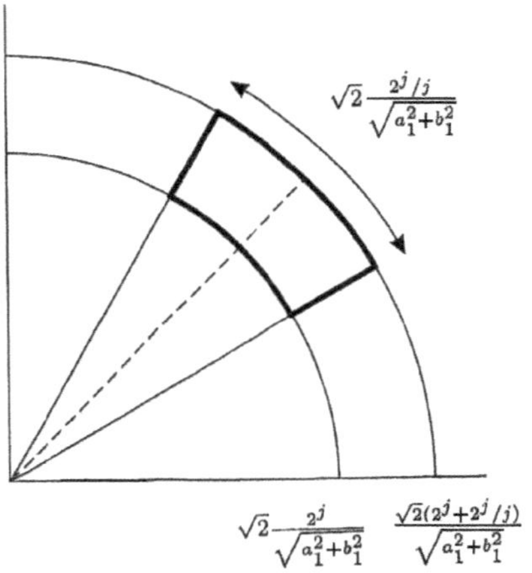

The condition $0 < |a_2/b_2 - 1| < 1/j$ yields $b_2 > j$ and, therefore,

$$\sqrt{2}\,\frac{2^j/j}{\sqrt{a_1^2+b_1^2}} \gtrsim 1,$$

which allows the number of lattice points inside that region to be estimated by its area. Thus

$$\sum_{r_j(n)\ge 2} r_j^2(n) \lesssim \sum_{\substack{a_1,b_1\\ 0<|a_1/b_1|<1/j}} \frac{4^j}{(a_1^2+b_1^2)j^2} \lesssim \sum_{b_1<2^j} \frac{4^j}{b_1 j^3} \lesssim \frac{4^j}{j^2},$$

which easily yields inequality (8.2).

Proof of Lemma 8.5. First we observe (see [24] for details) that if we have two different decompositions of $n \equiv 0 \bmod (4)$ as a sum of two squares, $n = a^2 + b^2 =$

$c^2 + d^2$, then there exist integers a_1, b_1, a_2 and b_2 such that

$$n = (a_1^2 + b_1^2)(a_2^2 + b_2^2),$$
$$\frac{a_1}{b_1} = \tan\left(\frac{\arctan(a/b) + \arctan(c/d)}{2}\right), \quad \frac{a_2}{b_2} = \tan\left(\frac{\arctan(a/b) - \arctan(c/d)}{2}\right).$$

Let $n = a_r^2 + b_r^2 = a_s^2 + b_s^2$ be two different representations of n as the sum of two squares satisfying

$$2^j \le a_r \le b_r \le 2^j + \frac{2^j}{j}, \quad 2^j \le a_s \le b_s \le 2^j + \frac{2^j}{j},$$
$$a_r \equiv b_r \equiv a_s \equiv b_s \equiv 0 \pmod 2.$$

By our previous argument, there exist a_1, b_1, a_2 and b_2 such that

$$n = (a_1^2 + b_1^2)(a_2^2 + b_2^2),$$
$$\frac{a_1}{b_1} = \tan\left(\frac{\arctan(a_r/b_r) + \arctan(a_s/b_s)}{2}\right),$$
$$\frac{a_2}{b_2} = \tan\left(\frac{\arctan(a_r/b_r) - \arctan(a_s/b_s)}{2}\right).$$

The conditions imposed on a_r, b_r, a_s and b_s easily yield that

$$\left|\frac{a_2}{b_2}\right| < \frac{1}{j} \quad \text{and} \quad \left|\frac{a_1}{b_1} - 1\right| < \frac{1}{j}.$$

On the other hand, both quantities are strictly greater than 0, because otherwise we could not have started with two different representations of n as a sum of two squares.

Each pair of different representations

$$\{n = a_r^2 + b_r^2 = a_s^2 + b_s^2,\ a_s \equiv b_s \equiv a_r \equiv b_r \equiv 0 \pmod 2\}$$

produces different angles and, consequently, different integers a_1, b_1, a_2 and b_2. Therefore $\binom{r_j(n)}{2}$ is bounded above by the number of solutions of the problem

$$n = (a_1^2 + b_1^2)(a_2^2 + b_2^2), \quad 0 < \left|\frac{a_1}{b_1}\right| < \frac{1}{j}, \quad 0 < \left|\frac{a_2}{b_2} - 1\right| < \frac{1}{j}.$$

The lemma follows by observing that if $r_j(n) \ge 2$ then $r_j^2(n) \le 4\binom{r_j(n)}{2}$. □

Remark 8.5. It is not difficult to see that no subsequence of the squares $\{n_k^2\}$ such that $n_k^2 = o(k^2 \log k)$ can have the property $B_2[\infty]$.

Furthermore, one can build sequences of square numbers growing very fast and which, nevertheless, are not $B_2[\infty]$-sequences.

8.3 Fourier series with square frequencies

The question about the L^p-character of Fourier series whose frequencies are square numbers has a satisfactory answer under the monotonicity conditions of the coefficients.

Proposition 8.7. *Suppose that $\{a_n\}$ is a monotonically decreasing sequence of non-negative real numbers such that $\sum_{n=0}^{\infty} a_n e^{2\pi i n^2 x}$ is the Fourier series of an integrable function f. Then $f \in L^p[0,1]$ for every p, $1 \le p < 4$, and satisfies the estimate*

$$\|f\|_p \lesssim \|f\|_1.$$

The proof of the above proposition will be based on the following fact.

Lemma 8.8. *Let f be continuous in $[0,1]$ and such that*

(i) $\|f\|_\infty \le N$ *and* $\|f\|_2 \le N^{1/2}$.
(ii) *If* $|x - p/q| \le 1/q^2$, *then*

$$|f(x)| \le C_0 \left\{ q^{1/k} + \frac{N}{q^{1/k}} \right\}, \quad \textit{for some constant } C_0.$$

Then f satisfies the $2k$-weak-type inequality

$$\mu\{x : |f(x)| > \alpha > 0\} \le C \frac{N^k}{\alpha^{2k}}$$

for every $\alpha > 0$, where the constant $C = C(C_0)$, is independent of α and N, and μ denotes the Lebesgue measure in $[0,1]$.

Proof. Without loss of generality, we may assume that $C_0 = 1$, because if $C_0 > 1$ we will just consider then the function f/C_0. It will be equivalent to show that for every $\alpha > 0$ we have

$$\mu(E_\alpha) = \mu\{|f(x)| \ge 2N^{1/2}\alpha\} \le \widetilde{C} \frac{1}{\alpha^{2k}},$$

with $\widetilde{C}$ independent of N and α. Observe that it is enough to prove this in the case $\frac{1}{2}N^{1/2} \ge \alpha \ge 1$, because if $\alpha < 1$ then the inequality trivializes, and if $\alpha \ge \frac{1}{2}N^{1/2}$ we have that $E_\alpha = \emptyset$. If

$$x = \cfrac{1}{x_1 + \cfrac{1}{\frac{1}{x_2 + \frac{1}{x_3 + \cdots}}}}$$

is the continuous fraction expansion of the irrational number $x \in E_\alpha$ with convergents $\{P_\nu/Q_\nu\}$, then the following holds:

$$Q_\nu^{1/k} + \frac{N}{Q_\nu^{1/k}} \ge 2N^{1/2}\alpha.$$

That is, for every ν,

$$\text{either} \quad Q_\nu^{1/k} \geq N^{1/2}\alpha, \quad \text{or} \quad N/Q_\nu^{1/k} \geq N^{1/2}\alpha.$$

We have two possible cases:

Case 1. $Q_1^{1/k} \geq N^{1/2}\alpha$, i.e., $Q_1 \geq N^{k/2}\alpha^k$.

Since $x \leq 1/Q_1$, we have that $x \in I_0 = [0, \frac{1}{N^{k/2}\alpha^k}]$ and

$$\mu(I_0) = \frac{1}{N^{k/2}\alpha^k} \leq \frac{1}{2^k\alpha^{2k}}.$$

Case 2. There exists $\nu \geq 1$ such that

$$\frac{N}{Q_\nu^{1/k}} \geq N^{1/2}\alpha, \quad \text{i.e.,} \quad Q_\nu \leq \frac{N^{k/2}}{\alpha^k},$$
$$\text{and} \quad Q_{\nu+1}^{1/k} \geq N^{1/2}\alpha, \quad \text{i.e.,} \quad Q_{\nu+1} \geq N^{k/2}\alpha^k.$$

We have

$$\left|x - \frac{P_\nu}{Q_\nu}\right| \leq \frac{1}{Q_\nu Q_{\nu+1}} \leq \frac{1}{Q_\nu}\frac{1}{N^{k/2}\alpha^k},$$

that is,

$$x \in \left(\frac{P_\nu}{Q_\nu} - \frac{1}{Q_\nu}\frac{1}{N^{k/2}\alpha^k}, \frac{P_\nu}{Q_\nu} + \frac{1}{Q_\nu}\frac{1}{N^{k/2}\alpha^k}\right).$$

Given integers r and s such that $r < s$, $\gcd(r, s) = 1$, $s \leq N^{k/2}/\alpha^k$, let us consider the interval

$$I_{r,s} = \left(\frac{r}{s} - \frac{1}{s}\frac{1}{N^{k/2}\alpha^k}, \frac{r}{s} + \frac{1}{s}\frac{1}{N^{k/2}\alpha^k}\right).$$

Then the previous observations give us the inclusion

$$E_\alpha \setminus \{\text{rationals}\} \subset I_0 \cup \left(\bigcup_{r,s} I_{r,s}\right).$$

Therefore,

$$\mu(E_\alpha) \leq \mu(I_0) + \sum_{\substack{r=1\\(r,s)=1}}^{s-1} \sum_{s=1}^{N^{k/2}/\alpha^k} \mu(I_{r,s}) \leq \frac{1}{\alpha^{2k}} + \sum_{s=1}^{N^{k/2}/\alpha^k} \sum_{r=1}^{s-1} \frac{2}{s}\frac{1}{N^{k/2}\alpha^k}$$
$$\leq \frac{1}{\alpha^{2k}} + \sum_{s=1}^{N^{k/2}/\alpha^k} \frac{2}{N^{k/2}\alpha^k} \leq \frac{3}{\alpha^{2k}}. \qquad \square$$

Next let us consider the trigonometric polynomials

$$S_N(x; y) = \sum_{n=1}^{N} e^{2\pi i(n^2x+ny)}$$

and its Carleson maximal function

$$C_N^*(x;y) = \sup_{1\le M\le N} |S_M(x;y)|.$$

We will use now Carleson's maximal theorem to continue the proof of Proposition 8.7 showing that, for every $y \in [0,1]$, $S_N(x;y)$ satisfies the hypotheses of Lemma 8.8 in the case $k=2$.

Lemma 8.9. *Let x have a rational approximation of the form $|x - p/q| \le 1/q^2$, $1 \le q \le N^2$, $\gcd(p,q) = 1$. Then*

$$C_N^*(x;y) \lesssim \left\{q^{1/2} + \frac{N}{q^{1/2}}\right\}.$$

This result was known to Hardy and Littlewood [187, 188], who proved it using the approximate functional equation for θ-functions. The following proof of E. Bombieri, see [173], emphasizes the relationship between number theory and harmonic analysis throughout the use of Carleson's theorem [175].

Proof. First let us observe that if $|x - x_0| \le \frac{1}{4N^2}$ and $|y - y_0| \le \frac{1}{4N}$, then we have the inequality

$$C_N^*(x;y) \le 100\, C_N^*(x_0;y_0).$$

This is because

$$\begin{aligned}
S_M(x;y) &= \sum_{n=1}^{M} e^{2\pi i(n^2x+ny)} = \sum_{n=1}^{M} e^{2\pi i(n^2x_0+ny_0)} \cdot e^{2\pi i(n^2(x-x_0)+n(y-y_0))} \\
&= \sum_{n=1}^{M} (S_n(x_0;y_0) - S_{n-1}(x_0;y_0))\, e^{2\pi i(n^2(x-x_0)+n(y-y_0))} \\
&= \sum_{n=1}^{M-1} S_n(x_0;y_0) \left(e^{2\pi i(n^2(x-x_0)+n(y-y_0))} - e^{2\pi i((n+1)^2(x-x_0)+(n+1)(y-y_0))}\right) \\
&\quad + S_M(x_0;y_0)\, e^{2\pi i(M^2(x-x_0)+M(y-y_0))}.
\end{aligned}$$

Taking the supremum in M, $1 \le M \le N$, we obtain

$$\begin{aligned}
C_N^*(x;y) &\le 2\, C_N^*(x_0;y_0) \left[1 + \sum_{n=1}^{N} (|x - x_0| 4\pi n + 4\pi |y - y_0|)\right] \\
&\le 100\, C_N^*(x_0;y_0).
\end{aligned}$$

Next we have, for all M, $1 \le M \le N$, the following inequality:

$$C_N^*(x;y) \le 2\, C_{2N}^*(x; y - 2Mx),$$

since for all $K \leq N$,

$$|S_K(x;y)| = \Big|\sum_{n=1}^{K} e^{2\pi i(n^2x+ny)}\Big| = \Big|\sum_{n=M+1}^{K+M} e^{2\pi i((n-M)^2x+(n-M)y)}\Big|$$
$$= \Big|\sum_{n=M+1}^{K+M} e^{2\pi i(n^2x+n(y-2Mx))}\Big| \leq 2\,C^*_{2N}(x;y-2Mx).$$

Rising the inequality to the square power and averaging over M, we get

$$|C^*_N(x;y)|^2 \leq \frac{4}{N}\sum_{M=1}^{N} |C^*_{2N}(x;y-2Mx)|^2$$
$$\leq \frac{4}{N}\sum_{n=1}^{M}(100)^2\, 8N \int_{I(2Mx-y)} |C^*_{2N}(x;y_0)|^2\, dy_0,$$

where $I(z) = \{w : |z-w| \leq \frac{1}{8N}\}$. An elementary calculation gives us an upper bound for the overlapping of the family of intervals $\{I_{(2Mx-y)}\}$:

$$\Big\|\sum_{M=1}^{N} \chi_{I_{(2Mx-y)}}\Big\|_\infty \lesssim \Big(\frac{N}{q}+\frac{q}{N}\Big),$$

and applying Carleson's maximal theorem, we obtain

$$C^*_N(x;y)^2 \lesssim \Big\{q+\frac{N^2}{q}\Big\}. \qquad \square$$

Proof of Proposition 8.7. Let us consider the maximal function

$$C^*_N(x) = \sup_{1\leq M\leq N} \Big|\sum_{n=1}^{M} e^{2\pi i n^2 x}\Big|.$$

Then Lemmas 8.8 and 8.9 give us the estimate

$$\mu(\{C^*_N(x) \geq 2N^{1/2}\alpha\}) \lesssim \frac{1}{\alpha^4},$$

and, in particular, we obtain

$$\|C^*_N\|_p \leq C_p N^{1/2}, \qquad \text{for } 2 \leq p < 4.$$

Given the Fourier series $f = \sum a_n e^{2\pi i n^2 x}$, we have the Littlewood–Paley equivalence of norms

$$\|f\|_p \simeq \Big\|\Big(\sum_k |\Delta_k(x)|^2\Big)^{1/2}\Big\|_p, \qquad 1 < p < \infty,$$

where

$$\Delta_k(x) = \sum_{n=2^k}^{2^{k+1}-1} a_n\, e^{2\pi i n^2 x}.$$

In particular, for $2 \le p < 4$ we have

$$\|f\|_p \lesssim \Big[\int_0^1 \Big(\sum_k |\Delta_k(x)|^2\Big)^{p/2}\Big]^{\frac{2}{p}\cdot\frac{1}{2}} \le \Big[\sum_k \Big(\int_0^1 |\Delta_k|^p\Big)^{2/p}\Big]^{1/2}$$
$$= \Big(\sum_k \|\Delta_k\|_p^2\Big)^{1/2}.$$

But

$$\Delta_k(x) = \sum_{n=2^k}^{2^{k+1}-1} a_n\, e^{2\pi i n^2 x} = \sum_{n=2^k}^{2^{k+1}-1} a_n\, [S_n(x) - S_{n-1}(x)]$$
$$= \sum_{n=2^k}^{2^{k+1}-1} (a_n - a_{n+1})\, S_n(x) + a_{2^{k+1}-1}\, S_{2^{k+1}-1}(x),$$

where $S_n(x)$ denotes the partial sums of f. Then, since the coefficients are monotonically decreasing, we get

$$|\Delta_k(x)| \le 2\, C^*_{2^{k+1}}(x) \cdot a_{2^k}.$$

Therefore,

$$\|f\|_p \lesssim \Big(\sum_k a_{2^k}^2 \cdot 2^k\Big)^{1/2} \lesssim \|f\|_2. \qquad \square$$

8.4 The fractal dimension of Riemann's graphs

In his thesis, Riemann introduces also several examples of continuous functions lacking derivatives in a dense set of points, a project culminated by Weierstrass, who produced the first known continuous and nowhere differentiable function using lacunary trigonometric series. It seems that Weierstrass was inspired by the erroneous claim, attributed to Riemann, that the following two continuous functions,

$$F(x) = \sum_{n=1}^{\infty} \frac{\sin(2\pi n^2 x)}{n^2} \quad \text{and} \quad G(x) = \sum_{n=1}^{\infty} \frac{\cos(2\pi n^2 x)}{n^2},$$

are nowhere differentiable.

These functions have been studied by several authors, and nowadays we have a rather complete knowledge of its points of differentiability. In the following, we shall be concerned about the Minkowski (or box counting) dimension of their graphs.

Remark 8.10. With a given small number $h = N^{-1}$, the unit interval $[0, 1)$ can be decomposed as a disjoint union of subintervals $I_j = [(j-1)\,h, jh)$, $j = 1, 2, \ldots, N$. The number of boxes of edge h needed to cover the graph of a continuous function $F\colon [0, 1) \to \mathbb{R}$ is

$$A_h = \sum_{j=1}^{N} \sup_{x,y\in I_j} |F(x) - F(y)| + o(h^{-1}).$$

It then follows easily that the fractal (box counting, or Minkowski) dimension of the graph of F is given by the following limit:

$$\dim(\mathrm{graph}(F)) = -\lim_{h\to 0^+} \frac{\log A_h}{\log h}.$$

As was already pointed (Chapter 3), the quantity above only makes sense when the limit exists. By that reason, it is convenient to define the lower and upper fractal dimensions, $\underline{\dim}$ and $\overline{\dim}$, replacing lim by $\underline{\lim} = \liminf$ and $\overline{\lim} = \limsup$, respectively.

Given $0 < \delta \le 1$, let us consider the families of continuous functions

$$F_\delta(x) = \sum_{n=1}^{\infty} \frac{\sin(2\pi n^2 x)}{n^{1+\delta}} \quad \text{and} \quad G_\delta(x) = \sum_{n=1}^{\infty} \frac{\cos(2\pi n^2 x)}{n^{1+\delta}}.$$

Theorem 8.11 ([176]). $\dim_{\mathrm{B}}(\mathrm{graph}(F_\delta)) = \dim_{\mathrm{B}}(\mathrm{graph}(G_\delta)) = 7/4 - \delta/2$.

Proof. With a fixed (big) integer N, let $J_{a/q}$ be the arc containing a/q in the Farey dissection of $(0, 1]$ of order $N^{1/2}$, that is

$$(0, 1] = \bigcup_{a,q} J_{a/q}, \quad \text{with} \quad J_{a/q} = \left(\frac{a'+a}{q'+q}, \frac{a+a''}{q+q''}\right),$$

and where $a'/q' < a/q < a''/q''$ are three consecutive fractions in the Farey sequence $F_{N^{1/2}}$.

It happens that

$$J_{a/q} \subset \left\{x \in (0, 1] : \left|x - \frac{a}{q}\right| < \frac{1}{qN^{1/2}}\right\}.$$

Next, we shall consider incomplete Gaussian sums

$$S_{n,m}(x) = \sum_{k=m+1}^{n} e^{2\pi i k^2 x}, \quad S_n(x) = S_{n,0}(x).$$

From the previous section, we know the estimate

$$|S_{n,m}(x)| \lesssim \frac{n}{\sqrt{q}} + \sqrt{q}$$

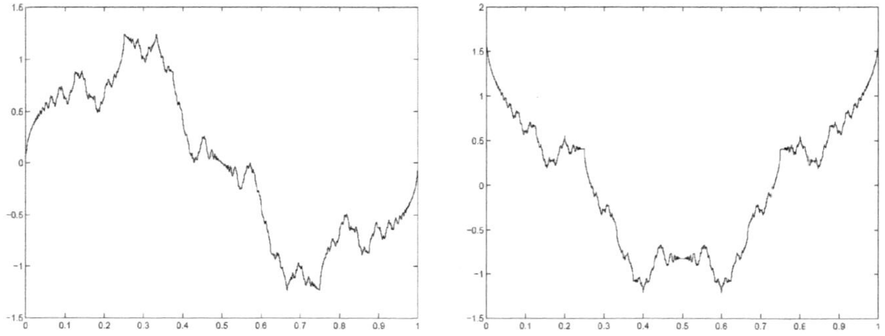

Fig. 8.1: Graphs of $F_1(x)$ (left) and $G_1(x)$ (right).

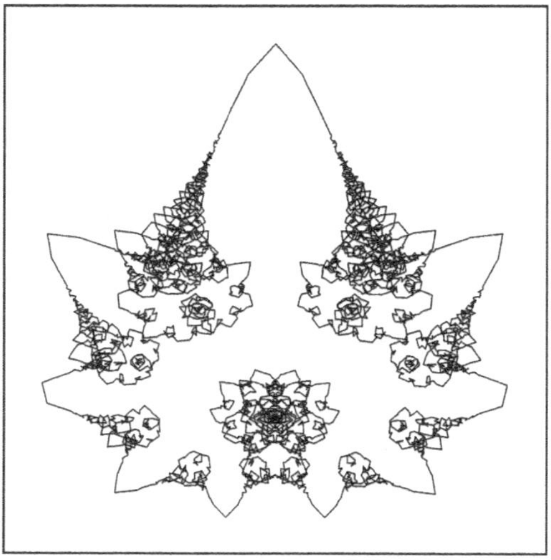

Fig. 8.2: Graph of $\sigma(t) = (F_{1/2}(t), G_{1/2}(t))$.

so long as $|x - a/q| \le 1/q^2$. For the complete sums, we have the explicit evaluation

$$\sum_{k=1}^{q} e^{2\pi i \frac{a}{q} k^2} = \varepsilon_q \sqrt{q} \left(\frac{a}{q}\right), \qquad \text{if } \gcd(a, q) = 1,$$

where $\left(\frac{a}{q}\right)$ denotes the Legendre symbol, and

$$\varepsilon_q = \begin{cases} 1 & \text{if } q \equiv 1 \bmod(4), \\ i & \text{if } q \equiv 3 \bmod(4), \\ 0 & \text{if } q \equiv 2 \bmod(4), \\ 1+i & \text{if } q \equiv 0 \bmod(4). \end{cases}$$

Given $n' = m + q[\frac{n-m}{q}]$, we have

$$S_{n,m}(a/q) = S_{n'-1,m}(a/q) + S_{n,n'}(a/q) = \varepsilon_q\left(\frac{a}{q}\right)\frac{n-m}{\sqrt{q}} + O(\sqrt{q}).$$

Let $\alpha(a/q; N)$ be the number of boxes of the mesh

$$\mathcal{M}_N = \left\{x = \frac{\ell}{N}, y = \frac{m}{n}; 0 \le \ell \le N\right\}$$

which are needed to cover the intersection graph$(F_\delta) \cap (J_{a/q} \times \mathbb{R})$, and let $\alpha(a/q; N, k)$ be the number of them contained inside the strip $I_k \times \mathbb{R}$, where $I_k = [k/N, (k+1)/N]$.

(i) We shall prove first the upper bound

$$\overline{\dim}_{\mathrm{B}}(\mathrm{graph}(F_\delta)) \le \frac{7}{4} - \frac{\delta}{2}.$$

Let M be a positive integer to be chosen later. From the definition of $\alpha(a/q; N, k)$, and after applying the mean value theorem, we get

$$\begin{aligned}\alpha\left(\frac{a}{q}; N, k\right) &\le 2 + N \sup_{x,y\in I_k\cap J_{a/q}} |F_\delta(x) - F_\delta(y)| \\ &\le 2 + 4\pi\left|\sum_{n\le M} n^{1-\delta} e^{2\pi i n^2 \xi}\right| + N\left|\sum_{n\ge M}\frac{1}{n^{1+\delta}}\left(e^{2\pi i n^2 x_0} - e^{2\pi i n^2 y_0}\right)\right|,\end{aligned}$$

where $\xi, x_0, y_0 \in I_k \cap J_{a/q}$. Partial summation yields

$$\begin{aligned}\alpha\left(\frac{a}{q}; N, k\right) &\lesssim 4\pi\left|\sum_{n\le M}[n^{1-\delta} - (n+1)^{1-\delta}]\, S_n(\xi)\right| \\ &\quad + |S_M(\xi)|\, M^{1-\delta} + (1+\delta) N \sum_{n\ge M}\frac{|S_{n,M}(x_0)| + |S_{n,M}(y_0)|}{n^{2+\delta}}.\end{aligned}$$

Now we choose $M = [\sqrt{N}]$, and from the known estimates for Gaussian sums, we get

$$\begin{aligned}\alpha\left(\frac{a}{q}; N, k\right) &\lesssim \left(\frac{M}{\sqrt{q}} + \sqrt{q}\right) + \left(\frac{M}{\sqrt{q}} + \sqrt{q}\right) M^{1-\delta} + N\left[\frac{1}{q^{1/2}}\cdot\frac{1}{N^{\frac{1}{2}(\frac{1}{2}+\delta)}}\right] \\ &\lesssim N^{1-\delta/2}\frac{1}{\sqrt{q}} + N^{1/2-\delta/2}\sqrt{q}.\end{aligned}$$

Next we observe that

$$\alpha\left(\frac{a}{q}; N\right) \lesssim \frac{|J_{a/q}|}{1/N}\left(N^{1-\delta/2}\frac{1}{\sqrt{q}} + N^{1-\delta/2}\sqrt{q}\right) \lesssim \frac{N^{3/2-\delta/2}}{q^{3/2}} + N^{1-\delta/2}\frac{1}{q^{1/2}}.$$

Therefore, the number of boxes of the mesh $\mathcal{M}_N$ needed to cover the graph is bounded by

$$\sum_{q\le N^{1/2}}\sum_{(a,q)=1}\left(\frac{N^{3/2-\delta/2}}{q^{3/2}}+N^{1-\delta/2}\frac{1}{q^{1/2}}\right)\lesssim\sum_{q\le N^{1/2}}\left(\frac{N^{3/2-\delta/2}}{q^{1/2}}+N^{1-\delta/2}q^{1/2}\right)$$
$$\lesssim N^{7/4-\delta/2},$$

implying the upper bound

$$\overline{\dim}_{\mathrm{B}}(\mathrm{graph}(F_\delta))\le\limsup_{N\to\infty}\frac{\log(N^{7/4-\delta/2})}{\log N}=\frac{7}{4}-\frac{\delta}{2}.$$

Obviously, the same proof also shows that

$$\overline{\dim}_{\mathrm{B}}(\mathrm{graph}(G_\delta))\le\frac{7}{4}-\frac{\delta}{2}.$$

(ii) We shall prove now the lower bounds

$$\underline{\dim}_{\mathrm{B}}(\mathrm{graph}(F_\delta))\ge\frac{7}{4}-\frac{\delta}{2}\quad\text{and}\quad\underline{\dim}_{\mathrm{B}}(\mathrm{graph}(G_\delta))\ge\frac{7}{4}-\frac{\delta}{2}.$$

Here we need to separate both cases. For F_δ, we choose P to be the set of prime numbers such that $p\equiv 3 \bmod (4)$ and $p\asymp\sqrt{N}$. For G_δ, we will proceed similarly, but changing to the primes $p\equiv 1 \bmod (4)$ for obvious reasons.

From the definition of Minkowski dimension, we have

$$\underline{\dim}_{\mathrm{B}}(\mathrm{graph}(F_\delta))\ge\liminf_{N\to\infty}\frac{\log\left(N\sum_{p\in P}\sum_{0\le a\le p}|F_\delta(a/p)-F_\delta(a/p+1/p^2)|\right)}{\log N}.$$

Next, we shall consider only the set $\mathcal{R}(p)$ of values of a which are quadratic residues mod (p). The corresponding sum over those a is greater than the absolute value of

$$\sum_{a\in\mathcal{R}(p)}\left(F_\delta\Big(\frac{a}{p}\Big)-F_\delta\Big(\frac{a}{p}+\frac{1}{p^2}\Big)\right)=\mathrm{Im}\left(\sum_n\sum_{r=1}^{(p-1)/2}\frac{e^{2\pi i r^2n^2/p^2}}{n^{1+\delta}}\left(1-e^{2\pi i n^2/p^2}\right)\right)$$
$$=\mathrm{Im}\left(\sum_n\frac{(-1+i\sqrt{p})\left(1-e^{2\pi i n^2/p^2}\right)}{2n^{1+\delta}}\right)$$
$$=\sqrt{p}\sum_n\frac{\sin^2(\pi n^2/q^2)}{n^{1+\delta}}+\sum_n\frac{\sin(2\pi n^2/q^2)}{2n^{1+\delta}}\gtrsim p^{1/2-\delta}.$$

The prime number theorem in arithmetic progressions yields

$$\underline{\dim}_{\mathrm{B}}(\mathrm{graph}(F_\delta))\ge\lim_{N\to\infty}\frac{\log\left(N\,N^{1/4-\delta/2}N^{1/2}/\log N^{1/2}\right)}{\log N}=\frac{7}{4}-\frac{\delta}{2}.\qquad\square$$

The extensions of Theorem 8.11 to power higher than two require several results from number theory.

The proof of many interesting propositions in analytic number theory can be sometimes reduced to get a non-trivial upper bound for an exponential sum. The first method to estimate these kind of sums was created in 1916 by Weyl, and it is completely based on the following lemma (see [196]).

Lemma 8.12 (Weyl's inequality). *If P is a polynomial of degree k and a/q is an irreducible fraction such that the leading coefficient of the polynomial P, say A, satisfies $|A - a/q| \le q^{-2}$, then*

$$\sum_{n\le N} e(P(n)) \lesssim \left(Nq^{-1/K} + N^{1-1/K} + N^{1-k/K}q^{1/K}\right) N^{\varepsilon}$$

for $K = 2^{k-1}$ and every $\varepsilon > 0$.

Using a method due to I. M. Vinogradov, it is possible to improve Weyl's inequality for large values of k. There are several kinds of results obtainable by Vinogradov's method, the following one is taken from Théorème 10.1 in [184] after some simplifications in the statement.

Lemma 8.13. *Let P, k, A and q be as in Weyl's inequality. Then, for $k \geqslant 9$ and $N \le q \le N^{k-2}$, we have the estimate*

$$\sum_{n\le N} e(P(n)) \lesssim N^{1-\delta},$$

where $\delta = (9k^2 \log k)^{-1}$.

In some contexts (e.g., studying sums of primes), exponential sums appear with coefficients that are only controlled in average, and so partial summation is not applicable. The large sieve inequality allows to obtain cancellation in some cases if we have a large enough number of these exponential sums. There are several formulations of the large sieve inequality depending on the context; perhaps the simplest is the following (see [172]).

Lemma 8.14 (Large sieve inequality). *Let $\delta > 0$ and $\{t_\nu\}_{\nu=1}^{R}$ be such that $0 \le t_1 < t_2 < \cdots < t_R < 1 - \delta$ and $t_{\nu+1} - t_\nu > \delta$. Then*

$$\sum_{\nu=1}^{R} \Big| \sum_{n\le N} a_n\, e(nt_\nu)\Big|^2 \le (N + \delta^{-1}) \sum_{n\le N} |a_n|^2$$

for arbitrary complex numbers a_n.

Let us now consider the family of functions

$$F_{\alpha,k}(x) = \sum_{n=1}^{\infty} \frac{c_n\, e(n^k x)}{n^\alpha}, \qquad \text{with} \qquad 0 < \underline{\lim}\, c_n \le \overline{\lim}\, c_n < \infty.$$

We can compute the fractal dimension of their graphs.

Theorem 8.15. *If* $(k+1)/2 \le \alpha \le k+1/2$, *then we have*

$$\dim(F_{\alpha,k}) = 2 + \frac{1}{2k} - \frac{\alpha}{k}.$$

Remark 8.16. Note that this result implies that the fractal dimension of $F_{\alpha,k}$ is non-integral for $(k+1)/2 \le \alpha < k+1/2$. On the other hand, if $\alpha > k+1/2$, $F_{\alpha,k}$ is absolutely continuous (note that $F'_{\alpha,k} \in L^2$), and the fractal dimension is trivially one. The range can be completed to $1 < \alpha \le k+1/2$ under the conjectural bound $\| \sum_{n\le N} c_n\, e(n^k x)\|_{2k} = O(N^{1/2+\varepsilon})$, which is true for $k=2$.

Proof of Theorem 8.15. Firstly we shall prove the upper bound

$$\overline{\dim}(F_{\alpha,k}) \le 2 + \frac{1}{2k} - \frac{\alpha}{k}. \tag{8.3}$$

Given $x, y \in I$, where I is a closed interval of length h, we separate the terms of $F_{\alpha,k}$ with $n > h^{-1/(k-1)}$, and we apply the mean value theorem to those with $n \le h^{-1/k}$. Then we obtain

$$|F_{\alpha,k}(x) - F_{\alpha,k}(y)| \lesssim h\,|S_1(\xi)| + |S_2(x)| + |S_2(y)| + h^{(\alpha-1)/(k-1)},$$

where

$$S_1(\xi) = \sum_{n\le h^{-1/k}} c_n\, n^{k-\alpha}\, e(n^k \xi) \quad \text{and} \quad S_2(t) = \sum_{h^{-1/k}<n\le h^{-1/(k-1)}} \frac{c_n\, e(n^k t)}{n^\alpha},$$

and $\xi \in I$. Taking $I = I_j$ (with I_j as in the definition of fractal dimension) and separating the contribution of odd and even j's to A_h (defined in Remark 8.10), we have

$$A_h \lesssim \sum_\nu |S_1(t_\nu)| + h^{-1} \sum_\nu |S_2(t'_\nu)| + h^{(\alpha-2k+1)/(k-1)} \tag{8.4}$$

where $h < t_{\nu+1} - t_\nu, t'_{\nu+1} - t'_\nu < 3h$.

Subdividing S_2 into dyadic intervals, we get

$$S_2(t'_\nu) \lesssim \sum_M \Big| \sum_{M<n\le 2M} \frac{c_n\, e(n^k t'_\nu)}{n^\alpha} \Big|,$$

where $M = 2^m h^{-1/k}$ with $h^{-1/k} \le M < h^{-1/(ic-1)}$. With a further subdivision of each dyadic interval $(M, 2M]$ into intervals of length $h^{-1}M^{1-k}$, we find M_1 and M_2, for each M, with $M < M_1 < M_2 \le 2M$ and $M_2 < M_1 + h^{-1}M^{1-k}$, such that

$$\sum_\nu |S_2(t'_\nu)| \lesssim \sum_M \frac{M}{h^{-1}\,M^{1-k}} \sum_\upsilon \Big| \sum_{M_1<n\le M_2} \frac{c_n\, e(n^k t'_\nu)}{n^\alpha} \Big|.$$

Let us denote $S_{2M}(t'_\nu)$ the innermost sum. Then, by (8.4) and Cauchy's inequality, we obtain

$$A_h^2 \lesssim h^{-1}\sum_\nu |S_1(t_\nu)|^2 + h^{-1}\sum_M M^{2k+\varepsilon}\sum_\nu |S_{2M}(t'_\nu)|^2 + h^{(2\alpha-4k+2)/(k-1)}. \tag{8.5}$$

The large sieve inequality implies

$$\sum |S_1(t_\nu)|^2 \lesssim (h^{-1}+h^{-1}) \sum_{n\le h^{-1/k}} |c_n|^2 n^{2k-2\alpha} \lesssim h^{-3+(2\alpha-1)/k-\varepsilon}$$

and

$$\sum |S_{2M}(t'_\nu)|^2 \lesssim (h^{-1}+h^{-1}) \sum_{M_1<n<M_2} |c_n|^2 M^{-2\alpha} \lesssim h^{-2}M^{1-k-2\alpha}.$$

Therefore, substituting in (8.5),

$$A_h^2 \lesssim h^{-4+(2\alpha-1)/k-\varepsilon},$$

and (8.3) follows.

Now we shall prove the lower bound

$$\underline{\dim}(F_{\alpha,k}) \geqslant 2 + \frac{1}{2k} - \frac{\alpha}{k}. \tag{8.6}$$

Let

$$f(x) = F_{\alpha,k}(x+h) - F_{\alpha,k}(x).$$

. By the definition of A_h and Hölder's inequality,

$$A_h \geqslant h^{-2}\|f\|_1 \geqslant h^{-2}\frac{\|f\|_2^3}{\|f\|_4^2}. \tag{8.7}$$

Parseval's identity proves

$$\|f\|_2^2 = \sum_n |c_n|^2 \frac{|1-e(n^k h)|^2}{n^{2\alpha}}.$$

Considering only the contribution of $1/(4h) < n^k < 3/(4h)$, we obtain

$$\|f\|_2^2 \gtrsim h^{(2\alpha-1)/k}.$$

On the other hand, subdividing into dyadic intervals we have that, for every $\varepsilon > 0$,

$$\|f\|_4^4 \lesssim \sup_N N^\varepsilon \sum_{N\le n<2N} c_n \frac{1-e(n^k h)}{n^\alpha} e(n^k x)\|_4^4$$

and hence, using again Parseval's identity,

$$\|f\|_4^4 \lesssim \sup_N N^{e-4\alpha} \sum_{N\le n<2N} \sum_{N\le m<2N} |r_k(n^k+m^k)|^2 |c_n|^2 |c_m|^2 |1-e(n^k h)|^2 \times |1-e(m^k h)|^2,$$

where r_k denotes the number of representations as a sum of two kth powers. It holds that $r_k(l) \lesssim l^e$ (because $l = m^k + n^k$ implies that $m+n$ divides l if k is odd, or $m^{k/2} + in^{k/2}$ divides l in $\mathbb{Z}[i]$ if k is even, and the number of divisors grows less than any positive power). Then

$$\begin{aligned}\|f\|_4^4 &\lesssim \sup_N N^{z-4\alpha}\Big(\sum_{N\le n<2N} |c_n|^2 |1-e(n^k h)|^2\Big)^2 \\ &\lesssim \sup_N N^{z-4\alpha} \min\left(N^2, N^{2+2k} h^{-2}\right) \lesssim h^{(4\alpha-2)/k-z}.\end{aligned}$$

Substituting the bounds for $\|f\|_2$ and $\|f\|_4$ in (8.7), we conclude

$$A_h \gtrsim h^{-2-1/2k+\alpha/k+\varepsilon}$$

for every $\varepsilon > 0$, which implies (8.6). □

8.5 Convergence and divergence of crystallographic partial sums

Take a function $f \in L^1(\mathbb{T})$ and consider its Fourier series

$$\sum_{k=-\infty}^{\infty} \hat{f}(k)\, e^{2\pi ikx}, \quad \text{where} \quad \hat{f}(k) = \int_0^1 f(x)\, e^{-2\pi ikx}\, \mathrm{d}x.$$

Here, $\mathbb{T}$ is identified with $[0,1)$ in the usual way. The traditional way of reconstructing the function f via its Fourier coefficients $(\hat{f}(k))_{k\in\mathbb{Z}}$ is to consider the N-partial sums

$$S_N f(x) = \sum_{|k|\le N} \hat{f}(k)\, e^{2\pi ikx}. \tag{8.8}$$

It is well known that there is convergence in norm: if $f \in L^p(\mathbb{T})$, for $1 < p < \infty$, then

$$S_N f \to f \quad \text{in } L^p(\mathbb{T}) \quad \text{as } N \to \infty.$$

And since Carleson [175] (and later Hunt [191]), it is also known that we have a.e.-convergence.

However, taking into account the interpretation of the Fourier coefficients as the diffraction pattern of a periodic distribution, it may seem to be more natural to pay attention to the coefficients that give us more information, that is, those of bigger magnitude, and to reconstruct the function ordering the Fourier coefficients in decreasing order of magnitude. That is, to consider, for each $\lambda > 0$, the partial sums

$$\tilde{S}_\lambda f(x) = \sum_{|\widehat{f}(k)|>\lambda} \widehat{f}(k)\, e^{2\pi i k x}, \tag{8.9}$$

and their limit when $\lambda \to 0$.

In 1996, see [192], T. W. Körner answered in the negative a question asked by L. Carleson and R. Coifman, proving the existence of a function $f \in L^2(\mathbb{T})$ such that

$$\limsup_{\lambda \to 0^+} \Big| \sum_{|\widehat{f}(k)|\geq\lambda} \widehat{f}(k)\, e^{2\pi i k x} \Big| = \infty \quad \text{for almost every } x \in \mathbb{T}. \tag{8.10}$$

This "bad" behaviour of Fourier series when summed with decreasing coefficients contrasts, for instance, with the case of other bases, such as wavelets, see [195].

Körner's proof is based on an ingenious modification of a construction due to Olevskii, and it also uses a probabilistic lemma of R. Salem and A. Zygmund. There is no explicit construction of a function with the required property.

A construction with Gaussian sums

With the use of Gaussian sums, one can produce explicit examples of functions $f \in L^p(\mathbb{T})$, with $p < 4/3$, such that (8.10) holds. Notice the range of p, compared with the result of Körner.

The construction is as follows. Fix $\varepsilon > 0$. For each $k \geq 0$, take

$$f_k(x) = \sum_{n=2^{2k}}^{2^{2k+2}-1} a_n \cos(2\pi n x), \quad \text{with } a_n = \begin{cases} \frac{1}{2^{k/2-k\varepsilon}}\left(1+\frac{1}{n}\right) & \text{if } n \text{ is a perfect square,} \\ \frac{1}{2^{k/2-k\varepsilon}}\left(1-\frac{1}{2^{2k}}\right) & \text{otherwise,} \end{cases} \tag{8.11}$$

and let

$$f(x) = 1 + 2\sum_{k=1}^{\infty} f_k(x). \tag{8.12}$$

One checks that

$$\|f_k\|_p \simeq 2^{k(\varepsilon+3/2-2/p)},$$

and so, taking ε small enough, we have that the function f defined in (8.12) belongs to $L^p(\mathbb{T})$ for each $p < 4/3$.

The aspect of the coefficients (the spectrum) of $f(x)$ across two consecutive blocks corresponding to $f_k(x)$ and $f_{k+1}(x)$ is depicted below.

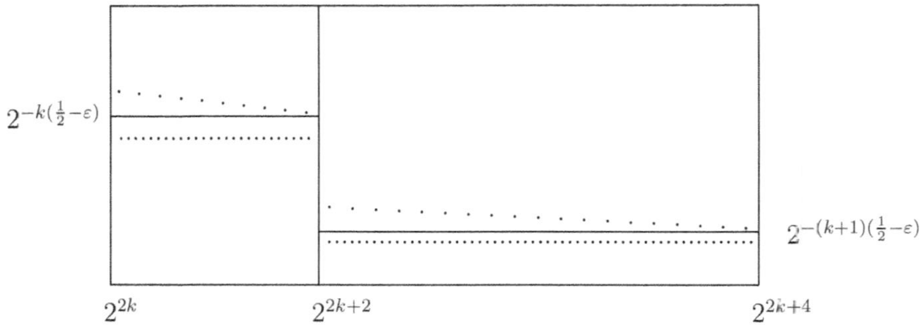

Observe that, for

$$b_k = \frac{1}{2^{k(1/2-\varepsilon)}}\left(1+\frac{1}{j}\right), \tag{8.13}$$

where $j = j(k)$ satisfies $2^{2k} \leq j < 2^{2k+2}$, and by the definition of the function f above, we have that

$$\sum_{|\hat{f}(\nu)|>b_k} \hat{f}(\nu)\, e^{2\pi i \nu x}$$
$$= \sum_{|\nu|\leq 2^{2k}} \hat{f}(\nu)\, e^{2\pi i \nu x} + \frac{1}{2^{k(1/2-\varepsilon)}} \sum_{2^{2k}\leq m^2<2^{2k}+j} \left(1+\frac{1}{m^2}\right) e^{2\pi i m^2 x}.$$

Thus, as $k \to \infty$, recalling the Carleson–Hunt convergence result, we get that, for almost all $x \in \mathbb{R}$,

$$\left|\sum_{|\hat{f}(\nu)|>b_k} \hat{f}(\nu)\, e^{2\pi i \nu x}\right| = \frac{1}{2^{k(1/2-\varepsilon)}}\left|\sum_{2^{2k}\leq m^2<2^{2k}+j} e^{2\pi i m^2 x}\right| + O(1). \tag{8.14}$$

Recall now the following facts.

Fact 8.17. *If* $|x-y| \leq 1/M^2$, *then*

$$\sup_{1\leq k\leq M}\left|\sum_{j=M}^{M+k} e^{2\pi i j^2 x}\right| \geq C \sup_{1\leq k\leq M}\left|\sum_{j=M}^{M+k} e^{2\pi i j^2 y}\right|.$$

Fact 8.18. *If* $\gcd(P,Q)=1$, *with* $M/2 \leq Q \leq M$ *and* Q *odd, then*

$$\sup_{1\leq k\leq M}\left|\sum_{j=M}^{M+k} e^{2\pi i j^2 P/Q}\right| \geq \sqrt{Q}.$$

Fact 8.17 follows from summation by parts, while Fact 8.18 is a consequence of Theorem 7.6.

We shall construct the sequence b_k in (8.13) in terms of x. Let $x \in [0,1]$ be an irrational number, and consider the sequence of convergents $\{P_m/Q_m\}$ of its continuous fraction expansion. They satisfy that $\gcd(P_m, Q_m) = 1$ and

$$\left|\frac{P_m}{Q_m} - x\right| < \frac{1}{Q_m^2}.$$

There is an infinite number of odd denominators Q_n.

The (sub)sequence $\{Q_\alpha\}$ of odd denominators determines dyadic blocks

$$2^{k_\alpha} \le Q_\alpha \le 2^{k_\alpha+1}. \tag{8.15}$$

With that sequence of indices $\{k = k_\alpha\}$, consider the numbers b_k of (8.13), where $j = j(k)$ is chosen so that

$$\left|\sum_{2^{2k} \le m^2 < 2^{2k+j}} e^{2\pi i m^2 x}\right| = \sup_{1\le \ell \le 2^k} \left|\sum_{s=2^k}^{2^k+\ell} e^{2\pi i s^2 x}\right|.$$

Using Fact 8.17 first, and then Fact 8.18, we get that

$$\sup_{1\le \ell \le 2^k} \left|\sum_{s=2^k}^{2^k+\ell} e^{2\pi i s^2 x}\right| \ge C \sup_{1\le \ell \le 2^k} \left|\sum_{s=2^k}^{2^k+\ell} e^{2\pi i s^2 P_\alpha/Q_\alpha}\right| \ge C\sqrt{Q_\alpha}.$$

Finally, with (8.14) and (8.15), we conclude that

$$\left|\sum_{|\hat f(\nu)|>b_k} \hat f(\nu)\, e^{2\pi i \nu x}\right| \ge C\, 2^{k\,\varepsilon} + O(1).$$

This proves that this function f satisfies the a.e.-divergence property in (8.12).

Convergence/divergence in norm

One also can show that L^p-convergence, $p < 2$, also fails for the partial sums $\tilde S_\lambda$ of (8.9). More concretely, we have the following.

Theorem 8.19. *For all p, $1 \le p < 2$, there is a function $f \in L^p(\mathbb{T})$ (explicitly constructed) such that*

$$\limsup_{\lambda\to 0^+} \|\tilde S_\lambda f\|_p = \infty.$$

The construction of this function is also of number-theoretical nature, and similar to the construction exhibited above, but with prime indices in the role of the perfect squares. See [182] for details.

One may argue that one reason for the failure of $\tilde S_\lambda f$ to converge is because the *uncertainty principle* has not been taken into account: it does not make sense to impose restrictions upon the size of $|\widehat f(k)|$ and not upon $|k|$ itself. Given a decreasing function ϕ, one can consider partial sums:

$$S_N^\phi f(x) = \sum_{\substack{|k| \le N \\ |\widehat{f}(k)| \ge \phi(N)}} \widehat{f}(k) e^{2\pi i k x}$$

Then our construction shows that different behaviours of S_N^ϕ, in terms of convergence or divergence in norm, depend upon the condition

$$N \cdot \phi^2(N) = o(1).$$

In particular, for $\phi(N) = N^{-\delta}$, $\delta > 0$, the convergence in the L^p-norm, $p < 2$, of

$$S_N^\delta f(x) = \sum_{\substack{|k| \le N \\ |\widehat{f}(k)| \ge N^{-\delta}}} \widehat{f}(k)\, e^{2\pi i k x}$$

requires an exponent $\delta > 1/2$, quite in line with the uncertainty principle. See [182].

Chapter 9
Weyl sums and atomic energy oscillations

The elements, if arranged according to their atomic weights, exhibit an apparent periodicity of properties.

Dmitri I. Mendeleev

9.1 Statement of objectives

Mathematicians are by nature reductionist and often dream, and also pretend, of being able to deduce rigorously the laws of Science from a small set of first principles. It is a plausible goal having a distinguished history, but very difficult to achieve. And it rarely excites physicists, because what mathematicians can rigorously prove is not usually a hot topic for them. Nevertheless, we can always quote E. Wigner's comment about the unreasonable effectiveness of mathematics as a fundamental instrument for the formulation and progress of physical theories.

We learned with Euclid that an axiom is a self-evident truth, but later discovered, first with Gauss, Bólyai and Lobachevski, and later with Gödel and the continuum hypothesis as a good example, that it may not be always the case. On the other hand, physicists are often willing to assume new principles that are far from being self-evident and with the only reason that they are extremely useful. That is the case of Schrödinger's equation and the ansatz that its wave function solutions provide accurate description of the physical properties of a given system.

This being a book about harmonic analysis and Fourier transform, it seems adequate to quote that the "uncertainty principle", a basic principle of quantum mechanics, has played an important role in several chapters: mainly in the statement that if a smooth function is adapted in configuration space to a parallelepiped (i.e., its derivatives on each directions decay accordingly with the corresponding side of the parallelepiped), then its Fourier transform is adapted to a dual parallelepiped in moment space, so that the product of their volumes must be bigger than an absolute constant (which with abuse of language may be called Planck's constant). But also in Chapter 6, when it was considered the impossibility, in communication theory, of the simultaneous confinement of a signal in both time and frequency. The interested reader may consult the references [199, 205, 214, 213], where the subject of the uncertainty principle in Fourier analysis and quantum mechanics has been developed.

A. Córdoba, *Suprematism in Harmonic Analysis*, Progress in Mathematics 356,
https://doi.org/10.1007/978-3-031-73737-4_9

However, without further ado, let us dive right into the midst of things, and state that in this chapter, we shall study the following sum, that plays a crucial role in the asymptotic analysis of non-relativistic atomic energies:

$$\Psi_Q(Z) = \sum_{l=1}^{l_{\mathrm{TF}}} \frac{2l+1}{\frac{1}{\pi}\int \left(V_{\mathrm{TF}}^Z(r) - \frac{l(l+1)}{r^2}\right)_+^{-1/2} \mathrm{d}r}\, \mu\left(\frac{1}{\pi}\int \left(V_{\mathrm{TF}}^Z(r) - \frac{l(l+1)}{r^2}\right)_+^{1/2} \mathrm{d}r\right),$$

where

- $\mu(x) = \mathrm{dist}\{x, \mathbb{Z}\}^2 - 1/12$,
- V_{TF}^Z is the Thomas–Fermi potential with charge Z, which satisfies the perfect scaling condition

(9.1) $$V_{\mathrm{TF}}^Z(r) = Z^{4/3}\, V_{\mathrm{TF}}^1(Z^{1/3} r),$$

  and we have

(9.2) $$V_{\mathrm{TF}}^1(r) = \frac{y(ar)}{r}, \qquad a = \left(\frac{3\pi}{2}\right)^{2/3},$$

  where y is the Thomas–Fermi function, solution of the Thomas–Fermi equation

(9.3) $$\begin{cases} y''(r) = y^{3/2}(r)/\sqrt{r}, \\ y(0) = 1, \\ \lim_{r\to+\infty} y(r) = 0, \end{cases}$$

- and l_{TF} is the greatest integer such that $V_{\mathrm{TF}}^Z(r) - l(l+1)/r^2$ is positive somewhere.

Here, we denote $(x)_+^{-1/2} = x^{-1/2}$ if $x > 0$, and $(x)_+^{-1/2} = 0$ otherwise.

To understand the role of the function $\Psi_Q(Z)$ in atomic physics, let us discuss briefly the Born–Oppenheimer model for a non-relativistic molecule.

9.2 The Born–Oppenheimer model

Consider

- M nuclei of positive charges $Z_1 e, \ldots, Z_M e$, placed at positions $y_1, \ldots, y_M \in \mathbb{R}^3$.
- $Z = (Z_1, \ldots, Z_M)$, $Y = (y_1, \ldots, y_M)$.
- N quantized electrons in positions $x_1, \ldots, x_N \in \mathbb{R}^3$.
- $-e$ = electron charge; m = electron mass; $\hbar$ = Planck's constant.

Following the Born–Oppenheimer model, a non-relativistic molecule is described by the Hamiltonian

$$H_{Y,Z,N} = -\sum_{j=1}^{N}\left(\frac{\hbar}{2m}\Delta_{x_j} + \sum_{k=1}^{M}\frac{Z_k e^2}{|x_j - y_k|}\right) + \sum_{1\le j<k\le N}\frac{e^2}{|x_j - x_k|} + \sum_{1\le m<n\le M}\frac{Z_n Z_m e^2}{|y_n - y_m|},$$

where the j-th Laplacian represents the quantized kinetic energy of the j-th electron, and we have also the Coulombian interactions: attraction between the nuclei and the electrons, and two repulsion terms corresponding to the nuclei, one of them, and the electrons, the other.

This Hamiltonian acts on

$$\mathcal{H} = \bigwedge_{j=1}^{N} L^2(\mathbb{R}^3 \otimes G), \quad G = \{+1, -1\}.$$

Here, the symbol $\bigwedge$ denotes the antisymmetric tensor product and encodes the Pauli exclusion principle (electron are fermions satisfying the statistics of Fermi–Dirac); while $+1$ (spin up) and -1 (spin down) denote the two kind of electronic spin.

The *ground state energy* is given by

$$E(Z) = \inf_N \inf_Y \inf_{\substack{\Psi \in \mathcal{H} \\ \|\Psi\|=1}} \langle \Psi, H_{Y,Z,N} \Psi \rangle.$$

To achieve the dream of explaining mathematically chemical phenomena, one would need a very precise calculation of these energies, a task that is several orders of difficulty beyond what can be obtained today.

In the following, we will confront the more "modest" goal of launching the first step to explain the periodic table, for which it will be enough to consider the Hamiltonian of neutral atoms.

9.3 Neutral atoms. Hartree–Fock and Thomas–Fermi theories

Consider a neutral atom with Z protons placed at the origin $y = 0$, and choose units such that $e^2 = 1$ and $\hbar/2m = 1$. Then

$$H_Z = -\sum_{j=1}^{Z} \left(\Delta_{x_j} + \frac{Z}{|x_j|} \right) + \sum_{1 \le j < k \le Z} \frac{1}{|x_j - x_k|}.$$

Although the spin variables are fundamental in the description of the electron behaviour, they are irrelevant from the point of view of understanding the mathematical model. Therefore, they can be omitted in the discussion, simplifying the structure of the Hilbert space of states to

$$\mathcal{H} = \bigwedge_{j=1}^{Z} L^2(\mathbb{R}^3) = \{\Psi(x_1, \ldots, x_Z), x_j \in \mathbb{R}^3 : \Psi \in L^2(\mathbb{R}^{3Z}), \Psi \text{ antisymmetric } \}.$$

Antisymmetry means that $\Psi(x_{\sigma(1)}, \ldots, x_{\sigma(Z)}) = (-1)^{\text{sign}(\sigma)} \Psi(x_1, \ldots, x_Z)$ for every permutation σ.

The electronic density is given by

$$\rho(x) = Z \int_{\mathbb{R}^{3(Z-1)}} |\Psi(x, x_2 \ldots, x_Z)|^2 \, dx_2 \cdots dx_Z,$$

and so,

$$\int_\Omega \rho(x) \, dx$$

is the expected number of electrons in $\Omega \subset \mathbb{R}^3$.

The ground state energy reads now

$$E(Z) = \inf_{\substack{\Psi \in \mathcal{H} \\ \|\Psi\|=1}} \langle \Psi, H_Z \Psi \rangle$$

Therefore, the task is to obtain the lowest eigenvalue of the Hamiltonian $\mathcal{H}_Z$ acting on antisymmetric square integrable functions in the 3Z-dimensional Euclidean space. But this is still a very hard problem, mainly due to the Coulombian interaction of the electrons.

From the earliest days of quantum mechanics it was known that solving explicitly the Schrödinger equation $H_Z \Psi = \lambda \Psi$ was an impossible task, and several approximated methods of solution were introduced.

If our Hamiltonian had the form

$$\widetilde{H}_Z = \sum_{k=1}^{Z} (-\Delta_{x_k} + V(x_k))$$

for a one electron potential V, then, by separation of variables, the problem would reduce from $3Z$ to 3 dimensions. Furthermore, exploiting the radial symmetry, it would become merely one-dimensional (i.e., an ordinary differential equation), and would be easily solved.

The main idea introduced to study atoms in quantum theory (physics and chemistry) is to approximate H_Z by $\widetilde{H}_Z$, where the potential $V(x)$ is chosen as follows: let us pretend for a moment that we know the electronic density ρ; then a good candidate for such one-electron potential would be the following:

$$V(x) = -\frac{Z}{|x|} + \int_{\mathbb{R}^3} \frac{\rho(y)}{|x-y|} \, dy.$$

It is easy to deduce that the ground state eigenfunction and the ground state electronic density for $\widetilde{H}_Z$ are given by

- $\Psi_{\mathrm{HF}}(x_1, \ldots, x_Z)$, the antisymmetrized product (Slater's determinant) of the eigenfunctions $\phi_1(x), \ldots, \phi_Z(x)$ for the three-dimensional Schrödinger operator $-\Delta + V(x)$.
- $\rho_{\mathrm{HF}}(x) = \sum_{k=1}^{Z} |\phi_k(x)|^2$.

The method (Hartree–Fock) is a great simplification of the original problem, because now we are confronting just a 3-dimensional calculation instead of a $3Z$-dimensional one. The system is closed asking the consistency condition (Hartree–Fock equation) that the obtained density equals the starting one:

$$\rho_{\mathrm{HF}} = \rho.$$

It is a remarkable and strange equation (neither an algebraic equation, a partial differential equation or an integral equation), and it is solved by iteration: starting with an initial density ρ_0, we produce the Hartree–Fock ρ_1, which taken as a new initial datum produces ρ_2, and so on. The limit density, in the hypothesis that it exists, would be the solution that we are looking for.

In this setting, it is easily deduced that the Hartree–Fock ground state is given, as was pointed above, by the antisymmetrized product of the eigenfunctions, and that the sum of the negative eigenvalues is the lowest possible energy of the Hamiltonian.

It is not obvious that the method will work and, in any case, the convergence will depend on the choice of an appropriate initial density. Here the Thomas–Fermi theory, also called statistical theory of atoms, came to help providing us with a good first-step approximations of both the density and the eigenvalue sum of a three-dimensional Schrödinger operator.

The Thomas–Fermi theory is a technique used to manage quantum mechanics for systems with many particles. There is no hope of understanding the usual wave function $\Psi(x_1, \ldots, x_Z)$ when Z is big. Instead, one tries to understand the electron density

$$\rho(x_1) = Z \int |\Psi(x_1, x')|^2 \, \mathrm{d}x',$$

which is a function of $x_1 \in \mathbb{R}^3$, and hope that important information about Ψ is contained in $\rho = \rho^Z_{\mathrm{TF}}$.

The Thomas–Fermi theory derives a PDE describing ρ^Z_{TF}, and guess the energy of the system from the knowledge of the electron density ρ^Z_{TF}, see [221]. It views the Quantum Mechanics as a classical gas filling the phase space and interacting with itself via the Coulomb repulsion and with the potential

$$V^Z_{\mathrm{TF}}(x) = -\frac{Z}{|x|} + \frac{1}{2} \int_{\mathbb{R}^3} \frac{\rho^Z_{\mathrm{TF}}(y)}{|x-y|} \, \mathrm{d}y,$$

which satisfies the Thomas–Fermi equations described in (9.1)–(9.3), and under the hypothesis of radial symmetry, those equations become a non-linear ordinary differential equation which can be solved, showing that the dependence on the parameter Z can be removed by a trivial scaling:

$$\rho^Z_{\mathrm{TF}}(x) = Z^2 \rho(Z^{1/3} x) \quad \text{and} \quad V^Z_{\mathrm{TF}}(x) = Z^{4/3} V(Z^{1/3} x),$$

where the density ρ and the potential V are some perfectly determined functions. These formulas have a natural interpretation in terms of the "atomic radius" $Z^{-1/3}$, and the average distance $Z^{-2/3}$ between the electrons.

A well established tradition in mathematics is to approximate non-linear equations by linear ones. The Thomas–Fermi and Hartree–Fock approaches involve approximating a linear system in a large number of variables by a non-linear system in a few variables, see [221].

Remark. A crucial step in the derivation of the Thomas–Fermi equation is the expression of the kinetic energy in terms of the electronic density function:

$$\mathcal{E}_{\mathrm{KE}} = C_{\mathrm{TF}} \int \rho^{5/3}(x)\,\mathrm{d}x,$$

so that the total energy is

$$\mathcal{E}_{\mathrm{TF}} = C_{\mathrm{TF}} \int \rho^{5/3}(x)\,\mathrm{d}x - \int V(x)\,\rho(x)\,\mathrm{d}x + \frac{1}{2}\iint \frac{\rho(x)\,\rho(y)}{|x-y|}\,\mathrm{d}x\,\mathrm{d}y.$$

See [221].

In order motivate the Thomas–Fermi ansatz about $\mathcal{E}_{\mathrm{KE}}$, and to establish an interesting connection with the lattice point problems of Chapter 7, in the following we shall describe a toy model of N free particles in a δ-box $B_\delta = [0,\delta] \times [0,\delta] \times [0,\delta]$ in $\mathbb{R}^3$. That is, we assume that they do not exert any force on each other, and then look for the ground state energy that these particles have, which is given by the wave function $\Psi(x_1,\ldots,x_N)$.

Observe that Ψ is supported on $B_\delta \times \cdots \times B_\delta = B_\delta^{(N)}$, and satisfies

$$\int_{B_\delta^{(N)}} |\Psi(x_1,\ldots,x_N)|^2 = 1, \qquad \mathcal{E}_{\mathrm{QE}} = \langle -\Delta\Psi, \Psi\rangle,$$

and $\rho(x_1) = N \int |\Psi(x_1,x')|^2\,\mathrm{d}x'$ is the particle density.

Let us also make the hypothesis that the wave function Ψ is antisymmetric (Pauli's exclusion principle), and then ask for the state of minimal energy, or ground state.

Let $f_1, f_2, \ldots$ be the eigenfunctions of $-\Delta$ under Dirichlet boundary conditions in B_δ, and let $\lambda_1, \lambda_2, \ldots$ be the corresponding eigenvalues. Since the Laplacian in $\mathbb{R}^{3N}$ splits by separation of variables, the eigenfunctions of our problem will be

$$\Psi_{m_1,\ldots,m_N}(x_1,\ldots,x_N) = f_{m_1}(x_1)\cdots f_{m_N}(x_N),$$

with eigenvalues $\lambda_{m_1} + \cdots + \lambda_{m_N}$.

Taking into account the antisymmetry, we must choose different m_j's if we want Ψ not to be identically equal to zero. Therefore, we arrive to the antisymmetric eigenfunctions

$$\Psi_{m_1,\ldots,m_N}(x_1,\ldots,x_N) = \frac{1}{N!}\sum_{\sigma\in S_N} \mathrm{sign}(\sigma)\, f_{m_1}(x_1)\cdots f_{m_N}(x_N),$$

where the sum is taken over all the elements σ of the symmetric group S_N.

But we know explicitly those eigenfunctions in the box B_δ; they are

$$\prod_{j=1}^{3} \sin\left(\frac{\pi \nu_j \cdot x_j}{\delta}\right), \qquad \text{where } \nu_j = (\nu_j^1, \nu_j^2, \nu_j^3) \in \mathbb{Z}^3,$$

and their corresponding eigenvalues are

$$\frac{\pi^2}{\delta^2}\left((\nu_j^1)^2 + (\nu_j^2)^2 + (\nu_j^3)^2\right) = \frac{\pi^2}{\delta^2}\,|\nu_j|^2.$$

To estimate the sum of the first eigenvalues, we take a sphere of radius r containing N-lattice points (except perhaps for a small error), and observe that

$$N \simeq \mathrm{vol}\{|\xi| < r : \xi_j \geq 0\} \simeq r^3.$$

Therefore, the energy is

$$\simeq \int_{|\xi|<r,\, \xi_j>0} \frac{\pi^2}{\delta^2}\,|\xi|^2\,\mathrm{d}\xi = C\,\frac{r^5}{\delta^2},$$

which we write in the form

$$C\left(\frac{N}{\delta^3}\right)^{5/3}\delta^3.$$

But N/δ^3 is the particle density in the box B_δ, and δ^3 is the volume. Therefore, dividing the space into small boxes B_δ^k, where $\rho = \rho_k$ is approximately constant, adding all those terms above, and taking limit when the number N of particles is big enough, we get the searched expression for the kinetic energy

$$C_{\mathrm{TF}} \int \rho^{5/3}(x)\,\mathrm{d}x.$$

See [221, 220] for a precise deduction of this formula and also for the expression of the Hartree–Fock equation in a more amenable form.

9.4 The ground state energy of the atom

The lines above are just a crude impressionistic view of the Hartree–Fock and Thomas–Fermi theories needed for the presentation of the main point here, that is the rigorous (mathematical theorems) results for the ground state energy of the atom, which can be summarized in the following (asymptotic) expansion:

$$E(Z) = C_{\mathrm{TF}}\,Z^{7/3} + C_{\mathrm{S}}\,Z^2 + C_{\mathrm{SD}}\,Z^{5/3} + \Psi_Q(Z) + \cdots$$

- The term $C_{\mathrm{TF}}\,Z^{7/3}$ was suggested in 1927 by Thomas–Fermi [212, 226], and proved in 1977 by Lieb–Simon [221].

- The term $C_s Z^2$ appeared first in numerical computations, was explained in 1952 by Scott [222], and proved rigorously by Hughes–Siedentop–Weikard in 1985-89 [224, 225, 218]. It reflects the fact that electrons near the nucleus behave as if they have not electronic interactions (screening effect).
- The term $C_{SD} Z^{5/3}$ was conjectured in 1930 by Dirac [203], and in 1981 by Schwinger [223]. It was proved by Fefferman–Seco [207] in 1990. It represents a small effect due to delicate electronic correlations.

The next term $\Psi_Q(Z)$ is oscillatory, and is given by the formula which was stated before:

$$\Psi_Q(Z) = \sum_{l=1}^{l_{TF}} \frac{2l+1}{\frac{1}{\pi}\int \mathcal{F}(r)^{-1/2}\,\mathrm{d}r} \cdot \mu\Big(\frac{1}{\pi}\int \mathcal{F}(r)^{1/2}\,\mathrm{d}r\Big),$$

where

$$\mathcal{F}(r) := \Big(V_{TF}^Z(r) - \frac{l(l+1)}{r^2}\Big)_+,$$

with the convention $(x)_+ = 0$ if $x \leq 0$, and where

$$V_{TF}^Z(r) = Z^{4/3}\, V(Z^{1/3} r)$$

is the Thomas–Fermi potential, and

$$l_{TF} = \{\text{greatest integer } l \text{ such that } V_{TF}^Z(r) - l(l+1)/r^2 > 0 \text{ somewhere}\}.$$

Finally,

$$\mu(x) = \operatorname{dist}(x, \mathbb{Z})^2 - \frac{1}{12}.$$

The graph of $\mu(x)$ is depicted below.

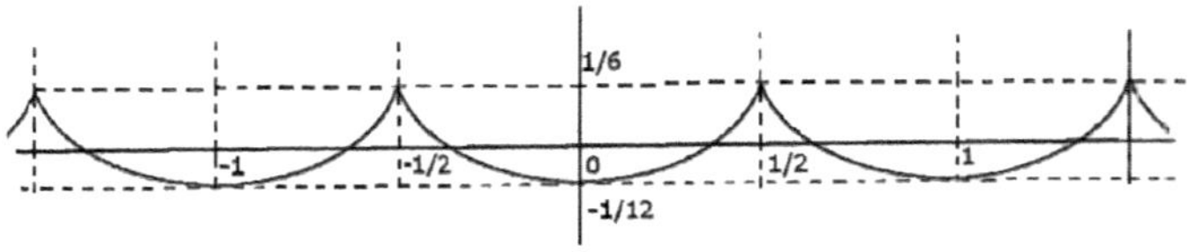

The oscillatory (almost periodic) character of Ψ_Q is very interesting, and there is a reasonable chance that it encodes the contribution to the energy of the electrons of valence. Our analysis of that term is contained in the following theorem.

Theorem 9.1.

$$-c_1 Z^{3/2} \leq \Psi_Q(Z) \leq c_1 Z^{3/2} \quad \text{and} \quad \Big(\frac{1}{Z}\int_1^Z |\Psi_Q(u)|^2\,\mathrm{d}u\Big)^{1/2} \geq c_2 Z^{3/2},$$

where $0 < c_1, c_2 < \infty$ *are universal constants.*

In particular, we have that

$$\text{Average of } |\Psi_Q| \approx Z^{3/2}.$$

The remainder of this chapter will be dedicated to presenting the details of the proof of Theorem 9.1. It is a lengthly argument that involves a series of similar calculations of oscillatory integrals employing techniques such as stationary phase, Van der Corput estimates, and Poisson summation formula. In a certain sense, as was pointed out in the introduction, it resides in the antipodal site of the mathematical haikus of Chapter 1.

For readers well versed in the mathematical theory of atoms, and familiar with the contents of references [208, 209, 218, 221], this chapter will provide a comprehensive exposition of the estimates of the intriguing oscillatory term appearing in the asymptotic development of the ground state energy. However, for those who may not be acquainted with the Hartree–Fock and Thomas–Fermi models, but posses an interest in the harmonic analysis of previous chapters, delving into the details of this proof can further stimulate the belief in the power of this field as a suitable tool for describing natural phenomena. Let us think also that it serves as an incentive to explore the rich and potent mathematics contained in the aforementioned references, and that perhaps it will stimulate future progress in the mathematical dream of establishing chemistry "more geometrico demonstrata", as Spinoza would say.

In our analysis, we first decode its expression emphasizing the analogies with some classical formulas of number theory, namely those appearing in the problem of counting lattice points inside regions of Euclidean space. Then we can make use of the tools introduced by number theorists to deal with trigonometric sums to achieve the proof of the theorem.

From the abstract mathematical perspective, sums such as Ψ_Q are quite old, the best known going back to Gauss, which is related to estimating the number of integral lattice points inside a convex region, as was considered in Chapter 7.

It is worth noting the close similarity between our problem and the circle problem, which comes from a refined analysis of the number of bound states of quantum free particles in a box.

The proof is organized as follows. First, after making some trivial modifications to the well-known stationary phase lemma (Section 9.5), we set out (in Section 9.6) to study sums of the type

$$S(\lambda) = \sum_{l=1}^{\lambda} f\Big(\frac{l}{\lambda}\Big)\mu\Big(\lambda\phi\Big(\frac{l}{\lambda}\Big)\Big),$$

where $\phi''(x) \geq c_0 > 0$, and μ is a periodic function of average 0. Examples of such sums are the following.

1. If $f \equiv 1$, $\mu(x) = e^{2\pi i x}$ and $\phi(x) = x^2$, we have the well-known Gauss sums modulo λ.
2. If $f \equiv 1$ and $\mu(x) = x - [x] - 1/2$, then S represents the error term in the lattice point problem for a curve ϕ dilated by λ.

While the first item above is well understood, the second remains very hard. In our case, we will have to deal only with functions μ whose Fourier coefficients decrease rather rapidly ($\hat{\mu}(n) \sim |n|^{-3/2+\varepsilon}$), and this allows a complete analysis of the sums via the usual method of Van der Corput (Poisson summation followed by stationary phase; see [215]), since all expressions turn out to be absolutely convergent. A little elementary number theory argument will be needed here to rule out the possibility of a small denominator problem, which gives rise to an error term whose size depends on whether a certain number is rational or irrational.

In Section 9.7, we apply the results of Section 9.6 to Ψ_Q, obtaining a new sum Ψ_0, a leading "dual" version of Ψ_Q, reminiscent of the Jacobi identity for the modular function. Sharp upper bounds for Ψ_Q are an easy consequence of this. However, obtaining the right regularity properties for the curve and amplitude involved in the formula for Ψ_Q turns out to be rather tedious.

In Section 9.8, we obtain lower bounds for Ψ_Q in the form of an Ω-result, by understanding how Ψ_0 behaves on average.

These results were first published in [200, 201], and the following pages are an almost verbatim transcription of the contents of those papers, which depend heavily upon the estimates proven in [208].

9.5 Stationary phase estimates

Let us begin with a review of stationary phase. Consider $f \in C_0^\infty(\mathbb{R})$. Then, if $t > 0$,

$$\int_{-\infty}^{\infty} e^{itx^2} f(x)\,\mathrm{d}x = e^{\pi i/4}\sqrt{\frac{\pi}{t}}\int_{-\infty}^{\infty} e^{-\pi^2 i\xi^2/t}\,\widehat{f}(\xi)\,\mathrm{d}\xi.$$

Using the identity

$$e^s = 1 + \int_0^1 e^{su}\, s\,\mathrm{d}u,$$

we deduce

$$\begin{aligned}\int_{-\infty}^{\infty} e^{itx^2} f(x)\,\mathrm{d}x &= e^{\pi i/4}\sqrt{\frac{\pi}{t}}\left(f(0) + \int_{-\infty}^{\infty}\widehat{f}(\xi)\int_0^1 \frac{-\pi^2 i\xi^2}{t}\, e^{-\pi^2 iu\xi^2/t}\,\mathrm{d}u\,\mathrm{d}\xi\right)\\ &= e^{\pi i/4}\sqrt{\frac{\pi}{t}}\left(f(0) + \frac{i}{4t}\int_{-\infty}^{\infty}\widehat{f''}(\xi)\int_0^1 e^{-\pi^2 iu\xi^2/t}\,\mathrm{d}u\,\mathrm{d}\xi\right)\\ &= e^{\pi i/4}\sqrt{\frac{\pi}{t}}\, f(0) + \frac{i}{4t}\int_{-\infty}^{\infty} f''(x)\int_0^1 e^{itx^2/u}\,\frac{\mathrm{d}u}{u^{1/2}}\,\mathrm{d}x\\ &= e^{\pi i/4}\sqrt{\frac{\pi}{t}}\, f(0) + \frac{i}{4t^{3/2}}\int_{-\infty}^{\infty} f''(x)\, g_t(x)\,\mathrm{d}x, \qquad (9.4)\end{aligned}$$

for

$$g_t(x) = \int_0^1 e^{itx^2/u}\left(\frac{t}{u}\right)^{1/2}\mathrm{d}u.$$

Note that $g_t(x) = t^{1/2} g_1(t^{1/2}x)$, and that

$$g_1(x) = i\, e^{ix^2/u} \frac{u^{3/2}}{x^2}\Big|_0^1 - \frac{3i}{2} \int_0^1 e^{ix^2/u} \frac{u^{1/2}}{x^2}\, du,$$

hence $|g_1(x)| \le 2/|x|^2$ and thus, g_1 is integrable. Furthermore,

$$\|g_t\|_1 = \|g_1\|_1 = O(1) \quad \text{and} \quad |g_t(x)| \le 2|x|^{-2} t^{-1/2}.$$

We also consider one-sided integrals of the form

$$\int_0^\infty e^{itx^2} f(x)\, dx.$$

Define

$$f^+(x) = \begin{cases} f(x), & \text{if } x \ge 0, \\ f(-x), & \text{if } x \le 0, \end{cases}$$

and consider $f_\varepsilon = f^+ * \varphi_\varepsilon$, for a suitable approximation φ_ε of the identity. Using our previous identity, we obtain

$$\begin{aligned} &\int_0^\infty e^{itx^2} f(x)\, dx \\ &= \frac{1}{2} \lim_{\varepsilon\to 0} \int_{-\infty}^\infty e^{itx^2} f_\varepsilon(x)\, dx = \frac{1}{2} e^{\pi i/4} \sqrt{\frac{\pi}{t}} f(0) + \frac{i}{8t^{3/2}} \lim_{\varepsilon\to 0} \int_{-\infty}^\infty f_\varepsilon''(x) g_t(x)\, dx \\ &= \frac{1}{2} e^{\pi i/4} \sqrt{\frac{\pi}{t}} f(0) + \frac{i}{4t^{3/2}} \int_0^\infty f''(x)\, g_t(x)\, dx. \end{aligned}$$

The last step follows since g_t is integrable and both functions g_t and f^+ are even.

Definition 9.2. Let ϕ be such that

$$|\phi^{(n)}(x)| \le C_n \quad (0 \le n \le 5), \quad c_0 = \inf |\phi''(x)| > 0,$$

for x in a certain interval which will be clear in our applications. We denote by

$$c_0^+ = \min\{1, c_0\} \quad \text{and} \quad B(\phi) = \left(\frac{1 + \|\phi\|_{C^5}}{c_0^+}\right)^{54}.$$

Lemma 9.3 (Stationary phase lemma). *Let $f(x) \in C_0^2(\mathbb{R})$ be such that*

$$|f'(x)| \le \begin{cases} N_1, & \text{if } |x| \le L, \\ N_2, & \text{if } |x| > L, \end{cases} \quad \text{and} \quad |f''(x)| \le \begin{cases} M_1, & \text{if } |x| \le L, \\ M_2, & \text{if } |x| > L, \end{cases}$$

where $N_1 \le N_2$ and $M_1 \le M_2$, and let $\phi(x)$ such that $\phi(0) = \phi'(0) = 0$, $|\phi^{(n)}(x)| \le C_n$ for $0 \le n \le 5$, and $\phi''(x) \ge c_0 > 0$ for all x in the support of f.

Then

$$\Big|\int_{-\infty}^{\infty} e^{it\phi(x)} f(x)\,\mathrm{d}x - \Big(\frac{2\pi}{|t|\,\phi''(0)}\Big)^{1/2} e^{\mathrm{sign}(t)\pi i/4} f(0)\Big|$$
$$\le A\,B(\phi)t^{-3/2}\Big(\|f\|_\infty + N_1 + \frac{N_2}{t^{1/2}L} + M_1 + \frac{M_2}{t^{1/2}L}\Big). \tag{9.5}$$

Similarly,

$$\Big|\int_{0}^{\infty} e^{it\phi(x)} f(x)\,\mathrm{d}x - \Big(\frac{\pi}{2|t|\,\phi''(0)}\Big)^{1/2} e^{\mathrm{sign}(t)\pi i/4} f(0)\Big|$$
$$\le A\,B(\phi)t^{-3/2}\Big(\|f\|_\infty + N_1 + \frac{N_2}{t^{1/2}L} + M_1 + \frac{M_2}{t^{1/2}L}\Big). \tag{9.6}$$

We also have the usual L-independent estimates

$$\Big|\int_{-\infty}^{\infty} e^{it\phi(x)} f(x)\,\mathrm{d}x - \Big(\frac{2\pi}{|t|\,\phi''(0)}\Big)^{1/2} e^{\mathrm{sign}(t)\pi i/4} f(0)\Big| \tag{9.7}$$
$$\le A\,B(\phi)t^{-3/2}\,(\|f\|_\infty + \|f'\|_1 + \|f''\|_1)$$

and

$$\Big|\int_{0}^{\infty} e^{it\phi(x)} f(x)\,\mathrm{d}x - \Big(\frac{\pi}{2|t|\,\phi''(0)}\Big)^{1/2} e^{\mathrm{sign}(t)\pi i/4} f(0)\Big| \tag{9.8}$$
$$\le A\,B(\phi)t^{-3/2}\,(\|f\|_\infty + \|f'\|_1 + \|f''\|_1),$$

where A is a universal constant, and $B(\phi)$ is as defined above for x in the support of f.

Here, $\mathrm{sign}(t)$ stands for the function which equals 1 if $t > 0$ and -1 if $t < 0$.

Proof. It will obviously be enough to consider the case $t > 0$. Consider the change of variables given by

$$u(x) = x\sqrt{\frac{\phi(x)}{x^2}}$$

and its inverse $z(u)$. We begin by obtaining regularity properties of u and z.

Let $k \ge 1$. In what follows, A_k will denote a collection of universal constants depending only on k. First, we consider $|x| \le 1$ and define

$$\phi_1(x) = x^{-2}\,\phi(x).$$

Since

$$\phi_1(x) = \int_0^1 \int_0^1 \phi''(stx)\,s\,\mathrm{d}t\,\mathrm{d}s,$$

we have

$$\|\phi_1\|_{C^k} \le A_k\,\|\phi\|_{C^{k+2}}.$$

Next, define

$$\phi_2(x) = \sqrt{\phi_1(x)}, \quad |x| \le 1,$$

and note that

$$\frac{d^k \phi_2(x)}{dx^k} = \frac{1}{\phi_1^{k-1/2}(x)} \sum_{\substack{1\cdot i_1+\cdots+p\cdot i_p=k \\ i_j \ge 0}} c^{(k)}_{i_1,\ldots,i_p}\, \phi_1'(x)^{i_1}\cdots\phi_1^{(p)}(x)^{i_p},$$

which can easily be checked by induction, and is a particular case of the well-known formula attributed to F. Faà di Bruno. As a result, we have

$$\|\phi_2\|_{C^k} \le A_k \frac{(1+\|\phi_1\|_{C^k})^k}{c_0^{k-1/2}} \le A_k \frac{(1+\|\phi\|_{C^{k+2}})^k}{c_0^{k-1/2}},$$

where we have used the fact that $\phi_1(x) \ge c_0/2$. Therefore, since $u(x) = x\phi_2(x)$, we conclude that

$$\left|\frac{d^k u(x)}{dx^k}\right| \le A_k \frac{(1+\|\phi\|_{C^{k+2}})^k}{(c_0^+)^{k-1/2}}, \quad \text{when } |x| \le 1.$$

When $|x| > 1$, we obviously have that

$$\frac{d^k u(x)}{dx^k} = \sum_{i=1}^{k} \phi^{1/2-i}(x) \sum_{\substack{1\cdot i_1+\cdots+p\cdot i_p=k \\ i_j \ge 0}} c^{i,k}_{i_1,\ldots,i_p}\, \phi'(x)^{i_1}\cdots\phi^{(p)}(x)^{i_1},$$

hence

$$|u^{(k)}(x)| \le A_k \frac{(1+\|\phi\|_{C^k})^k}{(c_0^+)^{k-1/2}}, \quad |x| > 1,$$

so altogether we obtain

$$\|u\|_{C^k} \le A_k \frac{(1+\|\phi\|_{C^{k+2}})^{k+2}}{(c_0^+)^{k-1/2}}, \quad |x| > 1,$$

Finally, since

$$z'(u(x))\, u'(x) = 1$$

and, for $k \ge 2$,

$$\frac{\mathrm{d}^k z}{\mathrm{d}u^k}(u(x))(u'(x))^k$$
$$= \sum_{p=1}^{k-1} K_{k,p} \frac{\mathrm{d}^p z}{\mathrm{d}u^p}(u(x)) \sum_{\substack{1\cdot i_1+\cdots+q\cdot i_q=k+1-p \\ i_j \ge 0}} c^{k,p}_{i_2,\ldots,i_q}\, u'(x)^{i_1}\cdots u^{(q)}(x)^{i_q},$$

and

$$|u'(x)| = \frac{1}{2}\left|\frac{\phi'(x)}{\sqrt{\phi(x)}}\right| \ge c, \quad c = \frac{c_0}{\sqrt{C_2/2}},$$

we obtain by induction that

$$\left|\frac{d^k z(u)}{du^k}\right| \le A_k \left(\frac{1+C_2}{c_0^+}\right)^{k^2} (1+\|u\|_{C^k})^{k^2}.$$

With our previous estimate for $\|u\|_{C^k}$, we then conclude that

$$\left|\frac{d^k z(u)}{du^k}\right| \le A_k \left(\frac{1+\|\phi\|_{C^{k+2}}}{c_0^+}\right)^{2k^3}.$$

Then,

$$\int_{-\infty}^{\infty} e^{it\phi(x)} f(x)\,\mathrm{d}x = \int_{-\infty}^{\infty} e^{itu^2} \tilde{f}(u)\,\mathrm{d}u, \quad \text{for} \quad \tilde{f}(u) = f(z(u))\, z'(u).$$

Note that

$$\tilde{f}(0) = f(0)\sqrt{2/\phi''(0)}.$$

Since $z'(u) \le c^{-1}$,

$$|f'(z(u))| \le \begin{cases} N_1, & \text{if } |u| \le cL, \\ N_2, & \text{otherwise,} \end{cases} \qquad |f''(z(u))| \le \begin{cases} M_1, & \text{if } |u| \le cL, \\ M_2, & \text{otherwise.} \end{cases}$$

As a result, using formula (9.4), we arrive at

$$\left|\int_{-\infty}^{\infty} e^{it\phi(x)} f(x)\,\mathrm{d}x - \left(\frac{2\pi}{t\phi''(0)}\right)^{1/2} e^{\pi i/4} f(0)\right| \le t^{-3/2}\left(|I_1| + |I_2| + |I_3|\right),$$

for

$$\begin{aligned} I_1 &= \int g_t(u)\, f(z(u))\, z'''(u)\,\mathrm{d}u, \\ I_2 &= \int g_t(u)\, f'(z(u))\, z'(u)\, z''(u)\,\mathrm{d}u, \\ I_3 &= \int g_t(u)\, f''(z(u))\, z'(u)^3\,\mathrm{d}u. \end{aligned}$$

Now,

$$|I_1| \le \|z\|_{C^3}\, \|f\|_\infty \int |g_t(u)|\,\mathrm{d}u \le A\left(\frac{1+\|\phi\|_{C^5}}{c_0^+}\right)^{54} \|f\|_\infty.$$

Next,

$$\begin{aligned} |I_2| &\le A\left(\frac{C_0}{\sqrt{c_0}}\right)\left(\frac{1+\|\phi\|_{C^4}}{c_0^+}\right)^{16}\left(N_1 \int_{|u|\le cL} |g_t(u)|\,\mathrm{d}u + 2N_2 \int_{|u|\ge cL} t^{-1/2} u^{-2}\,\mathrm{d}u\right) \\ &\le A\left(\frac{1+\|\phi\|_{C^5}}{c_0^+}\right)^{54}\left(N_1\, \|g_1\|_1 + \frac{2N_2}{t^{1/2} L}\right). \end{aligned}$$

Finally,

$$
\begin{aligned}
|I_3| &\leq M_1 c^{-3} \int_{|u|\leq cL} |g_t(x)|\,\mathrm{d}x + 2M_2 c^{-3} \int_{|u|\geq cL} t^{-1/2} u^{-2}\,\mathrm{d}u \\
&\leq M_1 c^{-3} \|g_1\|_1 + \frac{2M_2}{c^3 t^{1/2} L},
\end{aligned}
$$

which proves the first claim in our lemma. The one-sided integral is estimated in the same way. The L-independent estimates are obtained in a similar manner, except that integrals I_2 and I_3 in this case are estimated directly by

$$
|I_2| \leq \frac{C_0}{\sqrt{c_0}} \left(\frac{1 + \|\phi\|_{C^4}}{c_0^+}\right)^{16} \|f'\|_\infty \|g_t\|_1 \quad \text{and} \quad |I_3| \leq c^{-3} \|f''\|_\infty \|g_t\|_1.
$$

The one-sided estimate in this case is also analogous. □

This lemma will be complemented with the following trivial results:

Lemma 9.4. *Let $f \in C_0^2(\mathbb{R})$, and let ϕ be such that $|\phi'(x)| \geq d$ for all x in the support of f. Then*

$$
\left|\int_{\mathbb{R}} e^{it\phi(x)} f(x)\,\mathrm{d}x\right| \leq t^{-1} \left(\frac{\|f'\|_1}{d} + \frac{\|f\phi''\|_1}{d^2}\right), \tag{9.9}
$$

$$
\left|\int_{\mathbb{R}} e^{it\phi(x)} f(x)\,\mathrm{d}x\right| \leq t^{-1} \left(\frac{\|f\phi''\|_1}{d^2}\right) + 4t^{-2} \left(\frac{\|f''\|_1}{d^2} + \frac{\|f'\phi''\|_1}{d^3}\right), \tag{9.10}
$$

and

$$
\left|\int_{\mathbb{R}} e^{it\phi(x)} f(x)\,\mathrm{d}x\right| \leq 10t^{-2} \left(\frac{\|f''\|_1}{d^2} + \frac{\|f'\phi''\|_1}{d^3} + \frac{\|f\phi'''\|_1}{d^3} + \frac{\|f'(\phi'')^2\|_1}{d^4}\right). \tag{9.11}
$$

Proof. Integration by parts yields

$$
\begin{aligned}
\int_{\mathbb{R}} e^{it\phi(x)} f(x)\,\mathrm{d}x &= -\frac{1}{it} \int_{\mathbb{R}} e^{it\phi(x)} \frac{d}{dx}\left(\frac{f(x)}{\phi'(x)}\right) \mathrm{d}x \\
&= \frac{1}{it} \int_{\mathbb{R}} e^{it\phi(x)} \frac{f'(x)}{\phi'(x)}\,\mathrm{d}x - \frac{1}{it} \int_{\mathbb{R}} e^{it\phi(x)} \frac{f(x)\phi''(x)}{\phi'(x)^2}\,\mathrm{d}x.
\end{aligned} \tag{9.12}
$$

This yields (9.9). For (9.10), we perform another integration by parts to the first integral above, which equals

$$
\frac{1}{t^2} \int_{\mathbb{R}} e^{it\phi(x)} \frac{f''(x)}{\phi'(x)^2}\,\mathrm{d}x - \frac{2}{t^2} \int_{\mathbb{R}} e^{it\phi(x)} \frac{f'(x)\phi''(x)}{\phi'(x)^3}\,\mathrm{d}x,
$$

which yields (9.10). For (9.11), we integrate by parts also the last integral in (9.12), which gives, as needed,

$$
-t^{-2} \int \left(\frac{f'(x)\phi''(x)}{\phi'(x)^3} + \frac{f(x)\phi'''(x)}{\phi'(x)^3} - 3\,\frac{f(x)\phi''(x)^2}{\phi'(x)^4}\right) e^{it\phi(x)}\,\mathrm{d}x. \quad \square
$$

Lemma 9.5. *Let $f \in C_0^2((a,b))$, and let ϕ be such that $\phi''(x) \geq c_0 > 0$, and $\phi'(x) \neq 0$ for $x \in [a,b]$. Then,*

$$\left|\int_a^b e^{it\phi(x)} f(x)\, dx\right| \leq t^{-1}|b-a|\left(\frac{\|f''\|_\infty}{c_0} + \frac{\|f\|_\infty \|\phi''\|_\infty}{c_0^2}\right).$$

Remark. The point in this result is that the estimate is independent of $\inf|\phi'(x)|$.

Proof. The function f vanishes at a at order 2, which implies

$$|f(x)| \leq \|f''\|_\infty |x-a|^2 \quad \text{and} \quad |f'(x)| \leq \|f''\|_\infty |x-a|.$$

So, $\phi'(x) \geq c_0(x-a)$. The lemma follows trivially by integration by parts, since

$$\int_a^b e^{it\phi(x)} f(x)\, dx = \frac{i}{t}\int_a^b \left(\frac{f'(x)}{\phi'(x)} - \frac{f(x)\,\phi''(x)}{\phi'(x)^2}\right) e^{it\phi(x)}\, dx. \qquad \square$$

The following is a trivial variant of the usual Van der Corput lemma.

Lemma 9.6. *Let f be differentiable in $[a,b]$, and let ϕ be such that $\phi''(x) \geq c_0 > 0$ for $x \in [a,b]$. Then*

$$\left|\int_a^b e^{it\phi(x)} f(x)\, dx\right| \leq 8t^{-1/2} c_0^{-1/2}(1 + \|f\|_\infty + \|f'\|_1).$$

Proof. Let

$$R = t^{-1/2} c_0^{1/2},$$

and consider

$$E_1 = \{x : |\phi'(x)| \geq R\} \quad \text{and} \quad E_2 = \{x : |\phi'(x)| < R\}.$$

It is obvious that E_1 has at most two components, and $|E_2| = 2R$. The contribution of the integral over E_2 is thus trivial. The integral over E_1, after integration by parts, equals

$$\left.\frac{f(x)}{it\phi'(x)}\right|_{\partial E_1} + i(I_1 - I_2)$$

for

$$I_1 = \int_{E_1} e^{it\phi(x)} \frac{f'(x)}{t\phi'(x)}\, dx \quad \text{and} \quad I_2 = \int_{E_1} e^{it\phi(x)} \frac{f(x)\phi''(x)}{t\phi'(x)^2}\, dx.$$

The boundary terms contribute with at most $4\|f\|_\infty/(tR)$, which is fine, and the I_i are trivially estimated by

$$|I_1| \leq \frac{\|f'\|_1}{tR} \quad \text{and} \quad |I_2| \leq \|f\|_\infty \int_{E_1} \frac{\phi''(x)}{t\phi'(x)^2}\, dx \leq \frac{4\|f\|_\infty}{tR},$$

which gives us the bound in the claim of the lemma. $\square$

9.6 The heart of the matter

In this section, we consider a function μ periodic with period 1, average 0, and Fourier coefficients satisfying

$$|\hat{\mu}(n)| \leq M\,|n|^{-\sigma}, \quad \sigma \geq 1.$$

Let us also assume that

$$\sum_{n\neq 0} \sqrt{|n|}\;|\hat{\mu}(n)| < +\infty. \tag{9.13}$$

The estimates will depend on M in a trivial way, but since for the applications we will be satisfied with $M = 10$, we will not bother to keep track of the dependence on M. In fact, we will be mostly interested in $\hat{\mu}(n) = |n|^{-s}$, with $s = \sigma + it$ and $\sigma \geq 1$, and for the applications to the energy asymptotics we will be dealing with

$$\mu(x) = \operatorname{dist}\{x, \mathbb{Z}\}^2 - \frac{1}{12}, \quad \hat{\mu}(n) = \frac{e^{-\pi i n}}{2\pi^2 n^2}.$$

However, the estimates will be independent of the value of the sum in (9.13).

Consider also ϕ smooth, defined on $[a, b]$, and satisfying the crucial non-degeneracy condition $-\phi''(x) \geq c_0 > 0$: of course, the same argument would work if we assumed $\phi''(x) > c_0$, with only a few signs being flipped, but we choose this sign in our non-degeneracy condition because it is exactly the one satisfied by the function ϕ in our application to the sum $\Psi_Q(Z)$.

We also assume the bounds

$$|\phi^{(n)}(x)| \leq C_n, \quad \text{for } 0 \leq n \leq 5 \text{ and } x \in [a, b],$$

where $|b - a|$ is bounded by a universal constant, and define

$$S(\lambda) = \sum_{l \in (\mathbb{Z}+\gamma)\cap[a\lambda, b\lambda]} f\Big(\frac{l}{\lambda}\Big)\,\mu\Big(\lambda\phi\Big(\frac{l}{\lambda}\Big)\Big),$$

where γ is a real number.

In our applications, we will be concerned with the following two situations: on the one hand, we will have functions f and ϕ independent of λ; this simplifies some estimates, but the amplitude function f does not vanish at the endpoint b, which gives origin to a certain diophantine analysis of the phase ϕ. On the other hand, we will have to deal with functions f and ϕ which depend on λ, which will force us to keep track of error terms in a careful way; furthermore, there is no obvious multiscale analysis in the problem and we thus have to analyze blow up manually. However, in this case the amplitude function is supported inside $[a, b]$, which avoids diophantine discussions.

We summarize both cases as follows.

Case I: $f \in C_0^\infty((a, b])$.

In this case, we shall impose that the bounds satisfied by ϕ and f are universal, i.e., independent of λ. The obvious singularity in the sum appearing around $l = b\lambda$ will give rise to a purely arithmetic behavior of the sum.

Case II: $f \in C_0^\infty((a, b))$.

In this case, the functions ϕ and f will depend on λ in the sense that the bounds satisfied by ϕ will grow (slowly) as a function of λ. We will thus keep track carefully of the dependence of our error bounds in terms of the regularity assumptions of f and ϕ. The absence of singularities in this case will make the study of the sums purely analytical.

We wish to understand the behavior of $S(\lambda)$ for large λ in both cases.

Case I

As mentioned above, f and ϕ will satisfy universal bounds for its derivatives of the type

$$\|\phi'(x)\|_{C^5} \le C, \quad -\phi''(x) \ge c_0, \quad \|f\|_{C^2} \le C$$

for constants C and c_0 independent of λ. As a consequence, we will not keep track of the dependence of constants on the regularity properties of either f or ϕ, and the constant C will be ubiquitously used to denote a universal constant depending on the regularity properties of f and ϕ as stated above. Another constant will play a role, though, which is $\phi'(b)$ in the case that it is a rational number p/q: in this case, some constants will depend on q, and this dependence will be made explicit.

Let $\varphi(x)$ be supported on $(-\infty, b)$, identically equal to 1 on $[a, b - \lambda^{-1/2-\varepsilon}]$, for $\varepsilon = 1/20$, and φ as smooth as possible. We denote by

$$I_\lambda = [b - \lambda^{-1/2-\varepsilon}, b]$$

the set where φ' is supported inside the interval $[a, b]$.

It is clear that

$$S(\lambda) = \sum_{l \in \mathbb{Z}+\gamma} f\Big(\frac{l}{\lambda}\Big)\,\mu\Big(\lambda\phi\Big(\frac{l}{\lambda}\Big)\Big)\,\varphi(\lambda^{-1}l) + O(\|f\|_\infty \lambda^{1/2-\varepsilon}),$$

and that in the new sum above, only finitely many terms are non-zero.

Moreover, $\mu(\lambda\phi(x/\lambda))(f\varphi)(\lambda^{-1}x)$ is a piecewise smooth function of compact support. We set

$$\varphi_f(x) = (f\varphi)(x),$$

which satisfies $\|\varphi_f\|_\infty \le \|f\|_\infty$, and

$$(9.14) \qquad |\varphi_f'(x)| \le \begin{cases} C, & \text{if } x \notin I_\lambda, \\ \lambda^{1/2+\varepsilon}, & \text{if } x \in I_\lambda, \end{cases} \qquad |\varphi_f''(x)| \le \begin{cases} C, & \text{if } x \notin I_\lambda, \\ \lambda^{1+2\varepsilon}, & \text{if } x \in I_\lambda, \end{cases}$$

The Poisson summation formula yields

$$\begin{aligned}
\sum_{l\in\mathbb{Z}+\gamma} \mu\Big(\lambda\phi\Big(\frac{l}{\lambda}\Big)\Big)\,\varphi_f(\lambda^{-1}l) &= \sum_{l\in\mathbb{Z}} e^{2\pi i l\gamma} \int_{-\infty}^{\infty} \mu\Big(\lambda\phi\Big(\frac{x}{\lambda}\Big)\Big)\,\varphi_f(\lambda^{-1}x)e^{-2\pi i x l}\,\mathrm{d}x \\
&= \sum_{l\in\mathbb{Z}} e^{2\pi i l\gamma} \int_{\lambda a}^{\lambda b} \mu\Big(\lambda\phi\Big(\frac{x}{\lambda}\Big)\Big)\,\varphi_f(\lambda^{-1}x)e^{-2\pi i x l}\,\mathrm{d}x \\
&= \sum_{\substack{l\in\mathbb{Z}\\ n\neq 0}} \hat{\mu}(n)\,e^{2\pi i l\gamma} \int_{\lambda a}^{\lambda b} e^{2\pi i(\lambda n\,\phi(x/\lambda)-xl)}\,\varphi_f(\lambda^{-1}x)\,\mathrm{d}x \\
&= \lambda \sum_{\substack{l\in\mathbb{Z}\\ n\neq 0}} \hat{\mu}(n)e^{2\pi i l\gamma} \int_{a}^{b} e^{2\pi i\lambda(n\,\phi(x)-xl)}\,\varphi_f(x)\,\mathrm{d}x.
\end{aligned}$$

We will show below that the sum is absolutely convergent, due to the fast decrease of $\hat{\mu}$ assumed in (9.13), and the fast decrease of the integrals; therefore, the infinite sum can be taken in any order we like. Define

$$I(n,l) = e^{2\pi i l\gamma} \int_{a}^{b} e^{2\pi i\lambda(n\,\phi(x)-xl)}\,\varphi_f(x)\,\mathrm{d}x.$$

For integers n and l, define $x_{n,l}$ as the unique point (when it exists) satisfying $\phi'(x_{n,l}) = l/n$. Note that

$$c_0^{-1}\left|\frac{l}{n} - \frac{l'}{n'}\right| \geq |x_{n,l} - x_{n',l'}| \geq \|\phi''\|_\infty^{-1}\left|\frac{l}{n} - \frac{l'}{n'}\right|.$$

Define also

$$\begin{aligned}
\theta(n,l) &= n\,\phi(x_{n,l}) - l\,x_{n,l}, \\
\sigma_\phi(n,l) &= \lambda^{1/2}\frac{1}{|n\,\phi''(x_{n,l})|^{1/2}}\,e^{-\operatorname{sign}(n)\pi i/4+2\pi i(\lambda\theta(n,l)+\gamma l)}.
\end{aligned}$$

We write σ_ϕ to point out that σ depends only on ϕ: the amplitude f does not appear.

We begin with the following crude estimate, which is a trivial consequence of Lemma 9.6.

Lemma 9.7.

$$|I(n,l)| \leq C\lambda^{-1/2}|n|^{-1/2}.$$

This already implies that only the terms appearing for small n play a role in our sum.

Theorem 9.8. *With the previous notation, we have*

$$S(\lambda) = \sum_{\substack{n\neq 0,\, l\in\mathbb{Z}\\ x_{n,l}\in(a,b]}} \hat{\mu}(n)\,\varepsilon_{n,l}\,f(x_{n,l})\,\sigma(n,l) + A(\lambda),$$

where $\varepsilon_{n,l} = 1$, *unless* $\phi'(b) = l/n$, *when* $\varepsilon_{n,l} = 1/2$, *and*

$$A(\lambda) = o(\lambda^{1/2}).$$

If $\phi'(b) = p/q$, *then we have*

$$A(\lambda) = O(\lambda^{1/2-\gamma}), \quad \gamma > 0,$$

where the O *depends on* q.

If, however, $\phi'(b)$ *is irrational, the o-term depends on the diophantine properties of* $\phi'(b)$.

In any case, $|A(\lambda)| \leq C\sqrt{\lambda}$, *for* C *which only depends on* $\|f\|_{C^2}$ *and* $B(\phi)$, *where* $B(\phi)$ *was introduced in Definition* 9.2.

Proof. Note that there are three types of pairs (n, l): those for which $\phi'(x) = l/n$ for some $x = x_{n,l} \in [a, b)$, those such that $\phi'(x)$ never equals l/n for any $x \in [a, b)$, and those (if any) for which l/n equals $\phi'(b)$ ($\phi'(a)$ will play no role here, since f vanishes to infinite order at a). We need to deal with these cases separately, and we thus write

$$S(\lambda) = S_1(\lambda) + S_2(\lambda) + S_3(\lambda),$$

where

$$S_1(\lambda) = \lambda \sum_{\phi'(a) \leq l/n < \phi'(b)} \hat{\mu}(n)\, I(n, l),$$

$$S_2(\lambda) = \lambda \sum_{l/n = \phi'(b)} \hat{\mu}(n)\, I(n, l),$$

$$S_3(\lambda) = \lambda \sum_{l/n \notin [\phi'(a), \phi'(b)]} \hat{\mu}(n)\, I(n, l).$$

Sum S_1. For every term in this sum, the integrand in $I(n, l)$ has a stationary point $x_{n,l}$. Our stationary phase analysis then shows, using (9.14), that

$$\begin{aligned} &|\lambda I(n, l) - f(x_{n,l})\, \sigma_{n,l}| \leq \lambda\, E_{n,l}, \\ \text{(9.15)} \qquad &\text{with} \quad E_{n,l} = C(\lambda |n|)^{-3/2} \left(1 + \min\left\{\lambda^{1+2\varepsilon}, \frac{\lambda^{1/2+2\varepsilon}}{|n|^{1/2}\operatorname{dist}\{x_{n,l}, I_\lambda\}}\right\}\right), \end{aligned}$$

where the minimum appears as the best of estimates (9.5) and (9.6) above.

The terms in S_1 will be grouped into three categories. First, those for which $x_{n,l}$ falls far from b, and second, those for which $x_{n,l}$ falls near b. Within the second class, we will have to consider separately those that appear only when n is large, and those with n small.

Fix n in the sum above. For each n, the number of terms in the sum in l is at most $C|n|$. And of those terms, the number of l for which $x_{n,l}$ falls within d of I_λ is bounded by at most $1 + C|n|(d + \lambda^{-1/2-\varepsilon})$. If, say, $\phi'(b) = p/q$ is rational, and we take $d > \lambda^{-1/2}$, we have

$$d \geq |b - x_{n,l}| \geq C_2^{-1} \left| \frac{l}{n} - \frac{p}{q} \right| \geq \frac{C_2^{-1}}{|nq|},$$

which means that, if

$$|n| < \|\phi\|_{C^2}^{-1} q^{-1} d^{-1},$$

then there are no l such that $x_{n,l}$ falls within d if I_λ. Similarly, if $\phi'(b)$ is irrational, the number of such l is at most 1 for

$$|n| < \|\phi\|_{C^2}^{-1} d^{-1}$$

if $d > \lambda^{-1/2-\varepsilon}$. We denote this unique l (when it exists) by $l_0(n, d, \lambda)$, and we denote by $n_0(d, \lambda)$ the smallest $|n|$ for which n has such an l_0.

After all this, we choose

$$d = \lambda^{-7\varepsilon}, \quad c^\sharp = \begin{cases} \|\phi\|_{C^2}^{-1}, & \text{if } \phi'(b) \text{ is irrational,} \\ \|\phi\|_{C^2}^{-1} q^{-1}, & \text{if } \phi'(b) = p/q, \end{cases}$$

and break up

$$\begin{aligned} \lambda^{-1} S_1(\lambda) = & \sum_{\substack{|n| \leq c^4 d^{-1} \\ d(x_{n,1}, I_\lambda) \geq d}} \hat{\mu}(n)\, I(n,l) + \sum_{\substack{|n| > c^4 d^{-1} \\ \text{all } l}} \hat{\mu}(n)\, I(n,l) \\ & + \sum_{c^1 d^{-1} > |n| \geq n_0(d,\lambda)} \hat{\mu}(n)\, I(n, l_0(n,d,\lambda)), \end{aligned} \tag{9.16}$$

where it is understood that if $\phi'(b)$ is rational, the sum in the third term above is null, and the sum in n also in the third term is extended only to those n with a corresponding $l_0(n, d, \lambda)$.

For the first term, we use (9.5) to obtain

$$E_{n,l} \leq C (\lambda |n|)^{-3/2} (1 + |n|^{-1/2} \lambda^{1/2+9\varepsilon})$$

Since $|n| \leq c^\sharp \lambda^{7\varepsilon}$, we obtain (recall $\varepsilon = 1/20$),

$$E_{n,l} \leq C |n|^{-2} \lambda^{-1+9\varepsilon}$$

and, using again $\varepsilon = 1/20$, we obtain

$$\sum_{\substack{|n| \leq c^1 d^{-1} \\ d(x_{n,l}, I_\lambda) \geq d}} \lambda \hat{\mu}(n)\, I(n,l) = \sum_{\substack{|n| \leq c^1 d^{-1} \\ d(x_{n,l}, I_\lambda) \geq d}} \hat{\mu}(n)\, f(x_{n,l})\, \sigma_{n,l} + O\Big(\lambda^{1/2-\varepsilon} \sum_{n \neq 0} \Big|\frac{\hat{\mu}(n)}{n}\Big|\Big).$$

For the second term in (9.16), we use the trivial estimate (9.6) in (9.15) to conclude that

$$E_{n,l} = O\big(\lambda^{-1/2+2\varepsilon}\,|n|^{-3/2}\big),$$

and since $|n| \geq c^{\sharp}\lambda^{7\varepsilon}$,

$$E_{n,l} = O_q\big(\lambda^{-1/2-\varepsilon}\,|n|^{-15/14}\big),$$

and we obtain

$$\sum_{|n|>c^1 d^{-1}} \lambda\,\hat{\mu}(n)\,I(n,l) = \sum_{|n|>c^1 d^{-1}} \hat{\mu}(n)\,f(x_{n,l})\,\sigma_{n,l} + O_q\Big(\lambda^{1/2-\varepsilon}\sum_{n\neq 0}\frac{|\hat{\mu}(n)|}{|n|^{15/14}}\Big).$$

Note that, here, we could simply have used Lemma 9.6 to conclude that both $I(n,l)$ and $\sigma_{n,l}$ give a negligible contribution, but this would have required, either to use the stronger assumption that $\sigma > 3/2$, or to obtain an error estimate which depends on the value of the sum (9.13).

Finally, for the third term, it is clear that

$$\lim_{\lambda\to\infty} n_0(\lambda^{-7\varepsilon},\lambda) = \infty,$$

which, using Lemma 9.7, implies that

$$\begin{aligned}\Big|\sum_{|n|\geq n_0(d,\lambda)} \hat{\mu}(n)\,I(n,l_0(n,d,\lambda))\Big| &\leq C \sum_{|n|\geq n_0(d,\lambda)} |\hat{\mu}(n)|(\lambda|n|)^{-1/2}\\ &\leq C\lambda^{-1/2}\,|n_0(d,\lambda)|^{-\sigma+1/2} = o(\lambda^{-1/2}).\end{aligned}$$

Similarly, observe that

$$\Big|\sum_{|n|\geq n_0(d,\lambda)} \hat{\mu}(n)\,f(x_{n,l})\,\sigma_{n,l_0(n,d,\lambda)}\Big| \leq C\lambda^{1/2}\,|n_0(d,\lambda)|^{-\sigma+1/2} = o(\lambda^{1/2}).$$

Therefore, we can conclude that

$$S_1(\lambda) = \sum_{l/n\in[\phi'(a),\phi'(b))} \hat{\mu}(n)\,f(x_{n,l})\,\sigma_{n,l} + o(\lambda^{1/2}).$$

Sum S_2. If $\phi'(b)$ is irrational, this sum is empty. We thus assume that $\phi'(b)$ is rational.

If we try to proceed as we did for S_1, we find that $E_{n,l}$ is too big, and this has no remedy. This is so because we would be comparing $I(n,l)$ with the wrong thing: it is not $f(x_{n,l})\,\sigma_{n,l}$ what we should look at, but $f(x_{n,l})\,\sigma_{n,l}/2$ instead. We proceed as follows.

Say $\phi'(b) = l/n$. We have, by (9.8), that

$$\begin{aligned} e^{2\pi i l\gamma} &\int_0^1 e^{2\pi i\lambda(n\,\phi(x)-lx)}\,\varphi_f(x)\,\mathrm{d}x \\ &= e^{2\pi i l\gamma}\int_0^1 e^{2\pi i\lambda(n\,\phi(x)-lx)}\,f(x)\,\mathrm{d}x + O(\lambda^{-1/2-\varepsilon}) \\ &= \frac{1}{2\lambda}\,f(x_{n,l})\,\sigma_{n,l} + O((|n|\,\lambda)^{-3/2}) + O(\lambda^{-1/2-\varepsilon}), \end{aligned}$$

which yields

$$S_2(\lambda) = \frac{1}{2}\sum_{l/n=\phi'(b)} \hat{\mu}(n)\,f(x_{n,l})\,\sigma_{n,l} + O(\lambda^{1/2-\varepsilon}).$$

Sum S_3. As for S_1, we deal separately with those l and n for which $\phi'(x) - l/n$ is small or large, and for those for which it is small, we distinguish between small and large n.

When $|\phi'(x) - l/n| > d$ for all $x \in [a,b]$, we use (9.11) to obtain

$$I(n,l) = O\Big(\frac{\lambda^{-3/2+\varepsilon}}{|n|^2\,d^2} + \frac{\lambda^{-2}}{|n|^2\,d^3} + \frac{\lambda^{-2}}{|n|^2\,d^4}\Big),$$

which implies

$$\Big|\sum_{\{(n,l):|\phi'(x)-l/n|>d\}} \hat{\mu}(n)\,I(n,l)\Big| \le C\lambda^{-3/2}\sum_{\text{all } n\neq 0}\Big|\frac{\hat{\mu}(n)}{n}\Big|\Big(\frac{\lambda^{\varepsilon}}{d}+\frac{1}{\lambda^{1/2}d^2}+\frac{1}{\lambda^{1/2}d^3}\Big).$$

If we now set $d = \lambda^{-7\varepsilon}$, we obtain

$$\sum_{\{(n,l):|\phi'(x)-l/n|>d\}} \hat{\mu}(n)\,I(n,l) = O(\lambda^{-1/2-\varepsilon}).$$

When $0 < |\phi'(b) - l/n| \le d$, for a fixed n there are at most $1 + |n|d$ terms in the sum. And, as before, if $\phi'(b)$ is rational and $|n| < d^{-1}$, then there are no l, and if $\phi'(b)$ is irrational, there is at most one such l, which, if it really existed, we would denote by $l_0(n,d,\lambda)$; we denote by $n_0(d,\lambda)$ the first $|n|$ for which n has such an l. Therefore we break up the remaining part of $S_{3'}$ given by $0 < |\phi'(b) - l/n| \le d$, into

$$\sum_{\substack{(n,l):|\phi'(x)-l/n|\le d\\ |n|>d^{-1}}} \hat{\mu}(n)\,I(n,l). \quad \text{and} \quad \sum_{|n|\ge n_0(d,\lambda)} \hat{\mu}(n)\,I(n,l_0(n,d,\lambda)).$$

The first sum above is trivially controlled by (9.11), which implies

$$I(n,l) = O(\lambda^{-2+7/5}\,|n|^{-2}),$$

hence

$$\sum_{\substack{\{(n,l):|\phi'(x)-l/n|>d\}\\ |n|>d^{-1}}} \hat{\mu}(n)\, I(n,l) = O(\lambda^{-1/2-\varepsilon}).$$

For the second term, note as before that $\lim_{\lambda\to\infty} n_0(d,\lambda) = \infty$, and thus, using Lemma 9.6, we get

$$\Big| \sum_{|n|\geq n_0(d,\lambda)} \hat{\mu}(n)\, I(n, l_0(n,d,\lambda)) \Big| \leq C \sum_{|n|\geq n_0(d,\lambda)} |\hat{\mu}(n)|\, (\lambda\,|n|)^{-1/2}$$
$$\leq C\lambda^{-1/2}\, n_0^{-\sigma+1/2} = o(\lambda^{-1/2}).$$

All this implies that

$$S_3(\lambda) = o(\lambda^{1/2}),$$

and the theorem follows. □

Case II

In this case, we will not need $\varphi(x)$, since f is compactly supported and smooth. In fact, $\mu(\lambda\phi(x/\lambda))f(\lambda^{-1}x)$ is a piecewise smooth function of compact support, and by the Poisson summation formula, as before,

$$\sum_{l\in\mathbb{Z}+\gamma} \mu\Big(\lambda\phi\Big(\frac{l}{\lambda}\Big)\Big)\, f(\lambda^{-1}l) = \lambda \sum_{\substack{l\in\mathbb{Z}\\ n\neq 0}} \hat{\mu}(n)\, e^{2\pi i l\gamma} \int_a^b e^{2\pi i\lambda(n\,\phi(x)-xl)}\, f(x)\, \mathrm{d}x.$$

Define, as before,

$$I(n,l) = e^{2\pi i l\gamma} \int_a^b e^{2\pi i\lambda(n\,\phi(x)-xl)}\, f(x)\, \mathrm{d}x$$

and $x_{n,l}$ as the unique point (if it did exist) satisfying $\phi'(x_{n,l}) = l/n$. Also as before, we have

$$c_0^{-1}\left|\frac{l}{n} - \frac{l'}{n'}\right| \geq |x_{n,l} - x_{n',l'}| \geq \|\phi''\|_\infty^{-1}\left|\frac{l}{n} - \frac{l'}{n'}\right|.$$

Define also $\theta(n,l)$ and $\sigma(n,l)$ exactly as in Case I.

Theorem 9.9. *With the previous notation, we have*

$$S(\lambda) = \sum_{\substack{n\neq 0\\ l\in\mathbb{Z}}} \hat{\mu}(n)\, \sigma(n,l) + O\big(B(\phi)\|f\|_{C^2}(1 + \|f'\|_\infty)\big).$$

Proof. In this case, now there are only two types of pairs (n,l): those for which $\phi'(x) = l/n$ for some $x = x_{n,l} \in [a,b]$, and those such that $\phi'(x)$ never equals l/n, and thus we write

$$S(\lambda) = S_1(\lambda) + S_2(\lambda)$$

where

$$S_1(\lambda) = \lambda \sum_{l/n \in [\phi'(a), \phi'(b)]} \hat{\mu}(n)\, I(n,l),$$

$$S_2(\lambda) = \lambda \sum_{l/n \notin [\phi'(a), \phi'(b)]} \hat{\mu}(n)\, I(n,l).$$

Sum S_1 We proceed as in the previous section:

$$|\lambda I(n,l) - f(x_{n,l})\sigma_{n,l}| \leq \lambda E_{n,l},$$

where, by (9.7), $E_{n,l}$ is given now by

$$E_{n,l} = C\, B(\phi)\, \|f\|_{C^2}\, (\lambda |n|)^{-3/2}.$$

For each n, the number of terms in the sum is at most $(3 + C_1)|n|$. Therefore, we can conclude that

$$S_1(\lambda) = \lambda \sum_{l/n \in [\phi'(a), \phi'(b)]} \hat{\mu}(n)\, f(x_{n,l})\, \sigma_{n,l} + O(B(\phi)\, \|f\|_{C^2}\, \lambda^{-1/2}).$$

Sum S_2. By Lemma 9.5, we have

$$|I(n,l)| \leq C\, \|f\|_{C^2}\, B(\phi)\, (\lambda |n|)^{-1},$$

which we use when $|l| \leq 2C_1^+ |n|^{1-\delta}$, for $\delta > 0$, to obtain

$$\sum_{\{(n,l) : |l| \leq 2C_1^+ |n|^{1-\delta}\}} |\hat{\mu}(n)\, I(n,l)| \leq C \lambda^{-1}\, \|f\|_{C^2}\, B(\phi)\, C_1^+ \sum_{n \neq 0} \frac{|\hat{\mu}(n)|}{|n|^{\delta}}.$$

Outside of this range, we have

$$\left|\phi'(x) - \frac{l}{n}\right| \geq \frac{1}{2}\left|\frac{l}{n}\right|, \qquad \text{for all } x \in [a,b].$$

Therefore, (9.10) implies

$$|I(n,l)| \leq \frac{C\, \|f\|_{C^2}\, B(\phi)}{\lambda^2 |n|^2} \left(\frac{n^2}{l^2} + \frac{|n|^3}{|l|^3} + \frac{n^4}{l^4}\right).$$

Therefore,

$$\sum_{\{(n,l) : |l| > 2C_1^+ |n|^{1-\delta}\}} |\hat{\mu}(n)\, I(n,l)| \leq C \lambda^{-2}\, \|f\|_{C^2}\, B(\phi) \sum_{n \neq 0} \frac{|\hat{\mu}(n)|}{|n|^{1-4\delta}}. \qquad \square$$

9.7 Energy asymptotics

We plan to apply our previous estimates to the function

$$\Psi_C(Z) = 2\pi Z^{4/3} \sum_{l\in(Z+1/2)\cap[1,a^{-1/2}Z^{1/3}\Omega_c]} \eta(lZ^{-1/3})\,\mu\big(Z^{1/3}\,\phi(lZ^{-1/3})\big),$$

where

$$(9.17)\qquad \mu(x) = \operatorname{dist}\{x,\mathbb{Z}\}^2 - \frac{1}{12},$$

$$(9.18)\qquad \phi(\Omega) = \frac{1}{\pi}\int \Big(V^1_{\mathrm{TF}}(r) - \frac{\Omega^2}{r^2}\Big)_+^{1/2} dr = \frac{a^{-1/2}}{\pi}\int \Big(\frac{y(r)}{r} - \frac{a\Omega^2}{r^2}\Big)_+^{1/2} dr,$$

$$(9.19)\qquad \eta(\Omega) = \frac{\Omega}{P(\Omega)},$$

$$(9.20)\quad P(\Omega) = \int \Big(V^1_{\mathrm{TF}}(r) - \frac{\Omega^2}{r^2}\Big)_+^{-1/2} dr = a^{-3/2}\int_{r_1(a^{1/2}\Omega)}^{r_2(a^{1/2}\Omega)} \Big(\frac{y(r)}{r} - \frac{a\Omega^2}{r^2}\Big)^{-1/2} dr.$$

Here, $r_i(\Omega)$ are the two points where $y(r)/r$ equals Ω^2/r^2 (see below), and Ω_c is the supremum of the Ω for which

$$\frac{y(r)}{r} - \frac{\Omega^2}{r^2}$$

is positive somewhere, and the Thomas–Fermi function $y(r)$ was introduced before.

The crucial result we need is the non-vanishing of the second derivative of ϕ. This was proved in [208]. Because of its vital importance, we display it explicitly.

Theorem 9.10. *There exists a number c_0 such that*

$$-\phi''(\Omega) \ge c_0 > 0, \qquad \text{for all } \Omega \in (0,\Omega_c).$$

We will first recall some known results (which appear, for example, in [208, 218, 219] which we will need here. After that, we will complement them with further properties of ϕ and P, some of which are taken from similar estimates appearing in [207, 208, 209, 210, 211].

Review of earlier results

If we set $u(r) = ry(r)$, then u has a unique maximum at $r = r_c$, where $r_c \sim 2.1$. We set $\Omega_c^2 = u(r_c)$. Then, u is increasing on $[0, r_c]$ and decreasing on $(r_c, +\infty)$. This is a crucial fact whose proof goes back to Sommerfeld, and can be found in [218].

Around 0, u satisfies the expansions

$$u(x) = \sum_{n=2}^{\infty} u_n x^{n/2}, \quad u_2 = 1, \quad u_3 = 0, \quad u_4 \sim -1.588.$$

Rigorous numerical bounds for u_4 can be found in [208]. However, it is easy to see analytically that $u_4 < 0$. We also have

$$\sum_{n=2}^{\infty} |u_n| \, \rho_0^n < +\infty, \qquad \text{for some } \rho_0 > 0,$$

and therefore, the function

$$f(z) = \sum_{n=2}^{\infty} u_n z^n$$

is analytic in a small neighborhood around 0.

Around infinity, we have the expansion

$$u(x) = \frac{144}{x^2} \sum_{n=0}^{\infty} b_n \, x^{-n\alpha/2},$$

where

$$b_0 = 1, \quad b_1 \sim -13, \quad \alpha = \frac{\sqrt{73} - 7}{2} \sim 0.772.$$

Again, rigorous numerical bounds for b_1 are found in [208], and it can be seen analytically that $b_1 < 0$. We also have that

$$\sum_{n=0}^{\infty} |b_n| \, \rho_1^n < +\infty, \qquad \text{for some } \rho_1 > 0,$$

and, as a result, we have

$$u(x) = \frac{144}{x^2} \, f(x^{-\alpha/2}) \tag{9.21}$$

for a function f analytic in a neighborhood of 0.

Given any $\Omega \in (0, \Omega_c)$, there exist two numbers, $r_1(\Omega) \leq r_2(\Omega)$, where u equals Ω^2. We then have the following (see [201]).

Lemma 9.11. *The following formula holds:*

$$\phi(\Omega) = \frac{a^{-1/2}}{\pi} \, F(a^{1/2}\Omega),$$

where

$$F(\Omega) = \int (u(x) - \Omega^2)_+^{1/2} \, \frac{\mathrm{d}x}{x}, \quad F'(\Omega) = -\Omega \int (u(x) - \Omega^2)_+^{-1/2} \, \frac{\mathrm{d}x}{x},$$

$$F''(\Omega) = - \lim_{\delta \to 0} \left(\int_{r_1(\Omega)+\delta}^{r_2(\Omega)-\delta} (u(x) - \Omega^2)^{-3/2} \, y(x) \, \mathrm{d}x + c(\Omega) \, \delta^{-1/2} \right),$$

and where $c(\Omega)$ is uniquely specified by requiring the finiteness of the limit. Moreover, if b is any number less than $r_2(\Omega)$, then

$$\frac{d^2}{d\Omega^2}\int_{r_1(\Omega)}^{b}(u(x)-\Omega^2)_+^{1/2}\,\frac{dx}{x}$$
$$=-\lim_{\delta\to 0}\Big(\int_{r_1(\Omega)+\delta}^{b}(u(x)-\Omega^2)^{-3/2}y(x)\,dx+c_1(\Omega)\,\delta^{-1/2}\Big),$$

again for a constant c_1 that makes the limit finite. The corresponding symmetric case also holds.

Furthermore, F can be extended as an analytic function to a complex neighborhood of $(0,\Omega_c]$. However, 0 is an essential singularity of ϕ (or F), and, moreover,

$$\lim_{\Omega\to 0}\phi''(\Omega)\,\Omega^\gamma=\kappa,\qquad \textit{with } \gamma=\frac{9-\sqrt{73}}{2}>0,$$

where κ is a strictly negative real number.

A consequence of this which is of importance to us is that although ϕ and ϕ' remain bounded as we approach 0, the second derivative blows up slowly, and the third will blow up much faster. In other words, ϕ does not satisfy sensible non-degenerate multiscale analysis bounds.

Further background results

Here we will obtain growth and regularity properties of the functions ϕ and P above. We define

$$g_\gamma(x)=\int_1^{x^{-2}}(t-1)^{-1/2}\,t^\gamma\,dt,\qquad \text{for } 0<x\le\frac{1}{2}\cdot$$

We begin listing several elementary results of calculus.

Lemma 9.12. *For $\gamma\in\mathbb{R}$, we have*

$$g_\gamma^{(k)}(x)\le C_k(|\gamma|+1)^{k-1}x^{-2\gamma-1-k},\qquad \textit{for } k\ge 1. \tag{9.22}$$

Furthermore, if $\gamma<-1/2$, then

$$\frac{1-2^{\gamma+1/2}}{|\gamma+1/2|}\le g_\gamma(x)\le 100+\frac{100}{|\gamma+1/2|}, \tag{9.23}$$

and if $\gamma>-1/2$, then

$$\frac{x^{-2\gamma-1}-1}{|\gamma+1/2|}\le g_\gamma(x)\le\Big(100+\frac{100}{|\gamma+1/2|}\Big)x^{-2\gamma-1}, \tag{9.24}$$

where $0<x\le 1/2$.

Proof. Estimate (9.22) is completely trivial. For (9.23), we use

$$\int_1^2 t^{-1/2+\gamma}\,dt \le g_\gamma(x) \le 4 + \int_2^\infty (t-1)^{-1/2+\gamma}\,dt.$$

For (9.24), we use the fact that $t-1 > t/2$ for $t > 2$ to write

$$\int_1^{x^{-2}} t^{\gamma-1/2}\,dt \le g_\gamma(x) \le 2^{|\gamma|+3} + 2\int_2^{x^{-2}} t^{\gamma-1/2}\,dt,$$

which implies (9.24) after using the fact that $2^{|\gamma|+1/2} \le 4x^{-2\gamma-1}$. □

Lemma 9.13. *Define*

$$f(\Omega) = (\Omega_\varepsilon^2 - \Omega^2)^{-1/2}, \quad \textit{for } \Omega > 0.$$

Then

$$|f^{(k)}(\Omega)| \le C_k\,\Omega_\varepsilon^{-1-k} \quad \textit{for } \Omega \le \Omega_\varepsilon/2 \textit{ and } k \ge 0.$$

Proof.

$$f(\Omega) = \Omega_\varepsilon^{-1}\,g\Big(\frac{\Omega}{\Omega_\varepsilon}\Big), \qquad \text{for } g(x) = (1-x^2)^{-1/2}.$$ □

Lemma 9.14. *Given $\beta > 0$, $\delta > 0$, $\Omega_\varepsilon > 0$, τ, d and w_n for $n = 0, 1, 2, \ldots$, let*

$$f(\Omega) = \sum_{n=0}^\infty w_n\,m_n(\Omega), \quad \textit{for } \Omega > 0, \quad \textit{with } m_n(\Omega) = \Omega^{2\gamma_n+d}\,g_{\gamma_n}\Big(\frac{\Omega}{\Omega_\varepsilon}\Big),$$

and where $\tau_n = \tau + \beta n$ and $|\gamma_n + 1/2| > \delta$ for all $n \ge 0$.

Assume that $\sum_n w_n z^n$ has a radius of convergence ρ, and that $\Omega_\varepsilon^{2\beta} \le \rho/2$. Then,

$$\Big|\frac{d^k f}{d\Omega^k}(\Omega)\Big| \le \begin{cases} C\,\Omega^{2\tau+d-k}, & \textit{if } \tau < -1/2, \\ C\,\Omega^{d-1-k}, & \textit{if } \tau > -1/2, \end{cases} \quad \textit{when } \Omega \le \Omega_\varepsilon/2,$$

for a certain constant C which depends on everything except Ω.

Proof. Let us consider first those n such that $\gamma_n > -1/2$. In this case, using Lemma 9.13, we obtain

$$\begin{aligned}\Big|\frac{d^k m_n}{d\Omega^k}(\Omega)\Big| &\le \sum_{l=0}^k C(k;\delta)\,(n+1)^{k-l}\,\Omega^{2\gamma_n+d-(k-l)}\,\Omega_\varepsilon^{-l}\,\Big|\frac{d^l g_{\gamma_n}(\Omega)}{d\Omega^l}\Big(\frac{\Omega}{\Omega_\varepsilon}\Big)\Big| \\ &\le C(k;\delta)\,(n+1)^k\,\Omega_\varepsilon^{2\gamma_n+1}\,\Omega^{d-1-k}.\end{aligned}$$

If, on the other hand, $\gamma_n < -1/2$, the $l = 0$ term above has to be estimated by

$$C(k;\delta)\,(n+1)^k\,\Omega^{2\gamma_n+d-k},$$

and we obtain

$$\left|\frac{d^k m_n}{d\Omega^k}(\Omega)\right| \le C(k;\delta)\,(n+1)^k\,\left(\Omega_\varepsilon^{2\gamma_n+1}\,\Omega^{d-1-k}+\Omega^{2\gamma_n+d-k}\right) \\ \le 2C(k;\delta)(n+1)^k\,\Omega^{2\gamma_n+d-k}$$

because $\Omega \le \Omega_\varepsilon$. Since we can only have $\gamma_n < -1/2$ for finitely many n, we conclude that

$$\sum_{n=0}^{\infty} w_n\, m_n^{(k)}(\Omega) = \frac{d^k f}{d\Omega^k}(\Omega),$$

where the sum converges absolutely, and we obtain the required estimate. □

This ends our presentation of calculus results. In what follows, we will develop the regularity bounds for ϕ first, and then P.

Lemma 9.15. *For constants C_n and $c > 0$ we have*

$$|\phi(t)| \le C_0, \quad |\phi'(t)| \le C_1, \quad \left|\frac{d^n\phi}{d\Omega^n}\right| \le C_n\,\Omega^{-n+1+\alpha} \quad (n \ge 2)$$

and

$$-\phi''(\Omega) \ge c\,\Omega^{-1+\alpha}.$$

Proof. As in Lemma 9.11, in order not to bother with the presence of the constant a, we will prove this result for the function F instead.

The inequalities for ϕ and ϕ' are obvious. For the higher derivatives, the bounds outside a neighborhood of 0 are a direct consequence of the analytic extension of F to a complex neighborhood of (0, which is Corollary 1.3 in [208]. We are thus the right of 0, given by $(0, \bar{\Omega}_\varepsilon)$, for a small universal number $\bar{\Omega}_\varepsilon$ to be picked up later in the proof.

Arguing as in formulas (4.1a)–(4.1c) in [208], using Lemma 9.11, we write

$$-F''(t) = I_1 + I_2 + I_3,$$

for

$$I_1 = \int_\alpha^\beta (u(r)-\Omega^2)^{-3/2}\,y(r)\,\mathrm{d}r,$$

$$I_2 = \lim_{\delta\to 0}\left(\int_{r_1(\Omega)+\delta}^{\alpha} (u(r)-\Omega^2)^{-3/2}\,y(r)\,\mathrm{d}r - G_1(\Omega)\,\delta^{-1/2}\right),$$

$$I_3 = \lim_{\delta\to 0}\left(\int_{\beta}^{r_2(\Omega)-\delta} (u(r)-\Omega^2)^{-3/2}\,y(r)\,\mathrm{d}r - G_2(\Omega)\,\delta^{-1/2}\right),$$

with G_i such that the limit is finite, and α and β any numbers such that $r_1(\Omega) < \alpha < \beta < r_2(\Omega)$. In practice, we will take α and β such that

$$u(\alpha) = u(\beta) = \Omega_\varepsilon^2$$

for Ω_ε a small number, and later, we will take $\bar{\Omega}_\varepsilon \ll \Omega_\varepsilon$.

First, $I_1(\Omega)$ is C^∞ in a neighborhood of 0, and therefore satisfies

$$\left|\frac{d^k I_1}{d\Omega^k}(\Omega)\right| \leq C_k(\Omega_\varepsilon), \qquad \text{for } \Omega \in (0, \Omega_\varepsilon),$$

no matter which Ω_ε we will end up choosing.

For I_2, we write $I_2 = d\tilde{I}_2/d\Omega$, where, by Lemma 9.11,

$$\tilde{I}_2(\Omega) = \Omega \int_{r_1(\Omega)}^{\alpha} (u(r) - \Omega^2)^{-1/2} \frac{\mathrm{d}r}{r}.$$

Let $r(t)$ be the inverse of u near 0, $u(r(t)) = t$, and set

$$w(t) = r'(t)/r(t).$$

Changing variables above, we obtain

$$\tilde{I}_2(\Omega) = \Omega \int_{\Omega^2}^{\Omega_\varepsilon^2} (t - \Omega^2)^{-1/2} w(t)\, \mathrm{d}t = \Omega^2 \int_1^{\Omega^{-2}\Omega_e^2} (t-1)^{-1/2} w(t\Omega^2)\, \mathrm{d}t,$$

which implies, after differentiation,

$$I_2(\Omega) = 2\Omega \int_1^{\Omega^{-2}\Omega_\varepsilon^2} (t-1)^{-1/2} h(t\Omega^2)\, \mathrm{d}t - 2\,(\Omega_\varepsilon^2 - \Omega^2)^{-1/2} w(\Omega_\varepsilon^2)\, \Omega_\varepsilon^2,$$

for

$$h(t) = tw'(t) + w(t).$$

Next, we recall that $u(r) = rf(r^{1/2})$ for f analytic around 0 and $f(0) = 1$. Therefore, $u^{1/2}(r) = r^{1/2}\tilde{f}(r^{1/2})$, for $\tilde{f}$ also analytic around 0, or $u^{1/2}(r) = \tilde{\tilde{f}}(r^{1/2})$, for $\tilde{\tilde{f}}$ also analytic around 0, $\tilde{\tilde{f}}(0) = 0$ and $\tilde{\tilde{f}}'(0) = 1$. Therefore, $\tilde{f}$ has an analytic inverse g, with $g(0) = 0$ and $g'(0) = 1$, and therefore, $r^{1/2} = u^{1/2}\tilde{g}(u^{1/2})$, for $\tilde{g}$ analytic and $\tilde{g}(0) = 1$. Squaring both sides, we obtain

$$r(t) = t\,\tilde{\tilde{g}}(t^{1/2})$$

for $\tilde{\tilde{g}}$ analytic around 0, $\tilde{\tilde{g}}(0) = 1$. As a consequence of this, we also have

$$w(t) = t^{-1}\, W(t^{1/2})$$

for W analytic around 0, and

$$h(t) = t^{-1}\, H(t^{1/2}).$$

It was shown in [208] that $H(0) = H'(0) = 0$, and $H''(0) = -2y'(0)$, which implies that in fact,

$$h(t) = f_h(t^{1/2}) \quad \text{and} \quad f_h(0) = -y'(0).$$

Therefore,

$$f_h(z) = \sum_{n=0}^{\infty} h_n z^n \quad \text{and} \quad \sum_{n=0}^{\infty} |h_n| \rho_2^n < +\infty,$$

for ρ_2 a small universal constant. (Recall the relation $u(r) = ry(r)$.)

We break up

$$I_2(\Omega) = f_1(\Omega) + f_2(\Omega),$$

for

$$f_1(\Omega) = 2\Omega \int_1^{\Omega^{-2}\Omega_\varepsilon^2} (t-1)^{-1/2} h(t\Omega^2)\, dt = \sum_{n=0}^{\infty} 2 h_n \Omega^{1+n} g_{n/2}\left(\frac{\Omega}{\Omega_\varepsilon}\right),$$

$$f_2(\Omega) = -2(\Omega_\varepsilon^2 - \Omega^2)^{-1/2} w(\Omega_\varepsilon^2)\, \Omega_\varepsilon^2.$$

Lemma 9.13 shows that $f_2^{(k)}(\Omega)$ is bounded for all $k \geq 0$ and $\Omega \leq \Omega_\varepsilon/2$ by a constant that may depend on Ω_ε. For f_2, we apply Lemma 9.14 with $d = 1$, $\tau = 0$ and $\beta = 1/2$ to obtain

$$\left|\frac{d^k f_2}{d\Omega^k}(\Omega)\right| \leq C(k, \Omega_\varepsilon)\, \Omega^{-k}.$$

We conclude the analysis of I_2 by observing that all the bounds we obtained are in agreement with the statement of the lemma.

We continue now with I_3. Denote by $r(t)$ the inverse function of $u(r)$, such that $u(r(t)) = t$. We proceed as in Section 4 in [208] to construct $w(t) = -r'(t)/r(t)$ and then set $h(t) = tw'(t) + w(t)$, which allows us to argue as before to obtain (equation (4.21.a) in [208])

$$I_3 = 2\Omega \int_1^{\Omega^{-2}\Omega_\varepsilon^2} (t-1)^{-1/2} h(t\Omega^2)\, dt - 2(\Omega_\varepsilon^2 - \Omega^2)^{-1/2} w(\Omega_\varepsilon^2)\, \Omega_\varepsilon^2.$$

By (9.21), we have that $r^2 u(r) = g(r^{-\alpha})$ for g analytic in a neighborhood of 0, with $g(0) = 1/144$. Therefore, setting $z = r^{-\alpha}$ and $u(r) = t$ we have $t = z^{2/\alpha} g(z)$, or $t^{\alpha/2} = \tilde{g}(z)$ for a new $\tilde{g}$ analytic in a small neighborhood of 0, with $\tilde{g}(0) = 0$ and $\tilde{g}'(0) \neq 0$. Thus, $\tilde{g}$ has an analytic inverse, f, with $f(0) = 0$, $f'(0) \neq 0$, and we have $z = f(t^{\alpha/2})$, or $z = t^{\alpha/2} \tilde{f}(t^{\alpha/2})$ for $\tilde{f}$ analytic around 0 and $\tilde{f}(0) \neq 0$. Therefore, $r(t) = z^{-1/\alpha} = t^{-1/2} v(t^{\alpha/2})$ for a new function v analytic around 0 which also satisfies $v(0) \neq 0$. Hence,

$$r'(t) = t^{-3/2} v_p(t^{\alpha/2}) \quad \text{and} \quad r''(t) = t^{-5/2} v_{pp}(t^{\alpha/2}), \tag{9.25}$$

for functions v_p and v_{pp} analytic in a small neighborhood around 0. Therefore,

$$h(t) = \frac{1}{4t} f_h(t^{\alpha/2}),$$

where, by (9.25), $f_h(z)$ is analytic in a small neighborhood around 0, $|z| < \rho_h$. It is observed in [208] that $f_h(0) = 0$ and $f_h'(0) > 0$ (equation (4.20) in [208]). This allows us to put

$$I_3(\Omega) = f_1(\Omega) + f_2(\Omega),$$

where

$$(9.26)\qquad \begin{aligned} f_1(\Omega) &= \sum_{n=1}^{\infty} 2\,h_n\,m_n(\Omega),\\ \text{with}\quad m_n(\Omega) &= \Omega^{-1+n\alpha} \int_1^{\Omega^{-2}\,\Omega_\varepsilon^2} (t-1)^{-1/2}\, t^{-1+n\alpha/2}\,\mathrm{d}t, \end{aligned}$$

and

$$f_2(\Omega) = -2\,(\Omega_\varepsilon^2 - \Omega^2)^{-1/2}\, w(\Omega_\varepsilon^2)\,\Omega_\varepsilon^2.$$

If we make sure that

$$(9.27)\qquad \Omega_\varepsilon^\alpha \le \frac{1}{2}\,\rho_h,$$

we can invoke Lemma 9.14, with $\tau = -1 + \alpha/2$, which is less than $-1/2$, $\beta = \alpha/2$ and $d = 1$ to obtain

$$\left|\frac{d^k f_1}{d\Omega^k}(\Omega)\right| \le C(\Omega_\varepsilon; k)\,\Omega^{-1+\alpha-k}.$$

This ends the proof of all upper bounds in the statement of the lemma. For the lower bound for $-\phi''$, we use the notation in (9.26) to write

$$f_1(\Omega) = 2\,h_1\,m_1(\Omega) + \tilde{f}_1(\Omega),$$

where

$$\tilde{f}_1(\Omega) = \sum_{n=2}^{\infty} 2\,h_n\,m_n(\Omega).$$

Applying Lemma 9.12 to the first term above with $\gamma = -1 + \alpha/2 < -1/2$, and Lemma 9.14 applied to $\tilde{f}_1(\Omega)$ with $\tau = -1 + \alpha > -1/2$, we obtain

$$f_1(\Omega) \ge c\,h_1\,\Omega^{-1+\alpha}, \qquad |\tilde{f}_1(\Omega)| \le C(\Omega_\varepsilon).$$

Since all other terms in the break-up of $-\phi''$ remain bounded as $\Omega \to 0$, we conclude that

$$-\phi''(\Omega) \ge c\,\Omega^{-1+\alpha}, \qquad \Omega \le \bar{\Omega}_\varepsilon,$$

for $\bar{\Omega}_\varepsilon \ll \Omega_\varepsilon$, as required. □

We now turn our attention to the function P, defined in (9.20). In this case, rather than introducing a new function that allows us to eliminate the bothersome constant a, we will simply proceed as if a did not appear in the definition of P. This simplification clearly does not change the result, except of course, that the details of the proof will not contain the a dependence.

The following result is a trivial adaptation of Lemma 1.2 in [208] for P instead of ϕ.

Lemma 9.16. *We have that $P \in C^\infty(0, \Omega_c)$. Furthermore, P admits an analytic extension to a neighborhood of Ω_c.*

Proof. Let

$$H(\delta, \Omega) = \int_{r_1(\Omega)+\delta}^{r_2(\Omega)-\delta} (u(r) - \Omega^2)^{-1/2}\, r \, \mathrm{d}r.$$

Consider the analytic change of variables given by

$$t(r) = \begin{cases} (\Omega_c^2 - u(r))^{1/2}, & \text{if } r \geq r_c, \\ -(\Omega_c^2 - u(r))^{1/2}, & \text{if } r \leq r_c. \end{cases} \tag{9.28}$$

Note that t is smooth and strictly increasing in the range $(0, +\infty)$. We can therefore consider its inverse, $r(t)$, and use it to rewrite

$$H(\delta, \Omega) = \int_{t_1(\delta,\Omega)}^{t_2(\delta,\Omega)} (D^2 - t^2)^{-1/2}\, w(t)\, \mathrm{d}t,$$

where

$$t_1 = t(r_1 + \delta), \quad t_2 = t(r_2 - \delta), \quad D^2 = \Omega_c^2 - \Omega^2, \quad w(t) = r'(t) r(t).$$

Note that w is smooth on $(-\Omega_c, \Omega_c)$, and that

$$t_1 = -D(1 + \tau_1(\delta)), \quad t_2 = D(1 + \tau_2(\delta)), \quad c\delta \leq |\tau_i| \leq C\delta \quad \text{for } i = 1, 2, \tag{9.29}$$

uniformly on compact subsets of $(-\Omega_c, \Omega_c)$, which implies that

$$H(\delta, \Omega) = \int_{D^{-1}t_1}^{D^{-1}t_2} (1 - t^2)^{-1/2}\, w(tD)\, \mathrm{d}t$$

converges as $\delta \to 0$ uniformly to the C^1 function

$$H(0, \Omega) = \int_{-1}^{1} (1 - t^2)^{-1/2}\, w(tD)\, \mathrm{d}t = P(\Omega). \tag{9.30}$$

To show analyticity around Ω_c, note that $v(t)$ is analytic around 0; thus, it admits a convergent power series expansion given by

$$w(t) = \sum_{n=0}^{\infty} w_n t^n, \quad |t| \leq \rho,$$

which implies

$$P(\Omega) = \Omega \sum_{n=0}^{\infty} w_n D^n \int_{-1}^{1} (1 - t^2)^{-1/2}\, t^n \, \mathrm{d}t.$$

The integral corresponding to the odd terms in the sum is 0, which implies that in fact,

$$P(\Omega) = \Omega \sum_{n=0}^{\infty} w_{2n}\, D^{2n} \int_{-1}^{1} (1-t^2)^{-1/2}\, t^{2n}\, dt,$$

which defines an analytic function of Ω around Ω_c, since D^2 now is analytic in Ω. □

Lemma 9.17. *For constants C_n and $c > 0$, we have*

$$\left|\frac{d^k P}{d\Omega^k}(\Omega)\right| \le C_k\, \Omega^{-3-k}, \quad |P(\Omega)| \ge c\,\Omega^{-3}.$$

Proof. Lemma 9.16 establishes our inequalities outside an arbitrarily small neighborhood of 0. For a neighborhood to the right of 0 given by $(0, \bar{\Omega}_\varepsilon)$ we proceed as before, setting

$$f(\Omega) = I_1 + I_2 + I_3,$$

for

$$I_1 = \int_a^b (u(r) - \Omega^2)^{-1/2}\, r\, dr,$$
$$I_2 = \int_{r_1(\Omega)}^{a} (u(r) - \Omega^2)^{-1/2}\, r\, dr,$$
$$I_3 = \int_b^{r_2(\Omega)} (u(r) - \Omega^2)^{-1/2}\, r\, dr,$$

with a and b any numbers such that $r_1(\Omega) < a < b < r_2(\Omega)$. We will take a and b such that $u(a) = u(b) = \Omega_\varepsilon^2$ for Ω_ε a small number to be picked later.

The term I_1 is C^∞ around 0, and thus satisfies all the required upper bounds. For I_2, denote by $r(t)$ the inverse function of $u(r)$ around 0, such that $u(r(t)) = t$, $r(t) \le r_c$, and set $w(t) = r'(t)r(t)$. By the same argument as before, we can see that

$$w(t) = t\, f_0(t^{1/2}), \quad \text{for} \quad f_0(z) = \sum_{n=0}^{\infty} w_n\, z^n$$

analytic for $|z| \le \rho_3$, ρ_3 a small universal number, and $f_0(0) \ne 0$.

Then,

$$I_2 = \int_{\Omega^2}^{\Omega_c^2} (t - \Omega^2)^{-1/2}\, w(t)\, dt = \Omega \int_1^{\Omega^{-2}\Omega_c^2} (t-1)^{-1/2}\, w(t\Omega^2)\, dt$$
$$= \sum_{n=0}^{\infty} w_n\, \Omega^{3+n}\, g_{1+n/2}\left(\frac{\Omega}{\Omega_\varepsilon}\right).$$

Thus, Lemma 9.14, for $\Omega_\varepsilon \le \rho_3/2$, $\tau = 1$, $\beta = 1/2$, $d = 1$, yields

$$\left|\frac{d^k I_2}{d\Omega^k}(\Omega)\right| \le C(k;\Omega_\varepsilon)\, \Omega^{-k}, \quad \text{for } \Omega \le \frac{1}{2}\,\Omega_\varepsilon.$$

For I_3, denote by $r(t)$ the inverse function of $u(r)$ around infinity, such that $u(r(t)) = t, r(t) \geq r_c$, and set $w(t) = -r'(t)r(t)$. By the same argument as before, we can see that

$$w(t) = t^{-2} f_1(t^{\alpha/2}), \qquad \text{for} \quad f_1(z) = \sum_{n=0}^{\infty} w_n z^n$$

analytic for $|z| \leq \rho_4$, ρ_4 a small universal number, and $f_1(0) \neq 0$.

Then,

$$\begin{aligned} I_3 &= \int_{\Omega^2}^{\Omega_e^2} (t - \Omega^2)^{-1/2} w(t)\,\mathrm{d}t = \Omega \int_1^{\Omega^{-2}\Omega_c^2} (t-1)^{-1/2} w(t\Omega^2)\,\mathrm{d}t \\ &= \Omega^{-3} \sum_{n=0}^{\infty} w_n \Omega^{\alpha n} \int_1^{\Omega^{-2}\Omega_e^2} (t-1)^{-1/2} t^{-2+\alpha n/2}\,\mathrm{d}t = \sum_{n=0}^{\infty} w_n m_n(\Omega), \end{aligned}$$

where

$$m_n(\Omega) = \Omega^{-3+\alpha n} g_{-2+\alpha n/2}\left(\frac{\Omega}{\Omega_\varepsilon}\right).$$

Thus, Lemma 9.14, for $\Omega_\varepsilon \leq \rho_4/4$, $\tau = -2$, $\beta = \alpha/2$, $d = 1$, yields

$$\left|\frac{d^k I_3}{d\Omega^k}(\Omega)\right| \leq C(k; \Omega_\varepsilon)\,\Omega^{-3-k}, \qquad \text{for } \Omega \leq \frac{1}{2}\Omega_\varepsilon.$$

For the lower bound, we write

$$I_3(\Omega) = w_n m_0(\Omega) + \tilde{I}_3(\Omega), \qquad \text{with} \quad \tilde{I}_3(\Omega) = \sum_{n=1}^{\infty} w_n m_n(\Omega).$$

Lemma 9.12 applied to m_0 and Lemma 9.14 applied to $\tilde{I}_3$ with $\tau = -2 + \alpha/2 < -1/2$, $\beta = \alpha/2$, $d = 1$, yield that

$$|m_0(\Omega)| \geq c\,\Omega^{-3}, \quad |\tilde{I}_3(\Omega)| \leq C(\Omega_\varepsilon)\Omega^{-3+\alpha}, \qquad \text{for } \Omega \leq \frac{1}{2}\Omega_\varepsilon.$$

Therefore, for Ω small enough, we obtain

$$I_3(\Omega) \geq c\,\Omega^{-3}, \quad \Omega \leq \bar{\Omega}_\varepsilon,$$

for a number $\bar{\Omega}_\varepsilon \ll \Omega_\varepsilon$. Since I_1 and I_2 remain bounded as $\Omega \to 0$, the lemma is proved. □

Corollary 9.18. *For constants C_k and c_0, we have*

$$\eta(\Omega) \geq c_0 \Omega^4, \quad \left|\frac{d^k \eta}{d\Omega^k}(\Omega)\right| \leq C_k\,\Omega^{4-k}, \quad k \geq 0.$$

We apply now our growth estimates for ϕ and η to show that Ψ_C is very much like Ψ_Q in the introduction.

Lemma 9.19.

$$|\Psi_Q(Z) - \Psi_C(Z)| \leq C Z^{4/3}.$$

Proof. Set

$$\tilde{l} = \frac{l + 1/2}{\lambda}, \qquad \lambda = Z^{1/3},$$

and note that

$$l(l+1)\lambda^{-2} = \tilde{l}^2 - \frac{1}{4\lambda^2}.$$

We then use (9.1) and (9.2) to conclude that

$$\Psi_Q = 2\pi Z^{4/3} \sum_{l=1}^{l_{\text{TF}}} \frac{\tilde{l}}{P\left(\sqrt{\tilde{l}^2 - 1/(4\lambda^2)}\right)} \mu\left(Z^{1/3}\phi\left(\sqrt{\tilde{l}^2 - \frac{1}{4\lambda^2}}\right)\right).$$

Define $l_{\max}$ as the largest l appearing in the sum defining Ψ_C. Note that

(1) each term appearing in the definition of either Ψ_Q above, or in Ψ_C, is bounded by a constant independent of Z,
(2) the sum in Ψ_Q is taken over integers, while the sum in Ψ_Q is taken over half-integers,
(3) the number of terms in either sum (l_{TF} for Ψ_Q, and $l_{\max} - 1/2$ for Ψ_C) may differ slightly because in general, $l_{\text{TF}} + 1/2 \neq l_{\max}$.

We show now that the number of terms in both sums differ by at most 1. $l_{\max}$ is the greatest element in $\mathbb{Z} + 1/2$ which is less than or equal to $a^{-1/2} Z^{1/3} \Omega_c$. Similarly, l_{TF} is the largest integer that satisfies

$$l_{\text{TF}}(l_{\text{TF}} + 1) = \left(l_{\text{TF}} + \frac{1}{2}\right)\sqrt{1 - \frac{1}{(2l_{\text{TF}} + 1)^2}} \leq a^{-1/2} Z^{1/3} \Omega_c.$$

An immediate consequence of this is that

$$c Z^{1/3} \leq l_{\text{TF}} \leq C Z^{1/3} \quad \text{and} \quad c Z^{1/3} \leq l_{\max} \leq C Z^{1/3}.$$

Then, for Z large enough,

$$l_{\text{TF}} + 1/2 \leq a^{-1/2} Z^{1/3} \Omega_c + 1,$$

which implies

$$l_{\text{TF}} + \frac{1}{2} \leq l_{\max} + 1. \tag{9.31}$$

On the other hand,

$$l_{\max} - l_{\mathrm{TF}} \leq a^{-1/2} Z^{1/3} \Omega_c - l_{\mathrm{TF}}.$$

Since we must have

$$l_{\mathrm{TF}} + \frac{3}{2} > \frac{a^{-1/2} Z^{1/3} \Omega_c}{\sqrt{1-(2l_{\mathrm{TF}}+3)^{-2}}},$$

which implies

$$a^{-1/2} Z^{1/3} \Omega_c - l_{\mathrm{TF}} - \frac{1}{2} \leq \frac{3}{2},$$

we conclude that $l_{\max} - 1/2 - l_{\mathrm{TF}} < 1$, or

$$l_{\max} - \frac{1}{2} \leq l_{\mathrm{TF}}. \tag{9.32}$$

Thus, using (9.31),

$$\left| l_{\max} - \frac{1}{2} - l_{\mathrm{TF}} \right| \leq 1 \tag{9.33}$$

for Z large enough.

As a consequence of this, if we rewrite

$$\Psi_C(Z) = 2\pi Z^{4/3} \sum_{l=1}^{l_{\max}-1/2} \eta\Big(\Big(l+\frac{1}{2}\Big) Z^{-1/3}\Big) \mu\Big(Z^{1/3} \phi\Big(\Big(l+\frac{1}{2}\Big) Z^{-1/3}\Big)\Big),$$

$$\tilde{\Psi}_Q(Z) = 2\pi Z^{4/3} \sum_{l=1}^{l_{\max}-1/2} \frac{\tilde{l}}{P\big(\sqrt{\tilde{l}^2 - 1/(4\lambda^2)}\big)} \mu\Big(Z^{1/3} \phi\Big(\sqrt{\tilde{l}^2 - \frac{1}{4\lambda^2}}\Big)\Big),$$

which makes sense by (9.32), and recalling that $l_{\max}$ in fact refers to a half-integer, we see that

$$|\Psi_Q(Z) - \tilde{\Psi}_Q(Z)| \leq C Z^{4/3}$$

since, after all, the difference is at most one term of size at most $Z^{4/3}$. Next, we compare Ψ_C and $\tilde{\Psi}_Q$ term by term. To this end, we observe that, since $\tilde{l} \geq Z^{-1/3}$, we have

$$\left| \frac{1}{P(\tilde{l})} - \frac{1}{P\big(\sqrt{\tilde{l}^2 - 1/(4\lambda^2)}\big)} \right| \leq \frac{\big|P(\tilde{l}) - P\big(\sqrt{\tilde{l}^2 - 1/(4\lambda^2)}\big)\big|}{P(\tilde{l})\, P\big(\sqrt{\tilde{l}^2 - 1/(4\lambda^2)}\big)}$$

$$\leq C \tilde{l}^6 \tilde{x}^{-4} \tilde{l}\Big(1 - \sqrt{1 - \frac{1}{4\tilde{l}^2\lambda^2}}\Big) \leq C\tilde{l}\lambda^{-2}, \qquad \text{for } \tilde{x} \in \big[\sqrt{\tilde{l}^2 - 1/(4\lambda^2)}, \tilde{l}\big],$$

which implies

$$\left| \eta\Big(\frac{l+1/2}{\lambda}\Big) - \frac{\tilde{l}}{P\big(\sqrt{\tilde{l}^2 - 1/(4\lambda^2)}\big)} \right| \leq \tilde{l} \left| \frac{1}{P(\tilde{l})} - \frac{1}{P\big(\tilde{l}^2 - 1/(4\lambda^2)\big)} \right| \leq C \tilde{l}^2 \lambda^{-2}.$$

Similarly, since ϕ' is bounded,

$$\left|\phi(\tilde{l}) - \phi\left(\sqrt{\tilde{l}^2 - \tfrac{1}{4\lambda^2}}\right)\right| \leq C\,\tilde{l}\left(1 - \sqrt{1 - \tfrac{1}{4\tilde{l}^2\lambda^{-2}}}\right) \leq C\,\tilde{l}^{-1}\,\lambda^{-2},$$

and since μ is Lipschitz, we conclude that

$$\left|\mu\left(\lambda\phi\left(\frac{l+1/2}{\lambda}\right)\right) - \mu\left(\lambda\phi\left(\sqrt{\tilde{l}^2 - \tfrac{1}{4\lambda^2}}\right)\right)\right| \leq C\,\tilde{l}^{-1}\,\lambda^{-1}.$$

Therefore, the terms indexed by l in $\tilde{\Psi}_Q$ and Ψ_C differ by at most

$$C\left(\lambda^{-2}\,\tilde{l}^2 + \sup_{x\in[(\tilde{l}^2-1/(4\lambda^2))^{1/2},\,\tilde{l}]} |\eta(x)|\,\lambda^{-1}\,\tilde{l}^{-1}\right),$$

which implies, as required, that

$$|\Psi_Q(Z) - \tilde{\Psi}_C(Z)| \leq C\,Z^{4/3}\sum_{l=1}^{l_{\mathrm{TF}}}(Z^{-4/3}\,l^2 + Z^{-4/3}\,l^3) \leq C\,Z^{4/3}. \qquad \square$$

In what follows, we will denote either of the sums Ψ_C or Ψ_Q simply by Ψ, since we now know that both are the same modulo errors of order $Z^{4/3}$.

Define $\Omega_{n,l}$ such that $\phi'(\Omega_{n,l}) = l/n$, and

$$\theta(n,l) = n\,\phi(\Omega_{n,l}) - l\,\Omega_{n,l}.$$

Theorem 9.20.

$$\Psi(Z) = \Psi_0(Z) + o(Z^{3/2}),$$

where

$$\Psi_0(Z) = 2\pi\,Z^{3/2}\sum_{n,l}\frac{\eta(\Omega_{n,l})\,\hat{\mu}(n)}{|n\,\phi''(\Omega_{n,l})|^{1/2}}\,e^{2\pi i Z^{1/3}\theta(n,l)+\pi i(l-\mathrm{sign}(n)/4)}.$$

Also, Ψ_0 satisfies the bound

$$|\Psi_0(Z)| \leq C\,Z^{3/2}, \quad \textit{for a constant } C.$$

Proof. Construct now a partition of unity given by $\{U_\nu, \theta_\nu\}$ for $\nu = 0, 1, \ldots,$ such that

$$U_\nu = [a_\nu, b_\nu]\,, \quad a_\nu = 2^{-\nu-2}a^{-1/2}\Omega_c, \quad b_\nu = 2^{-\nu}a^{-1/2}\Omega_c,$$

$$d_\nu \overset{\text{def}}{=} b_\nu - a_\nu, \quad d_\nu \sim a_\nu \sim b_\nu,$$

$$\sum_\nu \theta_\nu(x) = \begin{cases} 1, & \text{if } 0 < x \leq \Omega_c, \\ 0, & \text{otherwise}\,, \end{cases} \quad \theta_\nu \in C_0^\infty, \quad \operatorname{supp}\theta_\nu \subset U_\nu \ \text{ for } \nu \geq 1,$$

$$\theta_0(x) \equiv 0 \ \text{ for } x \notin [a_0, b_0], \quad \textit{and} \quad \left|\frac{d^k\theta_\nu}{dx^k}(x)\right| \leq C_k\,d_\nu^{-k},$$

for universal constants C_k independent of ν. Clearly, we have

$$\Psi_C(Z) = 2\pi Z^{4/3} \sum_{\nu=0}^{\infty} S_\nu(Z^{1/3}), \tag{9.34}$$

for

$$S_\nu = \sum_{l\in\mathbb{Z}+1/2} (\eta\theta_\nu)\Big(\frac{l}{\lambda}\Big)\,\mu\Big(\lambda\phi\Big(\frac{l}{\lambda}\Big)\Big), \quad \lambda = Z^{1/3}.$$

Note that we have

$$\begin{cases} |\phi(x)| \le C, \quad |\phi'(x)| \le C, \quad c\,d_\nu^{-1+\alpha} \le |\phi''(x)| \le C\,d_\nu^{-1-\alpha}, \\ \Big|\dfrac{d^k\phi}{d\Omega^k}(x)\Big| \le C\,d_\nu^{1-k+\alpha} \quad (k \ge 2), \\ \Big|\dfrac{d^k(\theta_\nu\eta)}{d\Omega^k}(x)\Big| \le C\,d_\nu^{4-k} \quad (k \ge 0), \end{cases} \tag{9.35}$$

for $x \in U_\nu$, which implies that, in each S_ν, we have

$$B(\phi) \le C\,a_{\nu_1}^{-400} \le C\,2^{400\nu}. \tag{9.36}$$

We consider

$$\nu_1 = \varepsilon_1 |\log_2 \lambda|, \quad \text{with } \varepsilon_1 = 10^{-3}.$$

For $\nu = 0$, we apply Theorem 9.8 to obtain

$$S_0 = \sum_{\substack{n\neq 0\\ l\in\mathbb{Z}}} \varepsilon_{n,l}\hat{\mu}(n)\,\theta_0(\Omega_{n,l})\,\eta(\Omega_{n,l})\,\sigma(n,l) + o(Z^{1/6}). \tag{9.37}$$

For $0 < \nu \le \nu_1$, we use (9.35) and Theorem 9.9 to obtain

$$S_\nu = \sum_{\substack{n\neq 0\\ l\in\mathbb{Z}}} \hat{\mu}(n)(\theta_\nu\eta)(\Omega_{n,l})\,\sigma(n,l) + O(2^{400\nu}). \tag{9.38}$$

If $\nu > \nu_1$, we argue directly as we did before Theorem 9.8 to obtain that

$$S_\nu = \lambda \sum_{\substack{n\neq 0\\ l\in\mathbb{Z}}} \hat{\mu}(n)\,I(n,l), \quad I(n,l) = \int (\theta_\nu\eta)(x)\,e^{2\pi i\lambda(n\,\phi(x)-lx)}\,dx.$$

In order to analyze $I(n,l)$, we use Lemma 9.6 to obtain

$$|I(n,l)| \le C(\lambda|n|)^{-1/2}\,(d_\nu^{-1+\alpha})^{-1/2}\,d_\nu^{3+\alpha},$$

which we will use when $|l| \le 4\|\phi'\|_\infty |n|$, to obtain

$$\Big| \sum_{\{|l| \le 4\|\phi'\|_\infty |n|\}} \hat{\mu}(n)\, I(n,l) \Big| \le C\lambda^{-1/2}\, d_\nu^{4+\alpha/2}.$$

When $|l| \ge 4\|\phi'\|_\infty |n|$, we have $|\phi'(x) - l/n| \ge |l|/(4|n|)$; thus, we apply (9.10) directly to $I(n,l)$ to obtain

$$(9.39)\quad |I(n,l)| \le (\lambda|n|)^{-1} \frac{d_\nu^{4+\alpha}}{(|l|/(4|n|))^2} + 4(\lambda|n|)^{-2} \Big(\frac{d_\nu^3}{(|l|/(4|n|))^2} + \frac{d_\nu^{3+\alpha}}{(|l|/(4|n|))^3} \Big),$$

which implies that, for n fixed,

$$\Big| \sum_{\{l: |l| \ge 4\|\phi'\|_\infty |n|\}} I(n,l) \Big| \le C\Big(\lambda^{-1} d_\nu^{4+\alpha} + \frac{\lambda^{-2}\, d_\nu^3}{|n|} + \frac{\lambda^{-2}\, d_\nu^{3+\alpha}}{|n|} \Big),$$

which finally implies

$$(9.40)\qquad \Big| \sum_{\{l: |l| \ge 4\|\phi'\|_\infty |n|\}} \hat{\mu}(n)\, I(n,l) \Big| \le \lambda^{-1}\, d_\nu^3.$$

Putting (9.39) and (9.40) together, we obtain $|S_\nu| \le \lambda^{1/2} d_\nu^3$, which implies

$$(9.41)\qquad \Big| \sum_{\nu > \nu_1} S_\nu \Big| \le \lambda^{1/2}\, 2^{-3\nu_1} = \lambda^{1/2 - 3\varepsilon_1}.$$

Also,

$$\Big| \sum_{\nu > \nu_1} \sum_{\{(n,l): \Omega_{n,l} \in U_\nu\}} \frac{\hat{\mu}(n)(\eta\theta_\nu)(\Omega_{n,l})}{|n\phi''(\Omega_{n,l})|^{1/2}}\, e^{2\pi i \lambda \theta(n,l) + \pi i (l - \mathrm{sign}(n)/4)} \Big|$$
$$\le C \sup_{x \in \bigcup_{\nu=\nu_1+1}^{\infty} U_\nu} |\eta(x)| = O(d_{\nu_1}^4) = O(\lambda^{-4\varepsilon_1}),$$

and putting (9.34), (9.37), (9.38) and (9.41) together, we obtain

$$\Psi(Z) = \Psi_0(Z) + o(Z^{3/2}).$$

□

9.8 Lower bounds

Theorem 9.20 told us two things: that Ψ has a leading expression as a trigonometric sum, and that the size of this trigonometric sum, and therefore of Ψ, is at most of order $Z^{3/2}$. The question remains whether this bound is sharp or not. In related problems, such as the lattice point problem, sharp upper and lower bounds on average have been known for over fifty years (see [198] and [217] for some developments).

This, in our context, would translate into the statement that indeed $Z^{3/2}$ is best possible.

The aim of this section is to derive such estimates on average for the function Ψ_0. Classical ideas will work effortlessly after we show that a certain number is not 0. This number can be viewed as a certain (analytic, not necessarily arithmetic) L-function evaluated at the point $s = 2$. Thus, it is not surprising that such a condition appears if one thinks, for example, about the lattice point problem for a parabola.

First, we will consider real values for Z, and then use this to study the case of interest, integer Z, as it relates to our original goal to understand the ground state energy of an atom of nuclear charge Z.

We begin by defining

$$a_{n,l} = \frac{\hat{\mu}(n)\,\eta(\Omega_{n,l})}{|n\,\phi''(x_{n,l})|^{1/2}}\, e^{\pi i(l-\operatorname{sign}(n)/4)}.$$

Let $\{\theta_\nu\}$ the set of all possible values of $\theta(n,l)$, selected such that $\theta_\nu \neq \theta_{\nu'}$ for $\nu \neq \nu'$. Define

$$L \overset{\text{def}}{=} L_{\eta,\phi} = \sum_{\nu} \Big| \sum_{\theta(n,l)=\theta_\nu} a_{n,l} \Big|^2$$

which Lemma 9.22 will prove to be non-zero; if, however, L were equal to 0, then it is easy to see that in fact we would have that

$$\Psi_0(Z) \equiv 0, \qquad \text{all } Z.$$

In the meantime, we will simply assume $L \neq 0$.

Define also

$$\begin{aligned} C^* &= 1 + \left(\frac{48\,\|\eta\|_\infty\,\|\phi'\|_\infty}{c_0 L}\right)^{2/3}, \\ c^\# &= \inf\left\{|\theta(n,l) - \theta(n',l')| : \theta(n,l) \neq \theta(n',l') \text{ and } |n|,|n'| \le C^*\right\}. \end{aligned}$$

Therefore, setting as usual $\lambda = Z^{1/3}$,

$$\begin{aligned} Z^{-3}\,|\Psi_0(Z)|^2 &= \sum_{n,n',l,l'} a_{n,l}\,\overline{a_{n',l'}}\, e^{2\pi i\lambda(\theta(n,l)-\theta(n',l'))} \\ &\ge \sum_{\substack{n,n',l,l' \\ \theta(n,l)=\theta(n',l')}} a_{n,l}\,\overline{a_{n',l'}} + \sum_{\theta(n,l)\neq\theta(n',l')} a_{n,l}\,\overline{a_{n',l'}}\, e^{2\pi i\lambda(\theta(n,l)-\theta(n',l'))}. \end{aligned}$$

The first term above is our L defined above. In the second term, we separate the terms for small $|n|, |n'|$, which we keep untouched, and the ones for which either n or n' are large, which we estimate using the fact that $|l| \le \|\phi'\|_\infty |n|$ and that $|l'| \leq \|\phi'\|_\infty |n'|$, to obtain

$$
\begin{aligned}
&\geq L - \sum_{\substack{|\theta(n,l)-\theta(n',l')|>0 \\ |n|,|n'|\leq C^*}} a_{n,l}\, \overline{a_{n',l'}}\, e^{2\pi i\lambda(\theta(n,l)-\theta(n',l'))} \\
&\quad - 2\, c_0^{-1}\, \|\eta\|_\infty^2\, \|\phi'\|_\infty^2 \sum_{\substack{|n|\geq C^* \\ \text{all } n'}} |nn'|^{-5/2} \\
&\geq L - \sum_{\substack{|\theta(n,l)-\theta(n',l')|>0 \\ |n|,|n'|\leq C^*}} a_{n,l}\, \overline{a_{n',l'}}\, e^{2\pi i\lambda(\theta(n,l)-\theta(n',l'))} \\
&\quad - 24\, c_0^{-1}\, \|\eta\|_\infty^2\, \|\phi'\|_\infty^2 (C^* - 1)^{-3/2},
\end{aligned}
$$

which, by our choice of C^*, implies

$$
Z^{-3}\, |\Psi_0(Z)|^2 \geq \frac{L}{2} - \sum_{\substack{|\theta(n,l)-\theta(n',l')|>0 \\ |n|,|n'|\leq C^*}} a_{n,l}\, \overline{a_{n',l'}}\, e^{2\pi i\lambda(\theta(n,l)-\theta(n',l'))}.
$$

Now we consider $Z_0 \geq 1$ and $Z \leq Z_0/2$, and prepare to integrate both sides from Z_0 to $Z_0 + Z$. For that, note that $dZ = 3\lambda^2 d\lambda$ and, if we set $\lambda_0 = Z_0^{1/3}$, then

$$
(Z_0 + Z)^{1/3} = \lambda_0 + \Lambda, \quad \Lambda \leq Z/Z_0^{2/3}.
$$

Also,

$$
\left| \int_a^b \lambda^2\, e^{i\lambda\theta}\, d\lambda \right| \leq \frac{3b^2}{\theta}, \quad a, b > 0.
$$

Therefore,

$$
\begin{aligned}
&\int_{Z_0}^{Z_0+Z} |\Psi_0(z)|^2 \frac{dz}{z^3} \\
&\quad \geq \frac{Z}{2} L - 36\, Z_0^{2/3}\, c_0^{-1}\, \|\eta\|_\infty^2\, \|\phi'\|_\infty^2 \sum_{\substack{|\theta(n,l)-\theta(n',l')|>0 \\ |n|,|n'|\leq C^*}} \frac{|nn'|^{-5/2}}{2\pi\, |\theta(n,l) - \theta(n',l')|} \\
&\quad \geq \frac{Z}{2} L - 36\, Z_0^{2/3} \frac{\|\eta\|_\infty^2\, \|\phi'\|_\infty^2}{c_0\, c^\sharp} \sum_{|n|,|n'|\leq C^*} |nn'|^{-5/2} \\
&\quad \geq \frac{Z}{2} L - Z_0^{2/3} \frac{18000\, \|\eta\|_\infty^2\, \|\phi'\|_\infty^2}{c_0\, c^\sharp}.
\end{aligned}
$$

As a consequence, taking $Z/Z_0^{2/3}$ large depending on L, c_0, $c^\sharp$, $\|\eta\|_\infty$ and $\|\phi'\|_\infty$, but still Z not larger than $Z_0/2$, we have

$$
\int_{Z_0}^{Z_0+Z} |\Psi_0(z)|^2 \frac{dz}{z^3} \geq \frac{Z}{4} L.
$$

Next, we turn to the non-vanishing of L. In preparation for the proof, set

$$\alpha_1 = \phi'(0) \quad \text{and} \quad \alpha_2 = \phi'(a^{-1/2}\Omega_c),$$

and note that $0 > \alpha_1 > \alpha_2$. Clearly, the (n, l) that enter in the sum for Ψ_0 are determined by the lattice points in $\mathbb{Z}^2$ which fall in the double-cone

$$\Gamma = \{(u, v) : \alpha_2 |u| \leq -|v| < \alpha_1 |u|, uv < 0\}.$$

Define $x_{u,v}$, for $(u, v) \in \Gamma$, as the unique point satisfying

$$\phi'(x_{u,v}) = \frac{v}{n} \cdot$$

Note that $x_{u,v}$ is strictly positive.

Then, we define

$$\theta(u, v) = u\,\phi(x_{u,v}) - v x_{u,v}, \quad (u, v) \in \Gamma.$$

The following is a trivial fact.

Lemma 9.21. *On Γ, we have*

$$\nabla\theta(u, v) = (\phi(x_{u,v}), -x_{u,v}).$$

A consequence of this trivial fact is the result we mentioned above.

Lemma 9.22. $L \neq 0$.

Proof. We will show that there is one θ_γ which only has one $(n, l) \in \Gamma$ such that $\theta(n, l) = \theta_\gamma$. This clearly shows that L is not 0.

Let n_0 be the smallest positive integer such that $(n, l) \in \Gamma$ for some l. In our case, this l is negative. Choose the largest (negative) such l, which we denote by l_0. That is, l_0 is the largest negative integer that satisfies $l < \alpha_1 n_0$. Then we claim that there is no other pair $(n', l') \in \Gamma$ such that $\theta(n_0, l_0) = \theta(n', l')$. Indeed, since $\theta(n, l)$ is strictly positive for $n > 0$, and strictly negative for $n < 0$, such n' would also have to be positive. It cannot equal n_0 because, since we should have $l' < l_0$, by the previous lemma we have

$$\theta(n_0, l') > \theta(n_0, l_0).$$

We must then have $n' > n_0$, which also implies

$$l' < \alpha_1 n' < \alpha_1 n_0 \quad (< 0),$$

thus showing that $l' \leq l$. But this is also impossible because, also by the previous lemma, there exists a pair (ξ, η) on the segment joining (n_0, l_0) to (n', l'), such that

$$\theta(n', l') - \theta(n_0, l_0) = \phi(x_{\xi,\eta})\,(n' - n_0) - x_{\xi,\eta}\,(l' - l_0)$$

and this last expression is then strictly positive. □

We summarize all this in the following lemma.

Lemma 9.23. *There are a small constant κ_0 and a large constant K such that*

$$\int_{Z_0}^{Z_0+Z} z^{-3}\,|\Psi_0(z)|^2\,dz \geq \kappa_0\,Z$$

whenever $Z \geq K\,Z_0^{2/3}$, and $Z_0 \geq K$.

Proof. Our previous calculations show this result in the case that $Z \geq C\,Z_0^{2/3}$ but $Z \leq Z/2$, for a certain large constant C. For the general case, break up

$$\int_{Z_0}^{Z_0+Z} z^{-3}\,|\Psi_0(z)|^2\,dz = \sum_{n=0}^{N}\int_{Z_n}^{Z_{n+1}} z^{-3}\,|\Psi_0(z)|^2\,dz,$$

where

$$Z_{n+1} = 1.01\,Z_n \quad \text{when } n = 0, \ldots, N-1; \quad Z_{N+1} = Z_0 + Z,$$

and N is chosen so that $1.01^{N+1}\,Z_0 \leq Z_0{+}Z < 1.01^{N+2}\,Z_0$. Our previous calculations would apply to each of the integrals in the sum provided

$$Z_n - Z_{n-1} \geq C\,Z_{n-1}^{2/3}, \quad (n = 1, \ldots, N); \quad Z + Z_0 \geq Z_N^{2/3},$$

which amounts to requiring $Z_0 \geq 10^6$. □

This mean value information can be used to obtain information about the oscillating behavior of Ψ_0, as follows. Let

$$I = [Z_0, Z_0 + \hat{C}\,Z_0^{2/3}] = \bigcup_j I_j, \quad \text{for} \quad I_j = [Z_0 + \hat{c}\,j\,Z_0^{2/3}, Z_0 + \hat{c}\,(j+1)\,Z_0^{2/3}],$$

where $\hat{C}$ is large depending on K, and $\hat{c}$ is small depending on κ_0.

Denote also, for any function f,

$$m_j(f) = \inf_{x \in I_j} |x^{-3/2}\,f(x)|.$$

Corollary 9.24. *Given any $\varepsilon > 0$ small depending on k_0, any $\hat{C}$ and $\hat{c}$ as above, with the extra requirement that $\hat{c}$ is small depending on ε, and Z_0 also large enough, there exists a constant $0 < \alpha < 1$ such that $m_j(\Psi_0) < \varepsilon$ for at most $\alpha\,\hat{C}/\hat{c}$ of the I_j.*

Proof. Put

$$Z^{-3/2}\,\Psi_0(Z) = F(Z) + E(Z)$$

such that $F(Z)$ contains only finitely many terms in the sum, and $|E(Z)|$ is always less than ε. In particular, we have that in order that $m_j(\Psi_0) < \varepsilon$ we must have

$m_j(F) < 2\varepsilon$. It will therefore be enough to count how many of the $m_j(F)$ stay below 2ε to obtain the conclusion of the lemma.

Since both ϕ and ϕ' are bounded, we have the trivial bound

$$|F'(Z)| \leq C_\varepsilon Z^{-2/3}$$

for some constant C_ε which depends on ε. Thus, if $m_j(\Psi_0) \leq \varepsilon$, we have $F(Z) \leq 2\varepsilon + \hat{c}\, C_\varepsilon$, which implies

$$\int_{I_j} |F(z)|^2 \,\mathrm{d}z \leq 16\, |I_j|\, (\varepsilon^2 + \hat{c}^2\, C_\varepsilon^2).$$

Therefore, if we denote by

$$M = \text{number of } j \text{ such that } m_j(\Psi_0) < \varepsilon,$$

we use the trivial bound $|F(z)| \leq C$ for all z, for a universal constant C, to get

$$\kappa_0\, \hat{C}\, Z_0^{2/3} \leq \int_I |F(z)|^2 \,\mathrm{d}z \leq \Big(\frac{\hat{C}}{\hat{c}} - M\Big)\, C\, \hat{c}\, Z_0^{2/3} + 16 M\, \hat{c}\, Z_0^{2/3}\, (\varepsilon^2 + \hat{c}^2\, C_\varepsilon^2).$$

This implies that

$$M \leq \alpha\, \frac{\hat{C}}{\hat{c}}, \qquad \text{for } \alpha = \frac{C - \kappa_0}{C - 16(\varepsilon^2 + \hat{c}^2\, C_\varepsilon^2)}.$$

By adjusting $\hat{c}$ depending on ε, it is easy to make $\alpha < 1$. □

A consequence of this corollary is the following, which shows that the size of $\Psi_0(Z)$ is at least $c\, Z^{3/2}$, for a small constant c, when we restrict our attention to infinitely many integer values of Z.

Corollary 9.25.

$$\limsup_{\substack{Z \to \infty \\ Z = 1,2,3,\ldots}} |Z^{-2/3}\, \Psi_0(Z)| \neq 0.$$

Proof. Apply the previous corollary to any $\varepsilon > 0$ small as required, and to any $\hat{C}$ large and $\hat{c}$ small also as needed, to conclude that there are infinitely many intervals of lengths going to infinity where $|x^{-3/2}\, \Psi_0(x)|$ is never smaller than ε. □

Bibliography

1. Davis, K. M. and Chang, Y.-C.: *Lectures on Bochner–Riesz means*. London Mathematical Society Lecture Note Series 114, Cambridge University Press, Cambridge, 1987.
2. De Guzmán, M.: *Real variable methods in Fourier analysis*. Notas de Matemática, North-Holland Publishing Co., Amsterdam-New York, 1981.
3. Englert, B.-G.: *Semiclassical theory of atoms*. Lecture Notes in Physics 300, Springer, Berlin, 1988.
4. García-Cuerva, J. and Rubio de Francia, J. L.: *Weighted norm inequalities and related topics*. North-Holland Mathematics Studies 116, North-Holland Publishing Co., Amsterdam, 1985.
5. Grafakos, L.: *Modern Fourier analysis*. Second edition. Graduate Texts in Mathematics 250, Springer, NY, 2009.
6. Grosswald, E.: *Topics from the theory of numbers*. Second edition, Birkhäuser, Boston, MA, 1984.
7. Gutzwiller, M. C.: *Chaos in classical and quantum mechanics*. Interdisciplinary Applied Mathematics 1, Springer, New York, 1990.
8. Hörmander, L.: *The analysis of linear partial differential operators I (Distribution theory and Fourier analysis), II (Differential operators with constant coefficients), III (Pseudo-differential operators) and IV (Fourier integral operators)*. Grundlehren der mathematischen Wissenschaften 256, 257, 274 and 275, Springer 1983-1985.
9. Iwaniec, H. and Kowalski, E.: *Analytic number theory*. American Mathematical Society Colloquium Publications 53, American Mathematical Society, Providence, RI, 2004.
10. Mattila, P.: *Geometry of sets and measures in Euclidean spaces*. Cambridge Studies in Advanced Mathematics 44, Cambridge University Press, Cambridge, 1995.
11. Senechal, M.: *Quasicrystals and geometry*. Cambridge University Press, Cambridge, 1995.
12. Stein, E. M.: *Singular integrals and differentiability properties of functions*. Princeton Mathematical Series 30, Princeton University Press, Princeton, NJ, 1970.
13. Stein, E. M.: *Harmonic analysis: real-variable methods, orthogonality, and oscillatory integrals*. Princeton Mathematical Series 43, Princeton University Press, Princeton, NJ, 1993.
14. Stein, E. M. and Weiss, G.: *Introduction to Fourier analysis on Euclidean spaces*. Princeton Mathematical Series 32, Princeton University Press, Princeton, NJ, 1971.
15. Tao, T.: *Additive combinatorics*. Cambridge Studies in Advanced Mathematics 105, Cambridge University Press, Cambridge, 2006.
16. Watson, G. N.: *A treatise on the theory of Bessel functions*. Cambridge University Press, Cambridge; The Macmillan Company, New York, 1944.
17. Zygmund, A.: *Trigonometric series: Volumes I and II*. Cambridge University Press, London-New York, 1968.

A. Córdoba, *Suprematism in Harmonic Analysis*, Progress in Mathematics 356,
https://doi.org/10.1007/978-3-031-73737-4

Bibliography for Chapter 1

18. Apéry, R.: Irrationalité de $\zeta(2)$ et $\zeta(3)$. *Astérisque* **61** (1979), 11–13.
19. Beukers, F.: A note on the irrationality of $\zeta(2)$ and $\zeta(3)$. *Bull. London Math. Soc.* **11** (1979), no. 3, 268–272.
20. Bourgain, J. and Rudnick, Z.: Restriction of toral eigenfunctions to hypersurfaces and nodal sets. *Geom. Funct. Anal.* **22** (2012), no. 4, 878–937.
21. Caffarelli, L.: Interior $W^{2,p}$ estimates for solutions of the Monge–Ampère equation. *Ann. of Math. (2)* **131** (1990), no. 1, 135–150.
22. Caffarelli, L. A. and Cabré, X.: *Fully nonlinear elliptic equations*. American Mathematical Society Colloquium Publications 43, American Mathematical Society, Providence, RI, 1995.
23. Cilleruelo, J. and Córdoba, A.: *La teoría de los números*. Editorial Mondadori, 1992.
24. Cilleruelo, J. and Córdoba, A.: Trigonometric polynomials and lattice points. *Proc. Amer. Math. Soc.* **115** (1992), no. 4, 899–905.
25. Cilleruelo, J. and Tenenbaum, G.: An overlapping theorem with applications. *Publ. Mat., Barc.* **51** (2007), Spec. Iss., 107–118.
26. Córdoba, A.: Disquisitio numerorum. *Gac. R. Soc. Mat. Esp.* **4** (2001), no. 1, 249–260.
27. Córdoba, A. and Córdoba, D.: A pointwise estimate for fractionary derivatives with applications to partial differential equations. *Proc. Natl. Acad. Sci. USA* **100** (2003), no. 26, 15316–15317.
28. Córdoba, A., Córdoba, D. and Gancedo, F.: Interface evolution: the Hele–Shaw and Muskat problems. *Ann. of Math. (2)* **173** (2011), no. 1, 477–542.
29. Córdoba, A. and Gallego, A.: Una propiedad de conjuntos convexos. In *Actas XI Reunión de Matemáticos Españoles*, Murcia, 1970.
30. Córdoba, A. and Martínez, Á. D.: A pointwise inequality for fractional Laplacians. *Adv. Math.* **280** (2015), 79–85.
31. Córdoba, A. and Ocáriz, J.: A note on generalized Laplacians and minimal surfaces. *Bull. Lond. Math. Soc.* **52** (2020), no. 1, 153–157.
32. De Guzmán, M.: *Differentiation of integrals in* $\mathbb{R}^n$. Lecture Notes in Mathematics 481, Springer, Berlin-New York, 1976.
33. Fefferman, C.: Whitney's extension problem for C^m. *Ann. of Math. (2)* **164** (2006), no. 1, 313–359.
34. Giusti, E.: *Minimal surfaces and functions of bounded variation*. Monographs in Mathematics 80, Birkhäuser, Basel, 1984.
35. John, F.: Extremum problems with inequalities as subsidiary conditions. In *Studies and essays presented to R. Courant on his 60th birthday, January 8, 1948*, pp. 187–204. Interscience Publishers, New York, 1948.
36. Kneser, M.: Summenmengen in lokalkompakten abelschen Gruppen. *Math. Z.* **66** (1956), 88–110.
37. Ramada, D. S.: Arithmetical applications of an identity for the Vandermonde determinant. *Acta Arith.* **130** (2007), no. 4, 351–359.
38. Rivoal, T.: La fonction zêta de Riemann prend une infinité de valeurs irrationnelles aux entiers impairs. *C. R. Acad. Sci. Paris Sér. I Math.* **331** (2000), no. 4, 267–270.
39. Siegel, C. L.: *Transcendental numbers*. Annals of Mathematics Studies 16, Princeton University Press, Princeton, NJ, 1949.

Bibliography for Chapter 2

40. Córdoba, A.: On the Vitali covering properties of a differentiation basis. *Studia Math.* **57** (1976), no. 1, 91–95.
41. Córdoba, A.: Maximal functions: a proof of a conjecture of A. Zygmund. *Bull. Amer. Math. Soc. (N.S.)* **1** (1979), no. 1, 255–257.
42. Córdoba, A.: Maximal functions, covering lemmas and Fourier multipliers. In *Harmonic analysis in Euclidean spaces (Proc. Sympos. Pure Math., Williams Coll., Williamstown, Mass., 1978), Part 1*, pp. 29–50. Proc. Sympos. Pure Math. 35, Part 1, American Mathematical Society, Providence, RI, 1979.
43. Córdoba, A. and Fefferman, R.: A geometric proof of the strong maximal theorem. *Ann. of Math. (2)* **102** (1975), no. 1, 95–100.
44. Córdoba, A. and Fefferman, R.: On differentiation of integrals. *Proc. Nat. Acad. Sci. U.S.A.* **74** (1977), no. 6, 2211–2213.
45. Córdoba, A. and Fefferman, R.: On the equivalence between the boundedness of certain classes of maximal and multiplier operators in Fourier analysis. *Proc. Nat. Acad. Sci. U.S.A.* **74** (1977), no. 2, 423–425.
46. De Guzmán, M.: *Real variable methods in Fourier analysis*. Notas de Matemática, North-Holland Publishing Co., Amsterdam-New York, 1981.
47. Fefferman, R.: Some weighted norm inequalities for Córdoba's maximal function. *Amer. J. Math.* **106** (1984), no. 5, 1261–1264.
48. Hardy, G. H. and Littlewood, J. E.: A maximal theorem with function-theoretic applications. *Acta Math.* **54** (1930), no. 1, 81–116.
49. Martínez, Á. D.: Córdoba's differentiation theorem: revisited. *Ann. Inst. Fourier (Grenoble)* **73** (2023), no. 2, 559–565.
50. Nagel, A., Stein, E. M. and Wainger, S.: Differentiation in lacunary directions. *Proc. Nat. Acad. Sci. U.S.A.* **75** (1978), no. 3, 1060–1062.
51. Rey, G.: Another counterexample to Zygmund's conjecture. *Proc. Amer. Math. Soc.* **148** (2020), no. 12, 5269–5275.
52. Soria, F.: Examples and counterexamples to a conjecture in the theory of differentiation of integrals. *Ann. of Math. (2)* **123** (1986), no. 1, 1–9.
53. Stein, E. M.: Note on the class $L \log L$. *Studia Math.* **32** (1969), 305–310.
54. Stein, E. M.: Some problems in harmonic analysis. In *Harmonic analysis in Euclidean spaces (Proc. Sympos. Pure Math., Williams Coll., Williamstown, Mass., 1978), Part 1*, pp. 3–20. Proc. Sympos. Pure Math. 35, Part 1, American Mathematical Society, Providence, RI, 1979.
55. Strömberg, J.-O.: Weak estimates on maximal functions with rectangles in certain directions. *Ark. Mat.* **15** (1977), no. 2, 229–240.
56. Vitali, G.: Sui gruppi di punti e sulle funzione de variable real. *Atti Acad. Sci. Torino* **43** (1908), 229–246.
57. Jessen, B., Marcinkievicz, J. and Zygmund, A.: Note on differentiability of multiple integrals. *Fund. Math.* **25** (1935), 217–234.

Bibliography for Chapter 3

58. Besicovitch, A. S.: Sur deux questions d'intégrabilité des fonctions. *J. Soc. Phys.-Math. (Perm')* **2** (1919), no. 1, 105–123.
59. Besicovitch, A. S.: On Kakeya's problem and a similar one. *Math. Z.* **27** (1928), no. 1, 312–320.
60. Besicovitch, A. S.: On differentiation of Lebesgue double integrals. *Fund. Math.* **25** (1935), 209–216.
61. Besicovitch, A. S.: A general form of the covering principle and relative differentiation of additive functions. *Proc. Cambridge Philos. Soc.* **41** (1945), 103–110.
62. Besicovitch, A. S.: The Kakeya problem. *Amer. Math. Monthly* **70** (1963), no. 7, 697-706.
63. Bourgain, J.: Besicovitch type maximal operators and applications to Fourier analysis. *Geom. Funct. Anal.* **1** (1991), no. 2, 147–187.
64. Bourgain, J.: On the dimension of Kakeya sets and related maximal inequalities. *Geom. Funct. Anal.* **9** (1999), no. 2, 256–282.
65. Córdoba, A.: The Kakeya maximal function and the spherical summation multipliers. *Amer. J. Math.* **99** (1977), no. 1, 1–22.
66. Córdoba, A.: The fat needle problem. *Bull. London Math. Soc.* **25** (1993), no. 1, 81–82.
67. Cunningham, F., Jr.: Three Kakeya problems. *Amer. Math. Monthly* **81** (1974), 582–592.
68. Davies, R. O.: Some remarks on the Kakeya problem. *Proc. Cambridge Philos. Soc.* **69** (1971), 417–421.
69. Goog, J. I. (editor): *The scientist speculates: an anthology of partly-baked ideas*. Heinemann (Pubs.) Ltd., London, 1962.
70. Kahane, J.-P.: Trois notes sur les ensembles parfaits linéaires. *Enseign. Math. (2)* **159** (1969), 185–192.
71. Kakeya, S.: Some problems on maxima and minima regarding ovals. *Tohoku Sci. Reports* **6** (1917), 71–88.
72. Katz, N. H.: Maximal operators over arbitrary sets of directions. *Duke Math. J.* **97** (1999), no. 1, 67–79.
73. Katz, N. H., Łaba, I. and Tao, T.: An improved bound on the Minkowski dimension of Besicovitch sets in $\mathbb{R}^3$. *Ann. of Math. (2)* **152** (2000), no. 2, 383–446.
74. Katz, N. and Tao, T.: Recent progress on the Kakeya conjecture. *Publ. Mat.* **46** (2002), 161–179.
75. Perron, O.: Über einen Satz von Besicovitsch. *Math. Z.* **28** (1928), no. 1, 383–386.
76. Rademacher, H.: On a theorem of Besicovitch. In *Studies in mathematical analysis and related topics; essays in honor of George Pólya*, pp. 294–296. Stanford University, Stanford, 1962.
77. Wolff, T.: An improved bound for Kakeya type maximal functions. *Rev. Mat. Iberoamericana* **11** (1995), no. 3, 651–674.
78. Wolff, T.: Recent work connected with the Kakeya problem. In *Prospects in mathematics: invited talks on the occasion of the 250th anniversary of Princeton University*, 129–162. American Mathematical Society, Providence, RI, 1999.

Bibliography for Chapter 4

79. Andersen, K. F. and John, R. T.: Weighted inequalities for vector-valued maximal functions and singular integrals. *Studia Math.* **69** (1980/81), no. 1, 19–31.
80. Benedek, A., Calderón, A. P. and Panzone, R.: Convolution operators on Banach space valued functions. *Proc. Nat. Acad. Sci. U.S.A.* **48** (1962), 356–365.
81. Calderón, A. P. and Zygmund, A.: On the existence of certain singular integrals. *Acta Math.* **88** (1952), 85–139.
82. Coifman, R. R. and Fefferman, C.: Weighted norm inequalities for maximal functions and singular integrals. *Studia Math.* **51** (1974), 241–250.
83. Córdoba, A.: Translation invariant operators. In *Fourier analysis (Proc. Sem., El Escorial, 1979)*, pp. 117–176. Asoc. Mat. Española 1, Asoc. Mat. Española, Madrid, 1980.
84. Córdoba, A. and Fefferman, C.: A weighted norm inequality for singular integrals. *Studia Math.* **57** (1976), no. 1, 97–101.
85. David, G. and Journé, J.-L.: A boundedness criterion for generalized Calderón–Zygmund operators. *Ann. of Math. (2)* **120** (1984), no. 2, 371–397.
86. De Leeuw, K.: On L_p multipliers. *Ann. of Math. (2)* **81** (1965), no. 2, 364–379.
87. Fefferman, C.: Inequalities for strongly singular convolution operators. *Acta Math.* **124** (1970), 9–36.
88. Fefferman, C. and Muckenhoupt, B.: Two nonequivalent conditions for weight functions. *Proc. Amer. Math. Soc.* **45** (1974), 99–104.
89. Fefferman, C. and Stein, E. M.: Some maximal inequalities. *Amer. J. Math.* **93** (1971), no. 1, 107–115.
90. Helson, H. and Szegö, G.: A problem in prediction theory. *Ann. Mat. Pura Appl.* **51** (1960), 107–138.
91. Hirschman, I. I., Jr.: On multiplier transformations. *Duke Math. J.* **26** (1959), 221–242.
92. Hörmander, L.: Estimates for translation invariant operators in L^p spaces. *Acta Math.* **104** (1960), 93–140.
93. Hunt, R., Muckenhoupt, B. and Wheeden, R.: Weighted norm inequalities for the conjugate function and Hilbert transform. *Trans. Amer. Math. Soc.* **176** (1973), 227–251.
94. Muckenhoupt, B.: Weighted norm inequalities for the Hardy maximal function. *Trans. Amer. Math. Soc.* **165** (1972), 207–226.
95. Rubio de Francia, J. L.: Vector-valued inequalities for operators in L^p-spaces. *Bull. London Math. Soc.* **12** (1980), no. 3, 211–215.
96. Rubio de Francia, J. L.: Factorization theory and A_p weights. *Amer. J. Math.* **106** (1984), no. 3, 533–547.

Bibliography for Chapter 5

97. Barceló, B.: *Análisis armónico geométrico: la restricción de la transformada de Fourier.* Ph.D. Thesis, Universidad Autónoma de Madrid, 1983.
98. Barceló, B.: On the restriction of the Fourier transform to a conical surface. *Trans. Amer. Math. Soc* **292** (1985), no. 1, 321–333.
99. Bourgain, J.: Besicovitch type maximal operators and applications to Fourier analysis. *Geom. Funct. Anal.* **1** (1991), no. 2, 147–187.
100. Bourgain, J.: L^p-estimates for oscillatory integrals in several variables. *Geom. Funct. Anal.* **1** (1991), no. 4, 321–374.
101. Carbery, A.: The boundedness of the maximal Bochner–Riesz operator on $L^4(\mathbb{R}^2)$. *Duke Math. J.* **50** (1983), no. 2, 409–416.
102. Carbery, A.: A weighted inequality for the maximal Bochner–Riesz operator on $\mathbb{R}^2$. *Trans. Amer. Math. Soc.* **287** (1985), no. 2, 673–680.
103. Carleson, L. and Sjölin, P.: Oscillatory integrals and a multiplier problem for the disc. *Studia Math.* **44** (1972), 287–299.
104. Christ, M.: On almost everywhere convergence of Bochner–Riesz means in higher dimensions. *Proc. Amer. Math. Soc.* **95** (1985), no. 1, 16–20.
105. Córdoba, A.: A radial multiplier and a related Kakeya maximal function. *Bull. Amer. Math. Soc.* **81** (1975), 428–430.
106. Córdoba, A.: The multiplier problem for the polygon. *Ann. of Math. (2)* **105** (1977), no. 3, 581–588.
107. Córdoba, A.: A note on Bochner–Riesz operators. *Duke Math. J.* **46** (1979), no. 3, 505–511.
108. Córdoba, A.: Multipliers of $\mathcal{F}(L^p)$. In *Euclidean harmonic analysis (Proc. Sem., Univ. Maryland, College Park, Md., 1979)*, pp. 162–177. Lecture Notes in Mathematics 779, Springer, Berlin, 1980.
109. Córdoba, A.: Geometric Fourier analysis. *Ann. Inst. Fourier (Grenoble)* **32** (1982), no. 3, vii, 215–226.
110. Córdoba, A. and López-Melero, B.: Spherical summation: a problem of E. M. Stein. *Ann. Inst. Fourier (Grenoble)* **31** (1981), no. 3, x, 147–152.
111. Córdoba, A., Nagel, A., Vance, J., Wainger, S. and Weinberg, D.: L^p bounds for Hilbert transforms along convex curves. *Invent. Math.* **83** (1986), no. 1, 59–71.
112. Córdoba, A. and Rogers, K.: Weighted estimates for conic Fourier multipliers. *Math. Z.* **278** (2014), 431–440.
113. Córdoba, A. and Rubio de Francia, J. L.: Estimates for Wainger's singular integrals along curves. *Rev. Mat. Iberoamericana* **2** (1986), no. 1-2, 105–117.
114. Drury, S. W.: Restrictions of Fourier transforms to curves. *Ann. Inst. Fourier (Grenoble)* **35** (1985), no. 1, 117–123.
115. Fefferman, C.: The multiplier problem for the ball. *Ann. of Math. (2)* **94** (1971), 330–336.
116. Fefferman, C.: A note on spherical summation multipliers. *Israel J. Math.* **15** (1973), 44–52.
117. Hörmander, L.: Oscillatory integrals and multipliers on FL^p. *Ark. Mat.* **11** (1973), no. 1-2, 1–11.
118. Igari, S.: Decomposition theorem and lacunary convergence of Riesz–Bochner means of Fourier transforms of two variables. *Tohoku Math. J. (2)* **33** (1981), no. 3, 413–419.
119. Journé, J.-L.: Calderón–Zygmund operators on product spaces. *Rev. Mat. Iberoamericana***1** (1985), no. 3, no. 3, 55–91.
120. Nagel, A., Stein, E. and Wainger, S.: Hilbert transforms and maximal functions related to variable curves In *Harmonic analysis in Euclidean spaces (Williamstown, 1978)*, pp. 95–98. Proc. Sympos. Pure Math. 35, Part 2, American Mathematical Society, Providence, RI, 1979.

121. Nagel, A. and Wainger, S.: L^2-boundedness of Hilbert transforms along surfaces and convolution operators homogeneous with respect to a multiple parameter group. *Amer. J. Math.* **99** (1977), no. 4, 761–785.

122. Sjölin, P.: Fourier multipliers and estimates for Fourier transforms of measures carried by smooth curves in $\mathbb{R}^2$. *Studia Math.* **51** (1974), 169–182.

123. Rubio de Francia, J. L.: A Littlewood–Paley inequality for arbitrary intervals. *Rev. Mat. Iberoamericana* **1** (1985), no. 2, 1–14.

124. Ruiz, A.: *Multiplicadores asociados a curvas en el plano y teoremas de restricción de la transformada de Fourier a curvas en* $\mathbb{R}^2$ *y* $\mathbb{R}^3$. Ph.D. Thesis, Universidad Autónoma de Madrid, 1980.

125. Sogge, C. D.: A sharp restriction theorem for degenerate curves in $\mathbb{R}^2$. *Amer. J. Math.* **109** (1987), no. 2, 223–228.

126. Stein, E. M. and Wainger, S.: Problems in harmonic analysis related to curvature. *Bull. Amer. Math. Soc.* **84** (1978), no. 6, 1239–1295.

127. Tao, T., Vargas, A. and Vega, L.: A bilinear approach to the restriction and Kakeya conjectures. *J. Amer. Math. Soc.* **11** (1998), no. 4, 967–1000.

128. Tomas, P. A.: A restriction theorem for the Fourier transform. *Bull. Amer. Math. Soc.* **81** (1975), 477–478.

129. Zygmund, A.: On Fourier coefficients and transforms of functions of two variables. *Studia Math.* **50** (1974), 189–201.

Bibliography for Chapter 6

130. Balodis, P. and Córdoba, A.: The convergence of multidimensional Fourier–Bessel series. *J. Anal. Math.* **77** (1999), 269–286.
131. Barceló, J. A. and Córdoba, A.: Band-limited functions: L^p-convergence. *Trans. Amer. Math. Soc.* **313** (1989), no. 2, 655–669.
132. Carbery, A., Hernández, E. and Soria, F.: The behaviour on radial functions of maximal operators along arbitrary directions and the Kakeya maximal operator. *Tohoku Math. J. (2)* **41** (1989), no. 4, 647–656.
133. Carbery, A., Romera, E. and Soria, F.: Radial weights and mixed norm inequalities for the disc multiplier. *J. Funct. Anal.* **109** (1992), no. 1, 52–75.
134. Córdoba, A.: The disc multiplier. *Duke Math. J.* **58** (1989), no. 1, 21–29.
135. Córdoba, A.: Singular integrals, maximal functions and Fourier restriction to spheres: the disk multiplier revisited. *Adv. Math.* **290** (2016), 208–235.
136. Flammer, C.: *Spheroidal wave functions*. Stanford University Press, Stanford, California, 1957.
137. Landau, H. J. and Pollak, H. O.: Prolate spheroidal wave functions, Fourier analysis and uncertainty II. *Bell Syst. Tech. J.* **40** (1961), no. 1, 65–84.
138. Landau, H. J. and Pollak, H. O.: Prolate spheroidal wave functions, Fourier analysis and uncertainty III. The dimension of the space of essentially time- and band-limited signals. *Bell Syst. Tech. J.l* **41** (1962), no. 4, 1295–1336.
139. Mockenhaupt, G.: On radial weights for the spherical summation operator. *J. Funct. Anal.* **91** (1990), no. 1, 174–181.
140. Rubio de Francia, J. L.: Transference principles for radial multipliers. *Duke Math. J.* **58** (1989), no. 1, 1–19.
141. Sawyer, E. T.: A characterization of a two-weight norm inequality for maximal operators. *Studia Math.* **75** (1982), no. 1, 1–11.
142. Slepian, D.: Prolate spheroidal wave functions, Fourier analysis and uncertainty IV. Extensions to many dimensions; generalized prolate spheroidal functions. *Bell Syst. Tech. J.* **43** (1964), no. 6, 3009–3058.
143. Slepian, D. and Pollak, H. O.: Prolate spheroidal wave functions, Fourier analysis and uncertainty I. *Bell Syst. Tech. J.* **40** (1961), no. 1, 43–63.
144. Stratton, J. A., Morse, P. M., Chu, L. J., Little, J. D. C. and Corbató, F. J.: *Spheroidal wave functions including tables of separation constants and coefficients*. Technology Press of MIT. and John Wiley, New York; Chapman & Hall, London, 1956.
145. Vega, L.: *El multiplicador de Schrödinger, la función maximal y los operadores de restricción*. Ph.D. Thesis, Universidad Autónoma de Madrid, 1988.
146. Vega, L.: Schrödinger equations: pointwise convergence to the initial data. *Proc. Amer. Math. Soc.* **102** (1988), no. 4, 874–878.
147. Whittaker, E. T. and Watson, G. N.: *A course of modern analysis*. Cambridge Mathematical Library, Cambridge University Press, Cambridge, 1996.

Bibliography for Chapter 7

148. Bombieri, E. and Iwaniec, H.: On the order of $\zeta(1/2+it)$. *Ann Scuola Norm. Sup. Pisa* **13** (1986), no. 3, 449–472.
149. Chamizo, F.: Lattice point counting and harmonic analysis. In *Proceedings of the "Segundas Jornadas de Teoría de Números"*, pp. 83–99. Bibl. Rev. Mat. Iberoamericana, Rev. Mat. Iberoamericana, Madrid, 2008.
150. Chamizo, F. and Córdoba, A.: One-dimensional crystals and quadratic residues. *J. Number Theory* **65** (1997), no. 1, 101–104.
151. Chamizo, F. and Córdoba, A.: Puntos del retículo. In *Margarita Mathematica, en memoria de Jose Javier (Chicho) Guadalupe*, 59–76. Servicio de Publicaciones de la Universidad de La Rioja, Logroño, Spain, 2001.
152. Chamizo, F. and Córdoba, A.: A path integral approach to lattice point problems. *J. Math. Pures Appl. (9)* **81** (2002), no. 10, 957–966.
153. Chamizo, F. and Iwaniec, H.: On the sphere problem. *Rev. Mat. Iberoamericana* **11** (1995), no. 2, 417–429.
154. Córdoba, A.: La formule sommatoire de Poisson. *C. R. Acad. Sci. Paris Sér. I Math.* **306** (1988), no. 8, 373–376.
155. Córdoba, A.: Dirac combs. *Lett. Math. Phys.* **17** (1989), no. 3, 191–196.
156. Córdoba, A.: Lattice points. *J. Fourier Anal. Appl.* **3** (1997), Suppl 1, 859–870. (Proceedings of the conference dedicated to Professor Miguel de Guzmán (El Escorial, 1996).)
157. Fletcher, R. and Griffiths, D. F.: The generalized eigenvalue problem for certain unsymmetric band matrices. *Linear Algebra and Appl.* **29** (1980), 139–149.
158. Giacovazzo, C.: The diffraction of X-rays by crystals. In *Fundamentals of Crystallography*, 141–234. International Union of Crystallography Texts on Crystallography, Oxford Univ. Press, 1995.
159. Graham, S. W. and Kolesnik, G.: *Van der Corput's method of exponential sums*. London Mathematical Society Lecture Note Series 126, Cambridge University Press, Cambridge, 1991.
160. Grünbaum, F. A. and Moore, C. C.: The use of higher-order invariants in the determination of generalized Patterson cyclotomic sets. *Acta Cryst. Sect. A* **51** (1995), no. 3, 310–323.
161. Hardy, G. H.: The average order of the arithmetical functions $P(x)$ and $\Delta(x)$. *Proc. London Math. Soc. (2)* **15** (1916), 192–213.
162. Hardy, G. H.: On the representation of a number as the sum of any number of squares, and in particular of five. *Trans. Amer. Math. Soc.* **21** (1920), no. 3, 255–284.
163. Heath-Brown, D. R.: Lattice points in the sphere. In *Number theory in progress, Vol. 2 (Zakopane-Kościelisko, 1997)*, pp. 883–892. De Gruyter, Berlin, 1999.
164. Huxley, M. N.: Exponential sums and lattice points. III. *Proc. London Math. Soc. (3)* **87** (2003), no. 3, 591–609.
165. Iwaniec, H. and Mozzochi, C. J.: On the divisor and circle problems. *J. Number Theory* **29** (1988), no. 1, 60–93.
166. Jarník, V.: Über die Gitterpunkte auf konvexen Kurven. *Math. Z.* **24** (1926), no. 1, 500–518.
167. Lev, N. and Olevskii, A.: Quasicrystals and Poisson's summation formula. *Invent. Math.* **200** (2015), no. 2, 585–606.
168. Odlyzko, A. M. and Poonen, B.: Zeros of polynomials with 0, 1 coefficients. *Enseign. Math. (2)* **39** (1993), no. 3-4, 317–348.
169. Pauling, P. and Shappell, M. D.: The crystal structure of bixbyite and the C-modification of the sesquioxides. *Z. Krystallogr.* **75** (1930), 128–142.
170. Rosenblatt, J.: Phase retrieval. *Comm. Math. Phys.* **95** (1984), no. 3, 317–343.

Bibliography for Chapter 8

171. Atkin, A. O. L.: On pseudo-squares. *Proc. London Math. Soc. (3)* **14a** (1965), 22–27.

172. Bombieri, E.: *Le grand crible dans la théorie analytique des nombres* Astérisque 18 (1974), 112 p.

173. Bombieri, E.: On Vinogradov's mean value theorem and Weyl sums. In *Automorphic forms and analytic number theory (Montreal, PQ, 1989)*, pp. 7–24. Univ. Montréal, Montreal, QC, 1990.

174. Bombieri, E. and Zannier, U.: A note on squares in arithmetic progressions. II. *Atti Accad. Naz. Lincei Cl. Sci. Fis. Mat. Natur. Rend. Lincei (9) Mat. Appl.* **13** (2002), no. 2, 69–75.

175. Carleson, L.: On convergence and growth of partial sums of Fourier series. *Acta Math.* **116** (1966), 135–157.

176. Chamizo, F. and Córdoba, A.: The fractal dimension of a family of Riemann's graphs. *C. R. Acad. Sci. Paris Sér. I Math.* **317** (1993), no. 3, 455–460.

177. Chamizo, F. and Córdoba, A.: Differentiability and dimension of some fractal Fourier series. *Adv. Math.* **142** (1999), no. 2, 335–354.

178. Chamizo, F., Córdoba, A. and Ubis, A.: Fourier series in BMO with number theoretical implications. *Math. Ann.* **376** (2020), no. 1-2, 457–473.

179. Cilleruelo, J. and Córdoba, A.: $B_2[\infty]$-sequences of square numbers. *Acta Arith.* **61** (1992), no. 3, 265–270.

180. Cilleruelo, J. and Granville, A.: Lattice points on circles, squares in arithmetic progressions and sumsets of squares. In *Additive combinatorics*, pp. 241–262. CRM Proc. Lecture Notes 43, American Mathematical Society, Providence, RI, 2007.

181. Córdoba, A.: Encounters at the interface between number theory and harmonic analysis. In *Proceedings of the "Segundas Jornadas de Teoría de Números"*, pp. 101–118. Bibl. Rev. Mat. Iberoamericana, Revista Matemática Iberoamericana, Madrid, 2008.

182. Córdoba, A. and Fernández, P.: Convergence and divergence of decreasing rearranged Fourier series. *SIAM J. Math. Anal.* **29** (1998), no. 5, 1129–1139.

183. Duistermaat, J. J.: Self-similarity of "Riemann's nondifferentiable function". *Nieuw Arch. Wisk. (4)* **9** (1991), no. 3, 303–337.

184. Ellison, W. J. and Mendès-France, P.: *Les nombres premiers*. Hermann, Paris, 1975.

185. Erdős, P. and Rényi, A.: Additive properties of random sequences of positive integers. *Acta Arith.* **6** (1960), 83–110.

186. Gerver, J.: The differentiability of the Riemann function at certain rational multiples of π. *Proc. Nat. Acad. Sci. U.S.A.* **62** (1969), 668–670.

187. Hardy, G. H. and Littlewood, J. E.: Some problems of diophantine approximation. I: The fractional part of $n^k\theta$. *Acta Math.* **37** (1914), 155–191.

188. Hardy, G. H. and Littlewood, J. E.: Some problems of diophantine approximation. II: The trigonometrical series associated with the elliptic θ-functions. *Acta Math.* **37** (1914), 193–239.

189. Halberstan, H. and Roth, K. F.: *Sequences*. The Clarendon Press, Oxford, 1966.

190. Holschneider, M. and Tchamitchian, P.: Pointwise analysis of Riemann's "nondifferentiable" function. *Invent. Math.* **105** (1991), no. 1, 157–175.

191. Hunt, R.: On the convergence of Fourier series. In *Orthogonal expansions and their continuous analogues (Proc. Conf., Edwardsville, Ill., 1967)*, pp. 235–255. Southern Illinois Univ. Press, Carbondale, 1968.

192. Körner, T. W.: Divergence of decreasing rearranged Fourier series. *Ann. of Math. (2)* **144** (1996), no. 1, 167–180.

193. Montgomery, H. L.: *Ten lectures on the interface between analytic number theory and harmonic analysis*. CBMS Regional Conference Series in Mathematics 84, Published for the Conference Board of the Mathematical Sciences, Washington, DC; by the American Mathematical Society, Providence, RI, 1994.

194. Riemann, B.: *Über die Darstellbarkeit einer Function durch eine trigonometrische Reihe*. Dieterich, Göttingen, 1867.

195. Tao, T. On the almost everywhere convergence of wavelet summation methods. *Appl. Comput. Harmon. Anal.* **3** (1996), no. 4, 384–387.

196. Vaughan, R. C.: *The Hardy–Littlewood method*. Cambridge Tracts in. Mathematics 80, Cambridge University Press, Cambridge, 1981.

Bibliography for Chapter 9

197. Arnold, V. I.: Mathematical methods of classical mechanics. Graduate Texts in Mathematics 60, Springer Science+Business Media, New York, 1978.
198. Bleher, P. M.: Distribution of energy levels of a quantum free particle on a surface of revolution. *Duke Math. J.* **74** (1994), no. 1, 45–93.
199. Córdoba, A. and Fefferman, C.: Wave packets and Fourier integral operators. *Commun. Partial. Differ. Equ.* **3** (1978), no. 11, 979–1005.
200. Córdoba, A., Fefferman, C. and Seco, L. Á.: A trigonometric sum relevant to the nonrelativistic theory of atoms. *Proc. Natl. Acad. Sci.* **91** (1994), no. 13, 5776–5778.
201. Córdoba, A., Fefferman, C. and Seco, L. Á.: Weyl sums and atomic energy oscillations. *Rev. Mat. Iberoam.* **11** (1995), no. 1, 165–226.
202. Córdoba, A., Fefferman, C. and Seco, L. Á.: A number-theoretic estimate for the Thomas–Fermi density. *Commun. Partial Differ. Equ.* **21** (1996), no. 7-8, 1087–1102.
203. Dirac, P. A. M.: Note on exchange phenomena in the Thomas atom. *Math. Proc. Cambridge Phil. Soc.* **26** (1930), no. 3, 376–385.
204. Englert, B. G. and Schwinger, J.: Atomic-binding-energy oscillations. *Phys. Rev. A Gen. Phys.* **32** (1985), no. 1, 47–63.
205. Fefferman, C.: The uncertainty principle. *Bull. Amer. Math. Soc. (N.S.)* **9** (1983), no. 2, 129–206.
206. Fefferman, C.: Atoms and analytic number theory. In *American Mathematical Society centennial publications, Vol. II (Providence, RI, 1988)*, pp. 27–36. American Mathematical Society, Providence, RI, 1992.
207. Fefferman, C. and Seco, L. Á.: On the energy of a large atom. *Bull. Amer. Math. Soc.* **23** (1990), 525–530.
208. Fefferman, C. and Seco, L. Á.: Aperiodicity of the Hamiltonian flow in the Thomas–Fermi potential. *Rev. Mat. Iberoam.* **9** (1993), no. 3, 409–551.
209. Fefferman, C. and Seco, L. Á.: On the Dirac and Schwinger corrections to the ground-state energy of an atom. *Adv. Math.* **107** (1994), no. 1, 1–185.
210. Fefferman, C. and Seco, L. Á.: The density in a one-dimensional potential. *Adv. Math.* **107** (1994), no. 2, 187–364.
211. Fefferman, C. and Seco, L. Á.: The eigenvalue sum of a one-dimensional potential. *Adv. Math.* **108** (1994), no. 2, 263–335.
212. Fermi, E.: Un metodo statistico per la determinazione di alcune priopriètà dell'atomo. *Rend. Acad. Naz. Lincei* **6** (1927), 602–607.
213. Folland, G. B.: Harmonic analysis in phase space. Annals of Mathematics Studies 122, Princeton University Press, Princeton, NJ, 1989.
214. Folland, G. B. Sitaram, A.: The uncertainty principle: A mathematical survey. *J. Fourier Anal. Appl.* **3** (1997), 207–238.
215. Graham, S. W. and Kolesnik, G.: Van der Corput's method of exponential sums. London Mathematical Society Lecture Note Series 126, Cambridge University Press, Cambridge, 1991.
216. Gutzwiller, M. C.: *Chaos in classical and quantum mechanics*. Interdisciplinary Applied Mathematics 1, Springer New York, NY, 1990.
217. Heath-Brown, D. R.: The distribution and moments of the error term in the Dirichlet divisor problem. *Acta Arith* **60** (1992), no 4, 389–415.
218. Hughes, W.: An atomic energy lower bound that agrees with Scott's correction. *Adv. Math.* **79** (1990), no. 2, 213–270.
219. Hille, E.: On the Thomas–Fermi equation. *Proc. Nat. Acad. Sci. USA* **62** (1969), no. 1 , 7–10.

220. Lieb, E. H.: Thomas–Fermi and related theories of atoms and molecules. *Rev. Mod. Phys.* 53 (1981), no. 4, 603–641.

221. Lieb, E. H. and Simon, B.: The Thomas–Fermi theory of atoms, molecules and solids. *Adv. Math.* 23 (1977), no. 1, 22–116.

222. Scott, J. M. C.: The binding energy of the Thomas–Fermi atom. *Philos. Mag.* **43** (1952), 859–867.

223. Schwinger, J.: Thomas–Fermi model: The second correction. *Phys. Rev. A* **24** (1981), 2353–2261.
https://doi.org/cmp/1104159980

224. Siedentop, H. and Weikard, R.: On the leading energy correction for the statistical model of the atom: interacting case *Commun. Math. Phys.* **112** (1987), 471–490.

225. Siedentop, H., Weikard, R. and Müller, A. M. K.: On the leading correction of the Thomas–Fermi model: Lower bound. *Invent Math* **97** (1989), 159–193.

226. Thomas, L. H.: The calculation of atomic fields. *Math. Proc. Camb. Philos. Soc.* **23** (1927), no. 5, 542–548.